Porous Materials for Energy and Environment Applications

Porous Materials for Energy and Environment Applications

Editors

Guangxu Lan
Yi-Nan Wu

MDPI

Basel • Beijing • Wuhan • Barcelona • Belgrade • Novi Sad • Cluj • Manchester

Editors

Guangxu Lan
Peking University
Beijing
China

Yi-Nan Wu
Tongji University
Shanghai
China

Editorial Office
MDPI AG
Grosspeteranlage 5
4052 Basel, Switzerland

This is a reprint of articles from the Topic published online in the open access journals *Materials* (ISSN 1996-1944), *Nanomaterials* (ISSN 2079-4991), *Separations* (ISSN 2297-8739), and *Molecules* (ISSN 1420-3049) (available at: https://www.mdpi.com/topics/0063PZ39RP).

For citation purposes, cite each article independently as indicated on the article page online and as indicated below:

Lastname, A.A.; Lastname, B.B. Article Title. *Journal Name* **Year**, *Volume Number*, Page Range.

ISBN 978-3-7258-2351-2 (Hbk)
ISBN 978-3-7258-2352-9 (PDF)
doi.org/10.3390/books978-3-7258-2352-9

Contents

materials

MDPI

Article

Efficient Adsorption Removal of Phosphate from Rural Domestic Sewage by Waste Eggshell-Modified Peanut Shell Biochar Adsorbent Materials

Cancan Xu [1,2], Rui Liu [2,*], Lvjun Chen [2,3] and Quanxi Wang [4,*]

1 School of Environmental and Geographical Sciences, Shanghai Normal University, Shanghai 200234, China; xucancan523127@126.com
2 Zhejiang Provincial Key Laboratory of Water Science and Technology, Department of Environment, Yangtze Delta Region Institute of Tsinghua University, Jiaxing 314006, China; chenlj@tsinghua.edu.cn
3 School of Environment, Tsinghua University, Beijing 100084, China
4 College of Life Science, Shanghai Normal University, Shanghai 200234, China
* Correspondence: liuruitsinghuazj@gmail.com (R.L.); wangqx@shnu.edu.cn (Q.W.); Tel.: +86-0573-82581603 (R.L.)

Abstract: In order to promote the improvement of the rural living environment, the treatment of rural domestic sewage has attracted much attention in China. Meanwhile, the rural regions' sewage discharge standards are becoming increasingly stringent. However, the standard compliance rate of total phosphorus (TP) is very low, and TP has become the main limiting pollutant for the water pollutants discharge standards of rural domestic sewage treatment facilities. In this study, waste eggshell (E) was employed as a calcium source, and waste peanut shell (C) was employed as a carbon source to prepare calcium-modified biochar adsorbent materials (E-C). The resulting E-C adsorbent materials demonstrated efficient phosphate (P) adsorption from aqueous solutions over the initial pH range of 6–9 and had adsorption selectivity. At an eggshell and peanut shell mass ratio of 1:1 and a pyrolysis temperature of 800 °C, the experimental maximum adsorption capacity was 191.1 mg/g. The pseudo second-order model and Langmuir model were best at describing the adsorption process. The dominant sorption mechanism for P is that $Ca(OH)_2$ is loaded on biochar with P to form $Ca_5(PO_4)_3OH$ precipitate. E-C was found to be very effective for the treatment of rural domestic sewage. The removal rate of TP in rural domestic sewage was 91–95.9%. After adsorption treatment, the discharge of TP in rural sewage met the second-grade (TP < 3 mg/L) and even first-grade (TP < 2 mg/L). This study provides an experimental basis for efficient P removal by E-C adsorbent materials and suggests possible applications in rural domestic sewage.

Keywords: phosphate adsorption; modified biochar adsorbent materials; rural domestic sewage; discharge standard

Citation: Xu, C.; Liu, R.; Chen, L.; Wang, Q. Efficient Adsorption Removal of Phosphate from Rural Domestic Sewage by Waste Eggshell-Modified Peanut Shell Biochar Adsorbent Materials. *Materials* **2023**, *16*, 5873. https://doi.org/10.3390/ma16175873

Academic Editors: Yi-Nan Wu and Guangxu Lan

Received: 21 July 2023
Revised: 22 August 2023
Accepted: 24 August 2023
Published: 28 August 2023

1. Introduction

China is a big agricultural country with nearly 500 million rural people. Due to population growth and the improvement in social and economic living standards, the increase in rural domestic sewage is becoming one of the main environmental problems in rural areas [1]. In recent years, in order to reduce pollutant emissions and improve the rural living environment, rural domestic sewage treatment has received more and more attention by the Chinese government. Meanwhile, rural areas' sewage discharge standards are becoming more and more stringent [2]; currently, thirty-one provinces and cities have issued rural sewage discharge standards in China.

The most conventional treatment methods for domestic sewage in rural regions consist of an operational sequence of aerobic/anoxic/aerobic processes. Anaerobic–anoxic–oxic (A^2O) is one of the most typical schemes in China. However, some reports have observed that the efficient removal of COD, NH_3-N, and total phosphorus (TP) is hard to realize [3]. In the study by Yang et al., surrounding Erhai Lake in Yunnan Province, centralized rural sewage treatment facilities had good removal efficiency. However, some centralized sewage treatment facilities could not meet the water pollution discharge standards, in particular, the average standard compliance rate of TP was only 4.26%. TP has become the main limiting pollutant for the discharge standards of rural domestic sewage treatment facilities. Therefore, improving TP treatment efficiency for meeting discharge standards in rural regions is an important target of future research [4].

Different biological, chemical, and physical techniques have been created and widely applied in previous studies to manage and treat the phosphorus (P) present in wastewater [5,6]. Among these methods, the adsorption method is the most attractive since it is economically feasible and simple in real application. In addition, the nutrient-loaded adsorbents can be further utilized as a soil conditioner and fertilizer [7,8]. Therefore, it is an important research topic to prepare cost-effective adsorbents with good adsorption performance. Recently, the modified biochar synthesized from calcium-rich waste achieved a good P adsorption effect. For example, when marble waste was employed as a Ca source, agricultural waste was used as a carbon source to synthesize modified-biochar composites to remove P from waste streams, and the maximum P removal capacity of the composites was 263.17 mg/g. The methods of producing modified biochar are economical, environmentally friendly and green [9]. Tobacco straw was employed as the carbon source, while oyster shell was used as the source of Ca to create a modified-biochar adsorbent with a maximum P adsorption capacity of 88.64 mg/g [10]. According to statistics, the crop straw output was 819.7 tons, of which peanut shell straw, wheat straw and corn straw accounted for 8.9%, 21.6% and 28.9% in China, respectively [11]. Most peanut shells are disposed of by landfill or incineration, resulting in an impact on the environment and the waste of resources. The preparation of biochar from peanut shells is a low-cost, sustainable, win–win and green production method that can reduce peanut shell's environmental pollution [12]. Pure biochar is effective at adsorbing organic pollutants and cations [13,14], but the P removal rate is low by biochar from unmodified peanut shell waste, and modified biochar by metal is usually chosen to achieve P adsorption [11,15]. Eggshells contain calcium carbonate (~94%) and organic matter (~6%) [16]. By 2018, the world had generated approximately 8.6 million metric tons of eggshell waste [17]. Additionally, eggshells have the potential to be a source of calcium for the creation of Ca-modified biochar, which has the ability to absorb P and then be used as a phosphate fertilizer [16].

The removal of P in wastewater from pig farm, cattle farm, pond and human urine by Ca-modified biochar has been reported [9,10]. So far, the reports on the P treatment of rural domestic sewage by Ca-modified biochar are limited. In this study, in order to improve TP treatment efficiency to meet discharge standards in rural regions, waste eggshell was used as a Ca source and peanut shell waste as a carbon (C) source to synthesize adsorbent material (E-C) by pyrolysis. Some adsorption models were used to observe the adsorption performance of P by E-C. The mechanisms of P adsorption on the prepared E-C were analyzed. The feasibility of E-C as an adsorbent for TP removal was also assessed using rural domestic sewage. This study combines agricultural waste recycling with phosphorus pollution control, and we provide a possibility scheme for effective phosphorus removal/recovery from rural domestic sewage.

2. Materials and Methods

2.1. Experimental Materials

The peanut shells were obtained locally in Jiaxing (China). Before usage, they were air-dried and then ground to pass through 40 mesh sieves (0.425 mm). Waste eggshells were obtained from a mess hall in Jiaxing. They were dried at 60 °C in an oven after being cleaned three times. Additionally, there were then ground to pass through 40 mesh sieves (0.425 mm) before use. All chemical reagents (such as KH_2PO_4, HCl, NaOH, KCl, KNO_3,

K_2SO_4 and $KHCO_3$) were of analytical grade and purchased from Aladdin Industrial Corporation (Shanghai, China).

2.2. Preparation of Adsorbents

The predetermined weights of the eggshell and peanut shell power were determined in a mass ratio of 1:1. Deionized water was then added to the mixture, and then the mixture was blended overnight with a magnetic mixer. The mixture was placed in a 105 °C oven to dry. The dried mixture was transferred to a quartz boat, and then the mixture was placed in the tube furnace and calcined for 2 h under a nitrogen atmosphere at 800 °C. The heating rate was set at 5 °C/min. A pure peanut shell prepared biochar was named BC.

2.3. Phosphorus Adsorption Experiments

In this study, all adsorption experiments were carried out in batches. E-C adsorbent materials were placed in a 100 mL glass vial with a stopper containing 40 mL of P solutions. The vials were oscillated (180 rpm) in a shaker bath (HZQ-F100 Taicang, China) at pH 7 and 25 °C. The E-C adsorbent dose was 250 mg/L. Then, the suspension was filtered through a 0.45 μm PES syringe filter, and the filtered solution was used to detect the phosphorus concentration. A total of 0.1 mol/L NaOH or HCl was used to adjust the initial pH values of the solutions. KH_2PO_4 was used as a phosphorus source in the solutions because the main form of phosphate in the range of 2.13–7.20 is $H_2PO_4^-$ [18]. All experiments were performed in triplicate.

To evaluate the equilibrium adsorption parameters, adsorption isotherm experiments were conducted as follows: E-C dose, 250 mg/L; initial P concentrations of 5, 10, 25, 50, 100, 150 and 200 mg/L; initial pH, 7.0; and adsorption time, 24 h.

Adsorption kinetic tests were conducted as follows: E-C dose, 250 mg/L; initial P concentration, 100 mg/L; initial pH, 7.0; and adsorption time, 1–1440 min.

The effect tests of common competing substances (Cl^-, NO_3^-, SO_4^{2-} and HCO_3^-) on P removal by E-C were evaluated as follows: E-C dose, 250 mg/L; initial phosphate concentration, 100 mg/L; Cl^-, NO_3^-, SO_4^{2-} and HCO_3^- concentrations, 100 and 1000 mg/L; and adsorption time, 24 h.

The effects of different initial pH values on P removal by E-C were evaluated as follows: E-C dose, 250 mg/L; initial P concentration, 100 mg/L; initial pH of 6.0, 7.0, 8.0 and 9.0; and adsorption time, 24 h.

Rural residential sewage from two centralized rural sewage treatment facilities in Jiaxing (Zhejiang province) were utilized to assess the E-C adsorbent's suitability for actual phosphorus sewage. The first centralized rural sewage treatment facility was the anaerobic–anoxic–oxic (A^2O) process (named as F1), and the design capacity was 50 t/d. The second centralized rural sewage treatment facility was an A^2O process (named as F2), and the design capacity was 20 t/d. Every three days, a sample of rural home sewage was taken from the secondary sedimentation tank. A total of 40 mL of rural home sewage was collected in 100 mL glass vials with stoppers containing 0.01 g of E-C adsorbent materials, and the vials were oscillated (at 180 rpm) in a shaker bath for 4 h at 25 °C.

All experiments were repeated three times. The equilibrium adsorption capacity (q_e, mg/g) and adsorption capacity at different times t (q_t, mg/g) was calculated as follows:

$$q_e = (C_0 - C_e)\ V/m \tag{1}$$

$$q_t = (C_0 - C_t)\ V/m \tag{2}$$

where C_0 (mg/L) represents the initial P concentration, C_e (mg/L) represents the P concentration at equilibrium, and C_t (mg/L) represents the P concentration at time t; V (L) is the volume of the solution; and m (g) is the adsorbent mass.

2.4. Characterization and Analytical Method

The concentration of phosphorus in solution was determined using ammonium molybdate spectral photometry using an ultraviolet spectrophotometer (SHIMADZU UV-2450, Kyoto, Japan) according to the standard method [11]. The surface elements and morphologies of adsorbents before and after adsorption of P were studied using scanning electron microscopy with energy dispersive X-ray spectroscopy (SEM-EDS) (Regulus-810, Hitachi, Tokyo, Japan). The diffraction patterns of adsorbents before and after P adsorption were studied using an X-ray diffractometer (XRD, Bruker D8 Advance X-ray diffractometer, Karlsruhe, Germany). An FTIR spectrometer (Thermo Scientific Nicolet IS50, Waltham, MA, USA) was used to study the characterization of the functional groups in the adsorbents before and after P adsorption. A fully automated physisorption instrument (JW-BK200B, Beijing, China) was used to detect the specific surface area, pore volume, and average pore size of the adsorbents.

3. Results and Discussion

3.1. Basic Physicochemical Properties of Biochar

By weighing the samples before and after pyrolysis at 800 °C, the resulting mass loss (49.3%) during the preparation of the E-C adsorbent material was achieved. The basic physicochemical properties and major elements of the E-C and BC adsorbent materials are shown in Table 1. BC showed a specific surface area of 9.203 m^2/g, average pore size of 2.95 nm, and pore volume of 0.083 cm^3/g; the corresponding values for E-C were 64.956 m^2/g, 6.46 nm and 0.105 cm^3/g, respectively.

Table 1. Specific surface area, pore volume, average pore size and major elements of E-C and BC.

Adsorbent	Physical Properties			Major Elements		
	Specific Surface Area (m^2/g)	Pore Volume (cm^3/g)	Average Pore Size (nm)	C (%)	O (%)	Ca (%)
E-C	64.956	0.105	6.46	33.23	45.1	21.3
BC	9.203	0.083	2.95	78.5	13.53	1.26

The eggshell decomposed into CO_2 and CaO during high-temperature pyrolysis [19]. The formation of CO_2 broadens the material's pore size as an activation substance, and it can therefore be used to produce activated carbon [20]. In addition, the adsorbent's pore volume and specific surface area may increase by adding eggshell to the biochar [21].

The pore volume and surface area of the E-C were higher than those of the BC. Biochar's adsorption performance largely depends on its porosity and specific surface area. The larger the adsorbent's specific surface area and higher the porosity, the more surface adsorption sites there will be and the higher the adsorption performance [10,22,23]. After 24 h of adsorption, the P adsorption capacity of BC was only 5.3 mg/g, but the E-C reached 191.1 mg/g. This showed that Ca from $CaCO_3$ in eggshells was successfully introduced into the E-C adsorbent materials.

3.2. Adsorption Kinetics

Figure 1 shows the kinetic behavior results of P onto the E-C adsorbent materials. Clearly, during the initial hour the adsorption processes of E-C for P was rapid, with the rate slowing down until reaching an equilibrium within 4 h. The adsorption process involves the initial rapid diffusion of ions from the solution onto the surface of the external adsorbent. This is then followed by a slow adsorption process through the diffusion of ions into the pores on the inner surfaces of the adsorbent [20].

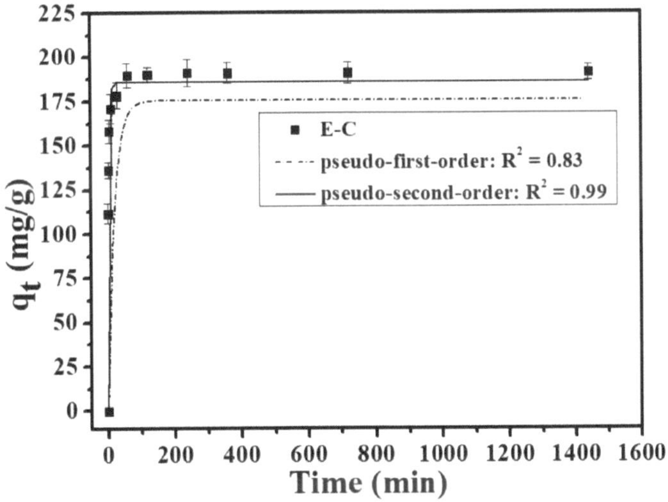

Figure 1. P adsorption kinetics onto E-C (dosage: 250 mg/L; temperature: 25 °C; initial P concentration: 100 mg/L; and pH = 7).

The pseudo-first-order model and pseudo-second-order model were used in the study of E-C adsorbent materials' adsorption kinetics for P in aqueous solutions:

$$\text{Pseudo-first-order kinetic equation: } q_t = q_e(1 - e^{-k_1 t}) \quad (3)$$

$$\text{Pseudo-second-order kinetic equation: } t/q_t = 1/k_2 q_e^2 + t/q_e \quad (4)$$

where q_t is the P adsorption amounts at time t, and q_e is the P adsorption amounts at equilibrium time; k_1 is the first-order kinetics' adsorption rate constant, and k_2 is the second-order kinetics' adsorption rate constant. The results showed that the coefficient of determination simulated with the pseudo-second-order model (R^2 = 0.99) is higher than the pseudo-first-order model (R^2 = 0.83) (Figure 1), which suggested that the pseudo-second-order kinetic model can well describe the adsorption kinetic process of P onto E-C, and a monolayer chemisorption occurred on the adsorbent surface. Based on the analysis of the pseudo-second-order theory and the chemisorption characteristics of E-C adsorbent materials, chemisorption was the main rate-limiting step of P adsorption by E-C [10,24].

3.3. Adsorption Isotherms

In this study, the isothermal experiments were carried out with different initial P concentrations (5, 10, 25, 50, 100, 150 and 200 mg/L). Langmuir and Freundlich adsorption isotherm models were used in the study of E-C adsorbent's adsorption isotherms for P in aqueous solutions:

$$\text{Langmuir: } q_e = K_L q_{max} C_e/(1 + K_L C_e) \quad (5)$$

$$\text{Freundlich: } q_e = K_F C_e^{1/n} \quad (6)$$

where q_e (mg/g) is the equilibrium adsorption amount, K_L (L/mg) is the Langmuir adsorption equilibrium constant, q_{max} (mg/g) represents the maximum adsorption capacity, C_e (mg/L) represents the P equilibrium concentration, K_F ($mg^{(1-1/n)}L^{1/n}/g$) represents the Freundlich adsorption constant representing the adsorption capacity of the adsorbent materials, and n represents an indication of linearity.

The fitted adsorption isotherms of P onto the E-C adsorbent materials are shown in Figure 2. The fitting effect of the Langmuir model on the experimental data is better than that of the Freundlich model in this study (Figure 2). The corresponding fitting parameters for the Langmuir model are as follows: q_{max} = 195.8 mg/g, K_L = 0.39 L/mg, and R^2 = 0.95. The corresponding fitting parameters for the Freundlich model are as follows: K_F = 47.3 $mg^{(1-1/n)}L^{1/n}/g$, 1/n = 0.32 L/mg, and R^2 = 0.91. Furthermore, the regression coefficient R^2 of the Langmuir model is higher than that of the Freundlich model. it is further confirmed that the Langmuir isotherm model adequately describes the experimental data of P adsorption by E-C, implying that the effective adsorption surface of E-C is homogeneous and a monolayer during the adsorption of P. this is consistent with the above adsorption kinetic results described by the pseudo-second-order model. Similar P adsorption behavior has been reported in other research of Ca-based adsorbents [10,25–28].

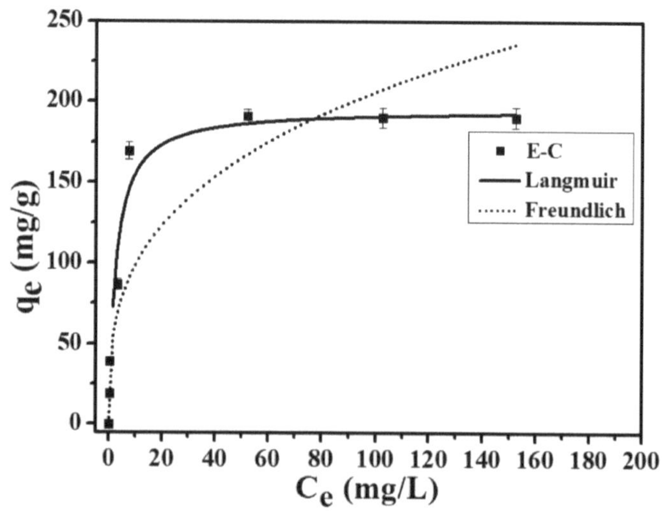

Figure 2. P adsorption isotherms onto the E-C (dosage: 250 mg/L; temperature: 25 °C; adsorption time: 24 h; and pH = 7).

The q_{max} of the E-C, through calculating the Langmuir isotherm, was 195.8 mg/g, which was consistent with the experimental maximum adsorption capacity of 191.1 mg/g, indicating that the E-C adsorbent materials in aqueous solution had excellent adsorption properties for removing P. The E-C adsorbent material's adsorption capacity for P was higher than that of the Ca-biochar adsorbent prepared from eggshell and peanut shells through ball milling and chemical impregnation methods (the maximum P adsorption capacity was 130.57 mg/g) reported by Liu et al. [28].

3.4. *Effect of Coexisting Anions*

In natural water and wastewater, some common anions are present that may compete with P for adsorption sites on the adsorbent [29]. Therefore, to further explore the adsorption ability of E-C, coexisting ions (i.e., KCl, $KHCO_3$, KNO_3 and K_2SO_4) were added into the adsorption system. These common anions have different levels of effect on the P adsorption capacity of the E-C adsorbent materials, following the order HCO_3^- > SO_4^{2-} > NO_3^- > Cl^- (Figure 3). The results showed that the anions NO_3^- and Cl^- had limited effect on the adsorption process of P; even for ion concentrations up to 1000 mg/L for Cl^- and NO_3^-, the adsorption capacity of P remained at 190.6 mg/g and 189.5 mg/g, respectively. The SO_4^{2-} anions had some effect on the adsorption process of P. As the concentration of SO_4^{2-} gradually increased from blank control to 100 mg/L and 1000 mg/L, the adsorption capacity of E-C decreased from 191.1 to 184.5 and 158.1 mg/g, respectively.

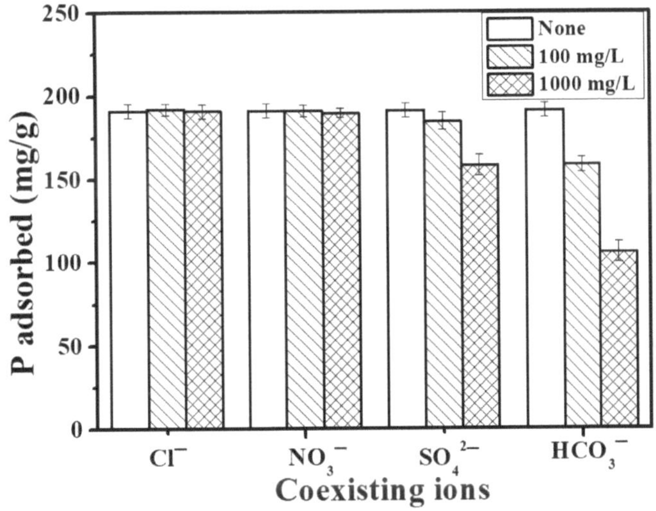

Figure 3. Effect of coexisting ions (Cl^-, NO_3^-, SO_4^{2-}, and HCO_3^-) on the P adsorption capacity of E-C (dosage: 250 mg/L; temperature: 25 °C; adsorption time: 24 h; initial P concentration: 100 mg/L).

On the other hand, HCO_3^- had a significant negative effect on the adsorption of P by E-C adsorbent materials. When the concentration of HCO_3^- gradually increased, P adsorption of the E-C adsorbent materials had a significant downward trend. HCO_3^- ionization in solution can generate CO_3^{2-}, and the adsorption sites on E-C are reduced because CO_3^{2-} will compete with P to bind Ca to produce a precipitate. Similar results have been reported in the study of other calcium-rich phosphate adsorbents [20,30]. However, when the concentration of HCO_3^- was as high as 1000 mg/L, P adsorption of E-C still remained at 105.8 mg/g. Thus, the E-C adsorbent materials have a higher adsorption selectivity for phosphate anions than for many other common anions. The above results show that the E-C adsorbent materials have significant potential in practical P-containing wastewater applications even in the presence of Cl^-, NO_3^-, SO_4^{2-}, and HCO_3^- common ions.

3.5. Influence of pH

The effect of pH is an important parameter impacting the absorption performance of a sorbent. Since the pH of rural domestic sewage is generally nearly 6–9, the effect of the initial solution pH (6–9) on the adsorption of P onto the E-C was examined. The P adsorption amount onto the E-C increased from 182.6 mg/g at pH = 6 to 191.1 mg/g at pH = 7 (Figure 4). Additionally, the P adsorption amount onto the E-C slightly decreased to 188.5 mg/g at pH = 8 and 186.7 mg/g at pH = 9 (Figure 4). Thus, the E-C adsorbent material showed a good adsorption effect on P under the initial pH range of 6–9.

In this study, calcium-active components contributed to the adsorption of P onto the E-C adsorbent material by strong chemical interaction mechanisms, resulting in the high adsorption capacity of the E-C for P over the pH range (6–9).

3.6. Adsorption Mechanisms

The adsorbent materials before and after P adsorption were analyzed by XRD, SEM and FTIR, and the adsorption mechanisms of P onto the E-C adsorption materials were discussed. Before and after P adsorption, the characteristic peaks of the BC had no significant change in the FTIR spectra analysis (Figure 5A). This was attributed to the fact that the ingredients in the BC did not react chemically with phosphates. The weak P adsorption property of the BC might be related to physical adsorption. Small amounts of P might be adsorbed into the pores of the BC. However, before and after P adsorption, the characteristic peaks of the E-C had a significant change in the FTIR spectra analysis (Figure 5B). The peaks at 872 cm^{-1} (C-O) and 1420 (C=C) were observed for the adsorbent before and after adsorption, suggesting that the C-O and C=C groups did not contribute to the adsorption of P. Before adsorption, the peak at 3641 cm^{-1} (O-H) was attributed to $Ca(OH)_2$ for the E-C [24]. After adsorption, the peak at 3641 cm^{-1} vanished, showing that the -OH took part in the chemical adsorption process of P. Furthermore, the peaks at 563, 601 and 1027 cm^{-1} (P-O) were significantly enhanced and attributed to phosphates, showing that P was adsorbed onto the E-C adsorbent materials. Similar results were reported by other researchers [9,20].

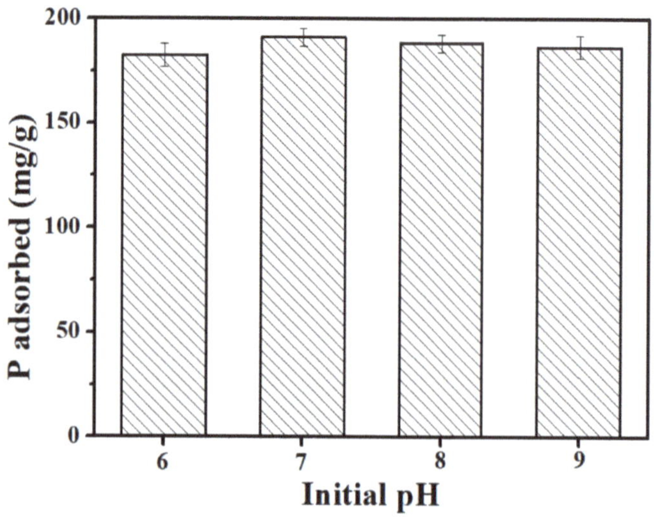

Figure 4. The effect of initial pH on the P adsorption capacity of the E-C (dosage: 250 mg/L; temperature: 25 °C; adsorption time: 24 h; and initial P concentration: 100 mg/L).

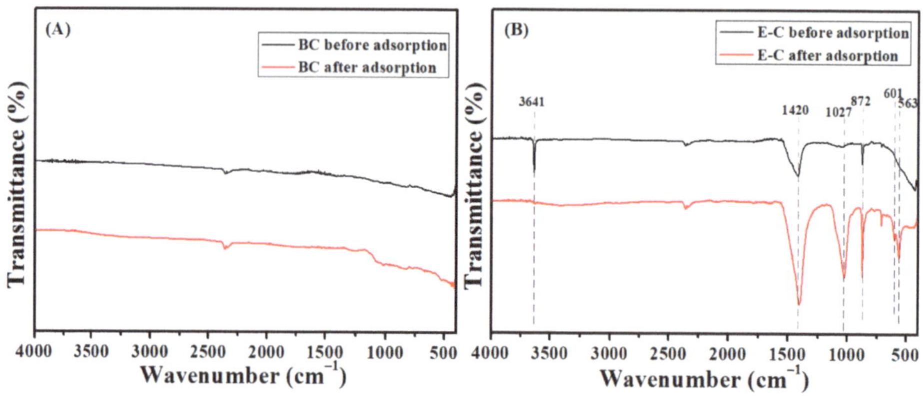

Figure 5. The FTIR spectra of BC (**A**) and E-C (**B**) before and after P adsorption.

The diffraction patterns of the E-C before and after P adsorption showed a significant change in the XRD spectra analysis (Figure 6). Before adsorption, the E-C had $CaCO_3$ characteristic peaks (PDF #05-0586, 2θ = 23°, 29.4°, 36°, 39.4°, 47.5°, 48.5°, and 57.4°) and $Ca(OH)_2$ characteristic peaks (PDF #04-0733, 2θ = 18.1°, 28.7°, 34.2°, 47.1°, 50.9°, 54.4°, 62.6°, 64.3°, 71.8°, and 84.8°). $Ca(OH)_2$ groups are the main active sites in the E-C adsorbent materials for P adsorption. $CaCO_3$ groups in eggshells were not fully decomposed at 800 °C.

In Lee's study, at a calcination temperature of 800 °C, the mineral composition of eggshell still contained 35.7% $CaCO_3$ [8].

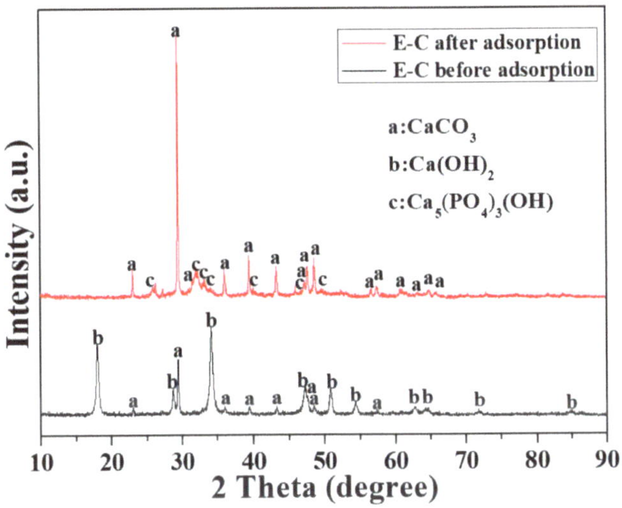

Figure 6. X-ray diffraction pattern of the E-C adsorbent materials before and after P adsorption.

After adsorption, $Ca(OH)_2$ diffraction peaks vanished, and the E-C had $Ca_5(PO_4)_3OH$ characteristic peaks (PDF #09-0432, 2θ = 25.9°, 29°, 32.2°, 39.8°, 46.7°, 49.5°, 53.2°, and 64.2°). The $Ca_5(PO_4)_3OH$, generated from the reaction of phosphate with $Ca(OH)_2$, was responsible for the removal of P from the water, consistent with the FTIR and kinetic findings. In addition, the $CaCO_3$ (PDF #05-0586) characteristic peaks still existed. It was pointed out that $CaCO_3$ did not participate in the chemical reaction with P. $CaCO_3$ might adsorb a small amount of P by physical adsorption.

Before and after P adsorption, the morphologies of the BC had no significant change though SEM images (Figure 7A,B). Before loading Ca, BC showed a porous structure, and SEM photography showed that the surface of the biochar was smooth, which might be related to the morphology of the biomass itself. The morphologies of the E-C adsorbent materials before and after P adsorption had a significant change when looking at the SEM images (Figure 7C,D). Before P adsorption, the surface of the E-C was relatively clean, but many impurity particles appeared on the surface. These particles were $CaCO_3$ and $Ca(OH)_2$. In addition, the E-C adsorbent materials had a porous structure before adsorption. After adsorption of P, a lot of flocculent precipitates appeared on the materials' surfaces, and the pores of the E-C were blocked. The above results show that during the adsorption of P onto the E-C adsorbent materials, a good deal of flocculent precipitates were generated on the surface of the E-C adsorbent materials and in the pores. Therefore, the porosity and effective active sites of the E-C were reduced, and finally the adsorption saturation of the E-C adsorbent materials was induced. From Figure 7E, after the adsorption of P, the phosphorus element was distributed on the E-C. Therefore, combined with Figure 6, it can be inferred that the precipitates on the E-C should be $Ca_5(PO_4)_3OH$.

According to the above results, the mechanisms of the E-C to adsorb P include physical adsorption, surface precipitation, and electrostatic attraction. The dominating sorption mechanism of the E-C for P is that $Ca(OH)_2$ reacts with phosphate to form $Ca_5(PO_4)_3OH$ precipitates, and this result is similar to other studies [9,25,31].

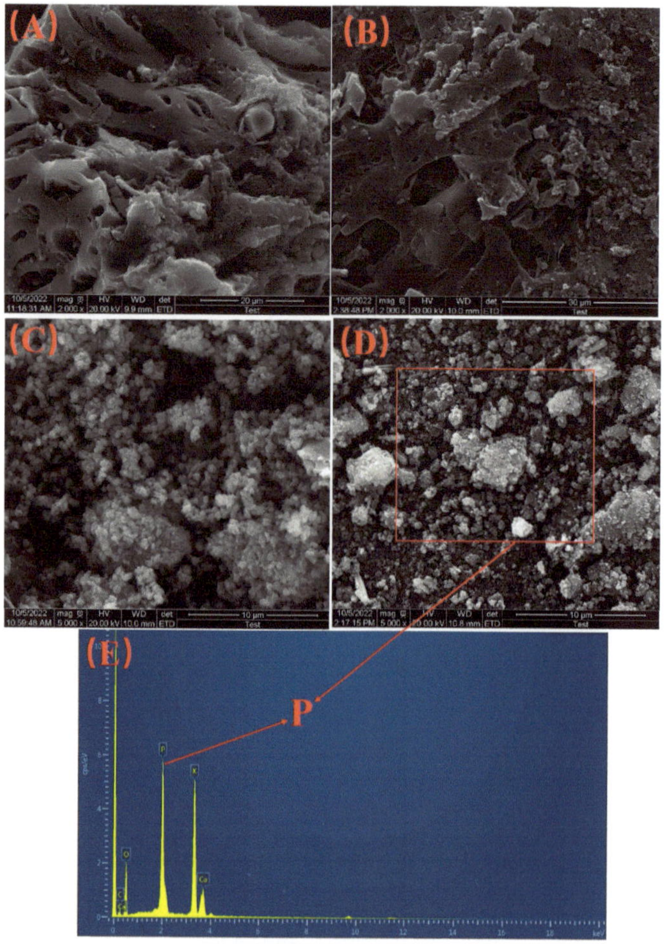

Figure 7. SEM photographs of the BC (**A**) before and (**B**) after P adsorption; E-C (**C**) before and (**D**) after P adsorption; (**E**) EDS image of the E-C after P adsorption.

3.7. Removal of Phosphate by E-C from Rural Domestic Sewage

It was observed from Figure 8 that the discharge of total phosphorus (TP) in rural domestic sewage of two centralized rural sewage treatment facilities in all sampling batches exceeded the second-grade Discharge Standard (DB33/973-2021) (TP < 3 mg/L) [32] in this study. The discharge of NH_3-N, COD and pH in rural sewage of F1 was 12.1–22.8 mg/L, 42–85 mg/L and 6.8–7.2, respectively. The discharge of NH_3-N, COD and pH in rural sewage of F2 was 6.8–19.1 mg/L, 35–76 mg/L and 6.7–7.2, respectively. The discharge of NH_3-N, COD and pH in rural sewage of two centralized rural sewage treatment facilities met the second-grade Discharge Standard (DB33/973-2021) (COD < 100 mg/L, NH_3-N < 25 mg/L, and pH 6~9) [32]. Phosphorus becomes the main limiting factor for reaching the discharge standard.

Figure 8 presents the results of phosphate adsorption by adding 0.01 g of E-C adsorbent materials to 40 mL of actual rural sewage. The TP content in the rural domestic sewage of F1 was 3.85–11.3 mg/L (Figure 8A), the TP content was 0.18–0.85 mg/L after adsorption, and the TP removal rate reached 92.5–95.2%. After adsorption treatment, the discharge of TP in rural sewage from all the sampling batches met the second-grade Discharge

Standard (DB33/973-2021) (TP < 3 mg/L), even meeting the first-grade Discharge Standard (DB33/973-2021) (TP < 2 mg/L) [32].

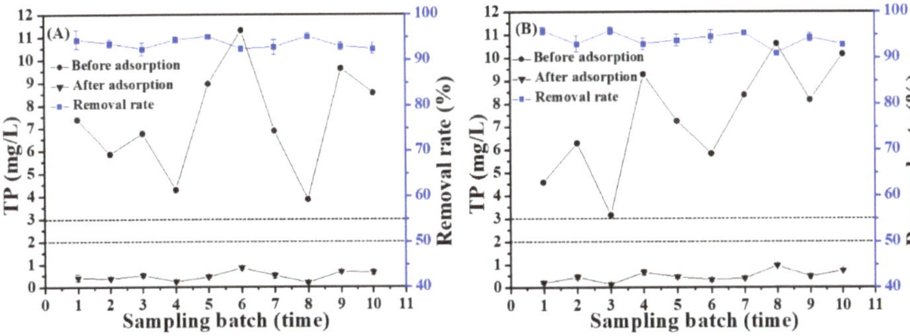

Figure 8. The removal rate of TP in rural domestic sewage by E-C, (**A**) F1 and (**B**) F2 (dosage: 250 mg/L; temperature: 25 °C; and adsorption time: 4 h).

The TP content in the rural domestic sewage of F2 was 3.15–10.6 mg/L (Figure 8B), the TP content was 0.13–0.95 mg/L after adsorption, and the TP removal rate reached 91–95.9%. After adsorption treatment, the discharge of TP in rural sewage from all the sampling batches met the second-grade Discharge Standard (DB33/973-2021) (TP < 3 mg/L), even meeting the first-grade Discharge Standard (DB33/973-2021) (TP < 2 mg/L).

Many complex components may be present in actual wastewater which can interfere with P removal by the adsorbent. In this study, the E-C adsorbent materials also exhibited a high P adsorption capacity from actual rural sewage. Therefore, the E-C adsorption materials have great potential in the treatment of large-scale rural domestic sewage containing P.

The recovery of adsorbents is an important economic parameter. However, the operation procedure of recycling P adsorbents is very complicated [21]. The main components of the E-C biochar materials are O, C, Ca and H; these elements are environmentally friendly, and the adsorbents have a mesoporous structure. After adsorption of P, a good deal of P-rich $Ca_5(PO4)_3(OH)$ is produced on the E-C adsorbent materials; furthermore,, it also has the physical properties of biochar. Therefore, the P-adsorbed E-C adsorbent materials may be further used as soil regulators and fertilizer. This method can realize the reuse of E-C and realizes the virtuous cycle of P resources in the ecosystem [9,20,30,33].

4. Conclusions

In this study, Ca-modified biochar (E-C) adsorbent materials were prepared utilizing eggshell and peanut shell waste as raw materials. The E-C adsorbent materials were used to remove P from aqueous solutions, and the results show that the prepared E-C adsorbent materials exhibited an excellent adsorption capacity over the initial pH range of 6–9 and adsorption selectivity. An experimental maximum adsorption capacity of 191.1 mg/g was obtained by the E-C sample that was prepared with a mass ratio of 1:1 and a pyrolysis temperature of 800 °C. The dominant sorption mechanism for phosphate was due to $Ca(OH)_2$ on the E-C adsorption materials reacting with phosphate to generate $Ca_5(PO_4)_3OH$ precipitate. E-C effectively removed P of rural domestic sewage. The removal rate of TP in rural domestic sewage was 91–95.9%. After adsorption treatment, the discharge of TP in rural sewage met second-grade Discharge Standard (DB33/973-2021) (TP < 3 mg/L), and even met the first-grade Discharge Standard (DB33/973-2021) (TP < 2 mg/L). This study provides the experimental basis for the efficient removal of phosphorus by E-C adsorbent materials, and it provides the possibility for its application in rural domestic sewage.

Author Contributions: C.X. conducted the analysis of the data, data modeling, and wrote the paper, R.L., L.C. and Q.W. contributed to the writing of the paper, R.L. contributed to design of the study, L.C contributed to interpretate data for the work, and Q.W. contributed to review it critically for important intellectual content. All authors have read and agreed to the published version of the manuscript.

Funding: This research was supported by the National Key Research and Development Program of China (No. 2020YFD1100103-2), Leading Talent of the Science & Technology Nova Program of Zhejiang (2020R52039), and Outstanding Innovative Team Supporting Plan of Jiaxing City.

Institutional Review Board Statement: Not applicable.

Informed Consent Statement: Not applicable.

Data Availability Statement: Not applicable.

Conflicts of Interest: The authors declare no conflict of interest.

References

1. Hu, S.A.; He, J.X. The willingness to household waste disposal practices of residents in rural China. *J. Mater. Cycles Waste* **2022**, *24*, 1124–1133. [CrossRef]
2. Wang, S.Y.; Hu, C.Q.; Cheng, F.K.; Lu, X.W. Performance of a combined low-consumption biotreatment system with cost-effective ecological treatment technology for rural domestic sewage treatment. *J. Water Process Eng.* **2023**, *51*, 103380. [CrossRef]
3. Gonzalez-Tineo, P.A.; Durán-Hinojosa, U.; Delgadillo-Mirquez, L.R.; Meza-Escalante, E.R.; Gortáres-Moroyoqui, P.; Ulloa-Mercado, R.G.; Serrano-Palacios, D. Performance improvement of an integrated anaerobic-aerobic hybrid reactor for the treatment of swine wastewater. *J. Water Process Eng.* **2020**, *34*, 101164. [CrossRef]
4. Yang, F.L.; Zhang, H.R.; Zhang, X.Z.; Zhang, Y.; Li, J.H.; Jin, F.M.; Zhou, B.X. Performance analysis and evaluation of the 146 rural decentralized wastewater treatment facilities surrounding the Erhai Lake. *J. Clean. Prod.* **2021**, *315*, 128159. [CrossRef]
5. Shi, Y.; Wang, X.L.; Qing, Z.X.; Song, Y.M.; Min, J.; Zhou, Y.N.; Du, J.; Wang, S.F. Using Iron Tailings for Phosphate Removal in Cemented Phosphogypsum (PG) Backfill. *Materials* **2022**, *15*, 8497. [CrossRef]
6. Soo, E.; Domoń, A.; Papciak, D.; Michel, M.M.; Cieniek, B.; Pajk, D. Characteristics of the Properties of Absodan Plus Sorbent and Its Ability to Remove Phosphates and Chromates from Aqueous Solutions. *Materials* **2022**, *15*, 3540. [CrossRef] [PubMed]
7. Jung, K.W.; Jeong, T.U.; Kang, H.J.; Chang, J.S.; Ahn, K.H. Preparation of modified-biochar from Laminaria japonica: Simultaneous optimization of aluminum electrode-based electro-modification and pyrolysis processes and its application for phosphate removal. *Bioresour. Technol.* **2016**, *214*, 548–557. [CrossRef]
8. Lee, J.I.; Kim, J.M.; Yoo, S.C.; Jho, E.H.; Lee, C.G.; Park, S.J. Restoring phosphorus from water to soil: Using calcined eggshells for P adsorption and subsequent application of the adsorbent as a P fertilizer. *Chemosphere* **2022**, *287*, 132267. [CrossRef]
9. Deng, W.D.; Zhang, D.Q.; Zheng, X.X.; Ye, X.Y.; Niu, X.J.; Lin, Z.; Fu, M.L.; Zhou, S.Q. Adsorption recovery of phosphate from waste streams by Ca/Mg biochar synthesis from marble waste, calcium-rich sepiolite and bagasse. *J. Clean. Prod.* **2021**, *288*, 125638. [CrossRef]
10. Feng, M.H.; Li, M.M.; Zhang, L.S.; Luo, Y.; Zhao, D.; Yuan, M.Y.; Zhang, K.Q.; Wang, F. Oyster Shell Modified Tobacco Straw Biochar: Efficient Phosphate Adsorption at Wide Range of pH Values. *Int. J. Environ. Res. Public Health* **2022**, *19*, 7227. [CrossRef]
11. Pan, W.L.; Xie, H.M.; Zhou, Y.; Wu, Q.Y.; Zhou, J.Q.; Guo, X. Simultaneous adsorption removal of organic and inorganic phosphorus from discharged circulating cooling water on biochar derived from agricultural waste. *J. Clean. Prod.* **2023**, *383*, 135496. [CrossRef]
12. Liu, X.Q.; Zhou, W.; Feng, L.; Wu, L.L.; Lv, J.L.; Du, W. Characteristics and mechanisms of phosphorous adsorption by peanut shell-derived biochar modified with magnesium chloride by ultrasonic-assisted impregnation. *ACS Omega* **2022**, *7*, 43102–43110. [CrossRef]
13. Li, X.Y.; Liu, T.L.X.; Han, X.; Li, Y.L.; Ma, X.L. Removal of heavy metals lead and ciprof-loxacin from farm wastewater using peanut shell biochar. *Environ. Technol. Inno.* **2023**, *30*, 103121. [CrossRef]
14. Kushwaha, R.; Singh, R.S.; Mohan, D. Comparative study for sorption of arsenic on peanut shell biochar and modified peanut shell biochar. *Bioresour. Technol.* **2023**, *375*, 128831. [CrossRef]
15. Qian, J.; Zhou, X.Y.; Cai, Q.S.; Zhao, J.J.; Huang, X.H. The study of optimal adsorption conditions of phosphate on Fe-modified biochar by response surface methodology. *Molecules* **2023**, *28*, 2323. [CrossRef] [PubMed]
16. Liu, D.D.; Hao, Z.K.; Chen, D.Q.; Jiang, L.P.; Li, T.Q.; Tian, B.; Yan, C.Q.; Luo, Y.; Chen, G.; Ai, H.F. Use of eggshell-catalyzed biochar adsorbents for Pb removal from aqueous solution. *ACS Omega* **2022**, *7*, 21808–21819. [CrossRef]
17. Waheed, M.; Yousaf, M.; Shehzad, A.; Inam-Ur-Raheem, M.; Kashif Iqbal Khan, M.; Rafiq Khan, M.; Ahmad, N.; Abdullah; Muhammad, A.R. Channelling eggshell waste to valuable and utilizable products: A comprehensive review. *Trends Food Sci. Tech.* **2020**, *106*, 78–90. [CrossRef]
18. Xie, F.Z.; Wu, F.C.; Liu, G.J.; Mu, Y.S.; Feng, C.L.; Wang, H.H.; Giesy, J.P. Removal of phosphate from eutrophic lakes through adsorption by in situ formation of magnesium hydroxide from diatomite. *Environ. Sci. Technol.* **2014**, *48*, 582–590. [CrossRef]

19. Panagiotou, E.; Kafa, N.; Koutsokeras, L.; Kouis, P.; Nikolaou, P.; Constantinides, G.; Vyrides, L. Turning calcined waste egg shells and wastewater to Brushite: Phosphorus adsorption from aqua media and anaerobic sludge leach water. *J. Clean. Prod.* **2018**, *78*, 419–428. [CrossRef]
20. Li, C.; Zhou, S.X.; Li, Q.Y.; Gao, G.M.; Zhang, L.J.; Zhang, S.; Huang, Y.; Ding, K.; Hu, X. Activation of sawdust with eggshells. *J. Anal. Appl. Pyrol.* **2023**, *171*, 105968. [CrossRef]
21. Liu, X.N.; Shen, F.; Qi, X.H. Adsorption recovery of phosphate from aqueous solution by CaO-biochar composites prepared from eggshell and rice straw. *Sci. Total Environ.* **2019**, *666*, 694–702. [CrossRef] [PubMed]
22. Bacelo, H.; Pintor, A.; Santos, S.; Boaventura, R.A.R.; Botelho, C.M.S. Performance and prospects of different adsorbents for phosphorus uptake and recovery from water. *Chem. Eng. J.* **2020**, *381*, 122566. [CrossRef]
23. Liu, M.C.; Wang, C.Z.; Guo, J.B.; Zhang, L.H. Removal of phosphate from wastewater by lanthanum modified bio-ceramisite. *J. Environ. Chem. Eng.* **2021**, *9*, 106123. [CrossRef]
24. Mitrogiannis, D.; Psychoyou, M.; Baziotis, I.; Inglezakis, V.J.; Koukouzas, N.; Tsoukalas, N.; Palles, D.; Kamitsos, E.; Oikonomou, G.; Markou, G. Removal of phosphate from aqueous solutions by adsorption onto $Ca(OH)_2$ treated natural clinoptilolite. *Chem. Eng. J.* **2017**, *320*, 510–522. [CrossRef]
25. Wang, S.D.; Kong, L.J.; Long, J.Y.; Su, M.H.; Diao, Z.H.; Chang, X.Y.; Chen, D.Y.; Song, G.; Shih, K. Adsorption of phosphorus by calcium-flour biochar: Isotherm, kinetic and transformation studies. *Chemosphere* **2018**, *195*, 666–672. [CrossRef]
26. Kong, L.J.; Han, M.N.; Shih, K.; Su, M.H.; Diao, Z.H.; Long, J.Y.; Chen, D.Y.; Hou, L.A.; Peng, Y. Nano-rod Ca-decorated sludge derived carbon for removal of phosphorus. *Environ. Pollut.* **2018**, *233*, 698–705. [CrossRef] [PubMed]
27. Sarker, P.; Liu, X.; Hata, N.; Takeshita, H.; Miyamura, H.; Maruo, M. Thermally modified bamboo-eggshell adsorbent for phosphate recovery and its sustainable application as fertilizer. *Environ. Pollut.* **2023**, *231*, 115992. [CrossRef]
28. Liu, X.Q.; Lv, J.L. Efficient Phosphate Removal from Wastewater by Ca-Laden Biochar Composites Prepared from Eggshell and Peanut Shells: A Comparison of Methods. *Sustainability* **2023**, *15*, 1778. [CrossRef]
29. Zhuo, S.N.; Dai, T.C.; Ren, H.Y.; Liu, B.F. Simultaneous adsorption of phosphate and tetracycline by calcium modified corn stover biochar: Performance and mechanism. *Bioresour. Technol.* **2022**, *359*, 127477. [CrossRef]
30. Liu, X.N.; Shen, F.; Smith Jr, R.L.; Qi, X.H. Black liquor-derived calcium-activated biochar for recovery of phosphate from aqueous solutions. *Bioresour. Technol.* **2019**, *294*, 122198. [CrossRef] [PubMed]
31. Rietra, R.; Hiemstra, T.; Riemsdijk, W. Interaction between calcium and phosphate adsorption on goethite. *Environ. Sci. Technol.* **2001**, *35*, 3369–3374. [CrossRef] [PubMed]
32. *DB33/973-2021*; Discharge Standard of Water Pollutants for Centralized Rural Sewage Treatment Facilities. The People's Government of Zhejiang Province: Hangzhou, China, 2021. (In Chinese)
33. Xu, Y.; Liao, H.; Zhang, J.; Lu, H.J.; He, X.H.; Zhang, Y.; Wu, Z.B.; Wang, H.Y.; Lu, M.H. A Novel Ca-Modified Biochar for Efficient Recovery of Phosphorus from Aqueous Solution and Its Application as a Phosphorus Biofertilizer. *Nanomaterials* **2022**, *12*, 2755. [CrossRef] [PubMed]

nanomaterials

Article

Multiscale Porous Carbon Materials by In Situ Growth of Metal–Organic Framework in the Micro-Channel of Delignified Wood for High-Performance Water Purification

Youngho Jeon [1], Dabum Kim [1], Suji Lee [1], Kangyun Lee [1], Youngsang Ko [1], Goomin Kwon [1], Jisoo Park [2], Ung-Jin Kim [1], Sung Yeon Hwang [1], Jeonghun Kim [2,*] and Jungmok You [1,*]

1 Department of Plant & Environmental New Resources, Kyung Hee University, 1732 Deogyeong-daero, Giheung-gu, Yongin-si 17104, Republic of Korea; crew75@khu.ac.kr (S.Y.H.)

2 Department of Chemical and Biomolecular Engineering, Yonsei University, 50 Yonsei-ro, Seodaemun-gu, Seoul 03722, Republic of Korea

* Correspondence: jhkim03@yonsei.ac.kr (J.K.); jmyou@khu.ac.kr (J.Y.)

Abstract: Porous carbon materials are suitable as highly efficient adsorbents for the treatment of organic pollutants in wastewater. In this study, we developed multiscale porous and heteroatom (O, N)-doped activated carbon aerogels (CAs) based on mesoporous zeolitic imidazolate framework-8 (ZIF-8) nanocrystals and wood using 2,2,6,6-tetramethylpiperidine-1-oxyl (TEMPO) oxidation, in situ synthesis, and carbonization/activation. The surface carboxyl groups in a TEMPO-oxidized wood (TW) can provide considerably large nucleation sites for ZIF-8. Consequently, ZIF-8, with excellent porosity, was successfully loaded into the TW via in situ growth to enhance the specific surface area and enable heteroatom doping. Thereafter, the ZIF-8-loaded TW was subjected to a direct carbonization/activation process, and the obtained activated CA, denoted as ZIF-8/TW-CA, exhibited a highly interconnected porous structure containing multiscale (micro, meso, and macro) pores. Additionally, the resultant ZIF-8/TW-CA exhibited a low density, high specific surface area, and excellent organic dye adsorption capacity of 56.0 mg cm^{-3}, 785.8 m^2 g^{-1}, and 169.4 mg g^{-1}, respectively. Given its sustainable, scalable, and low-cost wood platform, the proposed high-performance CA is expected to enable the substantial expansion of strategies for environmental protection, energy storage, and catalysis.

Keywords: multiscale porous carbon; tempo oxidation; water purification; wood microchannel

Citation: Jeon, Y.; Kim, D.; Lee, S.; Lee, K.; Ko, Y.; Kwon, G.; Park, J.; Kim, U.-J.; Hwang, S.Y.; Kim, J.; et al. Multiscale Porous Carbon Materials by In Situ Growth of Metal–Organic Framework in the Micro-Channel of Delignified Wood for High-Performance Water Purification. *Nanomaterials* **2023**, *13*, 2695. https://doi.org/10.3390/nano13192695

Academic Editor: Marco Stoller

Received: 11 September 2023
Revised: 27 September 2023
Accepted: 27 September 2023
Published: 3 October 2023

1. Introduction

With the rapid development of human society, the world has become faced with severe environmental pollution issues caused by hazardous organic chemicals, such as dyestuffs, oils, and organic solvents [1–3]. Particularly, industrial waste affects biodiversity, ecosystems, and human health through soil and water contamination. Organic dye chemicals are widely used in essential industries, such as dye, plastic, textile, paper, pharmaceutical, and food [4–9], and although most toxic organic dyes are filtered from industrial waste using water purification technology, their leakage into water bodies, even at low concentrations, can lead to serious environmental problems and hazards to human health. Particularly, their non-biodegradability in water, thermal stability, and use in everyday life increase their harmful effects [10–12]. Thus, wastewater treatment is essential to protect ecosystems and humans from harmful water pollutants. To prevent an increase in the environmental hazards caused by organic waste, various wastewater treatment approaches, such as coagulation, filtration, precipitation, adsorption, ion exchange, and advanced oxidation processes, have been employed for the removal of organic dyes from water systems [13–20]. Among them, the most common technique is adsorption based on the use of adsorbents (e.g., zeolite, clay, and silica) with high specific surface areas [21–26]. Adsorption is a relatively simple, effective, and economical technique (based on materials and operation cost and the minimal production of secondary pollutants). Thus, it is highly desirable to develop a high-performance absorbent for treating organic pollutants.

Carbon aerogels (CAs), which exhibit a large surface area, low density, and high porosity, have recently attracted considerable attention in environmental treatment, energy storage, catalysis, and sensors fields [27–30]. Over the past decades, CAs have been fabricated using (1) nanocarbon units (carbon nanotubes, carbon nanofibers, and graphene nanosheets) and (2) polymerized organic gels derived from precursors [31–35]. However, despite their intrinsic benefits, their high manufacturing cost and complex manufacturing processes, which hinder their large-scale commercial production, have limited their further application. Furthermore, most precursors used for the fabrication of CAs are considered to be carcinogenic to humans. To address these limitations, several recent studies have reported the high adsorption capability of biomass-derived CAs for organic pollutants [36–39]. For example, Yamauchi et al. reported the large-scale synthesis of biomass polymer-derived CAs using a simple and sustainable method as a high-performance adsorbent [40]. They demonstrated that biomass material is a suitable building block for CAs owing to its natural abundance, sustainability, and cost-effectiveness. Additionally, they demonstrated the enhanced surface area and mechanical properties of nanoparticle-embedded CAs, thereby expanding the applications of CAs.

Wood, one of the most abundant natural biomasses, exhibits a hierarchical structure with parallel hollow tubes comprising cellulose nanofibers and an amorphous matrix of lignin and hemicellulose [41–43]. As a biomass material, wood exhibits multiscale porous structures ranging from the macro to nano levels [44]. Macroscopic pores, such as cell lumen, resin channels, and pit apertures, as well as mesoporous pores, such as pores in the cell wall, can provide large specific surface areas and strong adsorption capacities. This hierarchically porous structure along with its natural abundance makes it an ideal template for the development of functional green nanocomposites [45]. Metal–organic frameworks (MOFs) are crystalline compounds, containing three-dimensional (3D) networks of metal ions and organic linkers, with tunable porosity [46,47]. Recently, MOFs have been extensively investigated for separation, storage, catalysis, and sensing applications, owing to their high specific surface area, large porosity (micropores/mesopores), and low density [48–57]. However, despite the innovative utilization of MOFs, most MOFs are produced in the form of crystals and powders, which limit their use in real-world applications. Recently, several studies have demonstrated the successful synthesis of wood/MOF composite adsorbents by the in situ synthesis of MOFs within the aligned wood channels [36,37,58,59]. For example, He et al. developed ZIF-67@wood aerogel via the in situ wet chemical deposition of ZIF-67 on the wood aerogel for the adsorptive purification of polluted wastewater [36]. However, the lack of continuous nucleation sites in natural wood templates results in a low mass loading and poor uniformity of MOFs, which significantly affect the adsorption efficiency of these adsorbents. To the best of our knowledge, there is only one report (Zhou et al. described a foam-like TEMPO-oxidized wood/MOF aerogels for efficient CO_2 capture) describing the use of a TEMPO-oxidized wood template for the in situ growth of MOFs [60].

Herein, we describe the development of high-performance activated CAs with a hierarchical pore structure ranging from the micro to meso- and macroscale by harnessing the advantages of wood and MOFs. To the best of our knowledge, this is the first report on a wood/ZIF-8-derived CA for the efficient removal of organic pollutants. In this study, a 2,2,6,6-tetramethylpiperidine-1-oxyl (TEMPO)-oxidized wood (TW) template with surface carboxyl groups was utilized as an important building block for the CAs. The surface carboxyl groups in the TW enabled the coordination of the cellulose network with multivalent metal ions, resulting in the efficient nucleation and in situ growth of MOFs within the wood template. Additionally, ZIF-8 with a high specific surface area and nitrogen content was synthesized in situ inside the TW, after which the wood was subjected to a freeze-drying process to obtain ZIF-8/TW-A. Subsequently, the ZIF-8/TW-A was subjected to a direct carbonization/activation process to obtain a highly interconnected multiscale porous and

O, N-doped CA (denoted by ZIF-8/TW-CA), where wood with a hierarchical cellular structure served as a template, and ZIF-8 provided micro and mesoporous networks. The resultant ZIF-8/TW-CA exhibited a low density, high specific surface area, and outstanding adsorption capacities for organic pollutants.

2. Materials and Methods

2.1. Materials

Balsa wood (Ochroma pyramidale) was purchased from Balsanara Co. (Chungju, Chungcheong-do, Republic of Korea). In addition, 6% sodium hypochlorite (NaClO) solution, methanol, *tert*-butanol, ethanol, *n*-hexane, chloroform, sodium hydroxide (NaOH), and hydrochloric acid (HCl) were purchased from Duksan Pure Chemicals Co. Ltd., Ansan, Gyeonggi-do, Republic of Korea. Zinc nitrate ($Zn(NO_3)_2{\cdot}6H_2O_6$), TEMPO, Sudan III, rhodamine B (RhB), phosphate-buffered saline (PBS), and sodium chlorite ($NaClO_2$) were purchased from Sigma-Aldrich, and 2-methyl imidazole (2-Melm) was purchased from TCI Co., Chou-ku, Tokyo, Japan).

2.2. Preparation of the TEMPO-Oxidized Wood

First, the balsa wood was cut into the desired size and immersed in 6% NaClO solution with vacuum impregnation for 24 h. After delignification, the delignified wood (DW) was washed with deionized water (DIW), and the lignin contents of the DW were measured using the Klason method. Briefly, the dry mass (0.3 g, M_0) was measured and treated with H_2SO_4 (72%, 3 mL) for 1 h at 30 °C. Thereafter, the mixtures were transferred to a beaker, diluted to 4% by mass of H_2SO_4 by adding 84 mL of DIW, and then boiled in an autoclave (121 °C, 1 h). After cooling, the mixtures were filtered using a glass filter membrane and washed with 250 mL of DIW water. The insoluble materials were dried and weighed (M_d), and the lignin content was calculated as follows: $(M_d/M_0) \times 100\%$. For TEMPO oxidation, the DW was immersed in TEMPO/NaClO/ $NaClO_2$ (0.016 g/1.25 mL/1.13 g) dissolved in PBS solution (100 mL, 0.1 M, pH 7.4) at 60 °C for 48 h. After the TEMPO oxidation, the TWs were washed in an ethanol/DIW mixture (1:1, *v*/*v*) and stored in methanol.

2.3. Preparation of ZIF-8/TW-A

ZIF-8/TW-A was synthesized according to a previously reported procedure [61]. First, the TW was immersed for 8 h with vacuum infiltration to achieve ion exchange between Zn and Na ions in $Zn(NO_3)_2{\cdot}6H_2O_6$ solution, which was prepared by dissolving 2.4 g of $Zn(NO_3)_2{\cdot}6H_2O$ in 20 g of methanol and 3 g of DIW. Subsequently, the 2-Melm solution (13.2 g of 2-Melm in 20 g of methanol and 3 g of DIW) was poured into the aforementioned solution. After 24 h, the ZIF-8/TW was washed with methanol, and the solvent was exchanged with tert-butanol. Thereafter, the ZIF-8/TW was subjected to freeze-drying.

2.4. Preparation of ZIF-8/TW-CA

First, the ZIF-8/TW-A composites were carbonized at 250 °C for 1 h in a tube furnace to remove water from the sample. Thereafter, the temperature was increased to 850 °C for 3 h in an N_2 atmosphere at a heating rate of 5 °C/min. After carbonization, the samples were further heated to 350 °C for 1 h in air at a heating rate of 5 °C/min. For comparison, ZIF-8/DW-CA samples were prepared without TEMPO oxidation, using the same process as that used for ZIF-8/TW-CA.

2.5. Characterizations

The morphology of the ZIF-8/wood-based CA was investigated using field emission scanning electron microscopy (FE-SEM; Carl Zeiss, model AURIGA, Oberkochen, Germany), and elemental analysis was performed using energy-dispersive X-ray spectroscopy (EDS). The thickness of the samples was measured using a digital caliper (Mitutoyo Co., model 2109S-10, Kawasaki, Japan). The argon adsorption isotherms were analyzed using a Micromeritics ASAP 2010 micropore system and calculated using the Brunauer–Emmett–

Teller (BET) and density functional theory (DFT) method. Fourier transform infrared (FT-IR) spectroscopy data were recorded from 4000 to 400 cm^{-1} using a Spectrum One System FT-IR spectrometer (Perkin-Elmer, Waltham, MA, USA). The chemical composition of the surface elements from 0 to 1200 eV was investigated using X-ray photoelectron spectroscopy (XPS; Thermo Electron model, K-Alpha, Waltham, MA, USA), and the crystallinity of the samples was analyzed using X-ray diffraction (XRD; Bruker, model D8 Advance, Billerica, USA) in the scanning range of 3−60°. Raman spectra were recorded using a spectrometer (QE-PRO, Ocean Optics, Inc., OR, USA) equipped with a 785 nm (22 mW) laser. Thermogravimetric analysis (TGA) was performed using a Discovery TGA (TA Instruments, New Castle, USA) under nitrogen atmosphere at a heating rate of 10 °C min^{-1}. Ultraviolet–visible (UV–VIS) curves were obtained using a UV–VIS spectrophotometer (Shimadzu, model UV-3600 Plus, Nishinokyo, Japan).

2.6. Adsorption Studies

The adsorption properties of ZIF-8/TW-CA were investigated via batch adsorption experiments with RhB. All experiments were performed by immersing 10 mg of the adsorbent samples in 10 mL of each concentration (10–1000 mg L^{-1}, pH 7) of RhB under stirring using a shaker at 80 rpm for 24 h. Thereafter, 1 M HCl and 1 M NaOH were added to the dye solutions to adjust the pH. The concentration of the solution after adsorption was analyzed using a UV–VIS spectrophotometer by measuring the prominent wavelength at 553 cm^{-1}. The adsorption spectra and standard calibration curves of the different concentrations of the RhB solution are illustrated in Figure S4. To evaluate the recyclability of the CA, the adsorbed adsorbent was rinsed thrice in a solution mixture containing ethanol and DIW (1:1, v/v) to desorb RhB. After washing, the sample was immersed in 10 mg L^{-1} of RhB again. The adsorption capacity (q_e) for RhB and the removal efficiency (%) were calculated as follows:

$$q_e = (C_0 - C_e)V/m, \tag{1}$$

$$Removal\ efficiency\ (\%) = (C_0 - C_e)/C_0 \times 100, \tag{2}$$

where q_e (mg g^{-1}) is the quantity of RhB adsorbed at equilibrium; C_0 (mg L^{-1}) is the initial concentration of RhB; C_e (mg L^{-1}) is the residual concentration of RhB solution after adsorption; and V (L) and m (g) are the volume of the adsorbate and mass of the adsorbent, respectively.

Adsorption Models

To understand the nonlinear forms of the kinetic adsorption and isotherm models, four adsorption models were adopted (isotherm model, nonlinear Langmuir (3) and nonlinear Freundlich (4); kinetic model, nonlinear pseudo-first-order (5) and nonlinear pseudo-second-order (6)). The equations and parameters of the adsorption models were expressed as follows: [62]

$$q_e = q_{max} K_L C_e \,/(1 + K_L C_e), \tag{3}$$

$$q_e = K_F C_e^{1/n}, \tag{4}$$

$$q_t = q_e\left(1 - e^{-k_1 t}\right), \tag{5}$$

$$q_t = q_e^2 K_2 t/(1 + q_e K_2 t). \tag{6}$$

3. Results

Figure 1a shows a schematic representation of the fabrication of ZIF-8/TW-CA through delignification, TEMPO oxidation, in situ ZIF-8 synthesis, and carbonization processes. First, pristine woods were pretreated via delignification to enhance the porosity and chemical accessibility of the woods (Figure 1a(i)) [63]. After the pretreatment, the yellowish color of the wood changed into white, indicating the delignification of the wood (Figure S1a,b). The successful delignification of the wood was evaluated using the Klason method and FT-IR (Figure S2). The results reveal a reduction of approximately 90% in the lignin content of the DW (2.42%) compared to that of pristine wood (21.5%) (Figure S2a). Furthermore, the decrease in the lignin-specific IR peaks at 1590, 1501, and 1452 cm^{-1}, which can be attributed to aromatic skeletal vibration, and that at 1230 cm^{-1}, corresponding to the guaiacyl ring with C–O stretching, confirmed the removal of lignin from the wood (Figure S2b). Additionally, the decrease in the intensity of the carbonyl group (C=O) at 1737 cm^{-1} indicated the removal of hemicellulose from the wood channel and middle lamella [64,65]. After delignification, we conducted TEMPO-mediated oxidation to selectively convert C6 primary hydroxyl groups to carboxyl groups on the surface of cellulose microfibrils (Figures 1a(ii) and S1c) to provide nucleation sites for the subsequent in situ ZIF-8 synthesis (Figure 1b). Conductometric titration revealed that the carboxyl group content in the TW was 1.68 $mmolg^{-1}$. After complete washing, the TWs were immersed in a zinc nitrate solution. The cell walls of the wood efficiently adsorbed Zn^{2+} via the electrostatic interactions between Zn^{2+} and the carboxylic groups. After methanol washing, the Zn^{2+}-loaded TWs were transferred into an organic ligand solution to achieve the nucleation and growth of ZIF-8 nanoparticles (Figure 1a(iii),b). Thereafter, the sample was freeze-dried to obtain ZIF-8/TW-A (Figure S1d), after which ZIF-8/TW-A was carbonized into ZIF-8/TW-CA under a N_2 atmosphere at 850 °C and further activated at 350 °C in air (Figures 1a(iv) and S1e). After carbonization, the air activation as a physical activation can effectively improve the surface area of the hierarchical porous structure and increase the number of oxygen-containing functional groups on the carbon surface in the ZIF-8/TW-CA [66].

The FE-SEM images (Figure 2a–c) indicate that the pristine wood and TW exhibited a typical natural wood structure with honeycomb-like cellular structures and vertical channels, confirming that pretreatments, such as delignification and TEMPO oxidation, did not result in the structural collapse of the wood. Additionally, mesopores (2–50 nm) and macropores (>50 nm) were observed at the aligned cellulose fibers in the cell wall of TW (Figure 2d,g), which may be attributed to the delignification and TEMPO oxidation. The extensive loss of lignin and its conversion into carboxyl groups during delignification and TEMPO oxidation enhanced the cell wall porosity and increased the specific surface area. The enhanced porosity and surface area will induce good chemical accessibility for the subsequent in situ ZIF-8 synthesis. Figure 2e,h demonstrate the good deposition of ZIF-8 on the entire surface of the cell walls inside the TW without large aggregation. After carbonization, ZIF-8/TW-A was converted into ZIF-8/TW-CA. Although the size of the ZIF-8-derived nanoporous carbon particles decreased because of the decomposition of organic ligands, the final ZIF-8/TW-CA adequately retained the original ZIF-8 shape (Figure 2f,i). The carbonization temperature is one of the critical parameters that determines the surface area and pore volume of the nanoporous carbon [67]. Additionally, after carbonization, the ZIF-8-embedded wood composites were converted into interconnected, heteroatom-doped, and porous CA by harnessing the advantages of ZIF-8 and wood [40].

Heteroatom-doped carbon is well known to exhibit excellent adsorption capabilities [68]. Compared to that of ZIF-8/TW-A, the volume of ZIF-8/TW-CA reduced by approximately 40%. Additionally, ZIF-8/TW-CA exhibited a low density of 56 mg cm^{-3}. The FT-IR results demonstrate the successful chemical transformation during the fabrication of ZIF-8/TW-CA (Figure 3a). A strong IR peak of C=O carbonyl groups was observed in the FT-IR spectrum of TW at 1603 cm^{-1}, confirming the TEMPO oxidation of pristine wood. Additionally, the representative IR peaks of neat ZIF-8 nanoparticles at 1584 cm^{-1}, which is assigned to the C=N stretching, and within 1350–1500 cm^{-1}, which is attributed to the ring stretching in imidazole, were observed in the FT-IR spectrum of ZIF-8/TW-A.

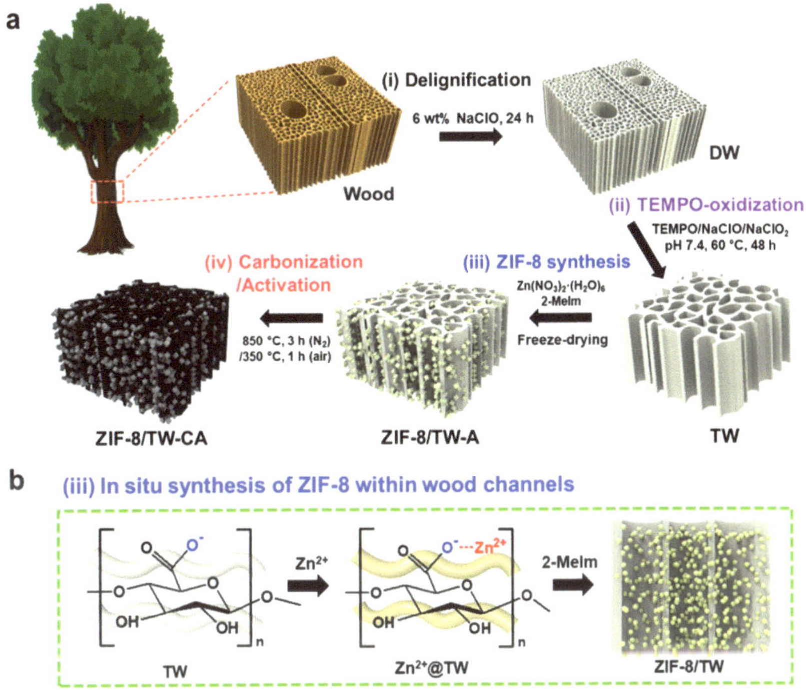

Figure 1. (**a**) Schematic representation of the fabrication of ZIF-8/TEMPO-oxidized wood carbon aerogel (ZIF-8/TW-CA): (i) delignification using NaClO solution (6 wt%, 24 h); (ii) TEMPO oxidation using TEMPO/NaClO/$NaClO_2$ solution (pH 7.4, 60 °C, 48 h) (purple); (iii) in situ synthesis of ZIF-8 using $Zn(NO_3)_2$, 2-MeIm solution, and freeze-drying (blue); (iv) carbonization at 850 °C (3 h, N_2 atmosphere) and activation at 350 °C (1 h, air) (red). (**b**) Illustration of the in situ process for obtaining ZIF-8/wood composites via TEMPO-oxidization.

This suggests the successful in situ growth of ZIF-8 in the TW. The XRD pattern further confirmed the embedding of ZIF-8 into the TW template (Figures 3b and S3), in which two broad diffraction peaks were observed at approximately 14.8°–16.8° and 22.5°, indicating the presence of cellulose I crystals in the wood. In addition, two broad diffraction peaks assigned to cellulose were observed in the XRD pattern of ZIF-8/TW-A, along with diffraction peaks at 7.2, 12.7, 16.3, and 18.4°, which can be attributed to the (110), (211), (310), and (222) crystal faces of ZIF-8, respectively (Figure S3a), demonstrating the successful growth of ZIF-8 crystals. These XRD patterns are consistent with the FT-IR results (Figure 3a). Further, two broad peaks were observed in the XRD pattern of ZIF-8/TW-CA at 25 and 43°, which are characteristic peaks of the (101) and (002) planes of carbon, respectively (Figure S3b) [40,60,66]. Next, we examined the chemical composition and environment of ZIF-8/TW-A and ZIF-8/TW-CA using XPS (Figure 3c). Compared to ZIF-8/TW-A (prior to carbonization), ZIF-8/TW-CA did not exhibit a Zn-related signal peak, suggesting the efficient removal of Zn content via carbonization. In addition, the XPS profiles revealed that ZIF-8/TW-CA comprised C, O, and N elements. Three peaks were observed in the high-resolution C 1s spectrum (Figure 3d) at 284.4, 285.4, and 287.2 eV, which can be attributed to the characteristic signals of C–C, C–O, and O–C=O bonds, respectively [66]. Additionally, three peaks were observed in the XPS O 1s spectrum (Figure 3e) at 530.8, 532.6, and 534.5 eV, which can be attributed to O–C=O, C=O, and C–O bonds, respectively [69]. Four peaks were observed in the N 1s spectrum (Figure 3f) at 398.3, 399.5, 400.5, and 403.2 eV, which can be attributed to pyridinic-N, pyrrolic-N, oxidized-N, and graphitic-N, respectively [40]. The introduction of heteroatoms (e.g., N, O) into the CA increased its adsorption ability for organic pollutants [70]. The XPS results of ZIF-8/TW-A and ZIF-8/TW-CA are with the EDS results (Figure 2j,k).

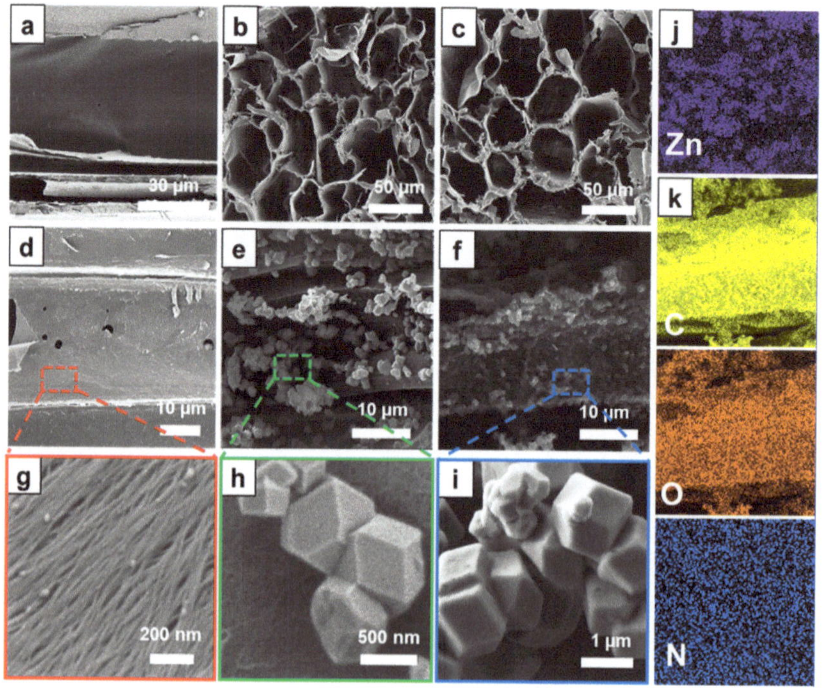

Figure 2. Field emission scanning electron microscopy (FE-SEM) images of (**a**,**b**) balsa wood, (**c**,**d**,**g**) tempo-oxidized wood (TW) (red), (**e**,**h**) ZIF-8/TW-A (green), and (**f**,**i**) ZIF-8/TW-CA (blue). Corresponding energy-dispersive spectroscopy (EDS) mapping images of (**j**) ZIF-8/TW-A and (**k**) ZIF-8/TW-CA.

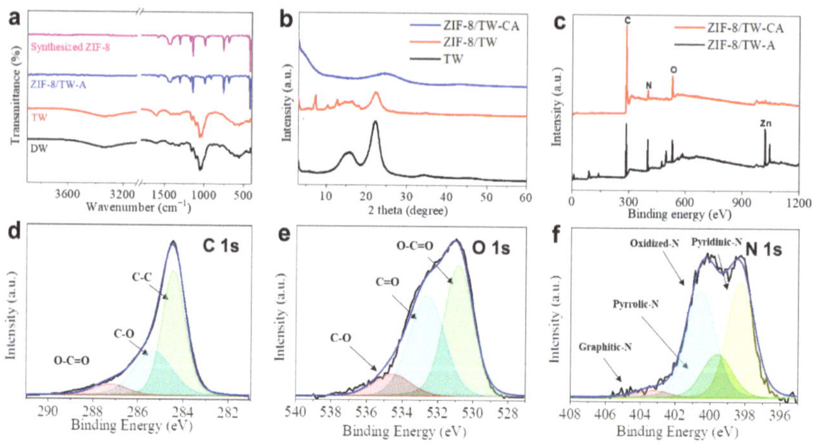

Figure 3. (**a**) Fourier transform infrared (FT-IR) spectra of delignified wood (DW), TW, ZIF-8/TW-A, and ZIF-8 nanoparticles. (**b**) X-ray diffraction (XRD) pattern of TW, ZIF-8/TW-A, and ZIF-8/TW-CA. (**c**) X-ray photoelectron spectroscopy (XPS) profiles of ZIF-8/TW-A and ZIF-8/TW-CA and high-resolution (**d**) C 1s, (**e**) O 1s, and (**f**) N 1s XPS profiles of ZIF-8/TW-CA.

The Raman spectra (Figure 4a) of ZIF-8/DW-CA and ZIF/TW-CA exhibited two characteristic carbon peaks at 1327 (D: disordered carbon) and 1588 (G: graphitic carbon). The ID/IG ratios of ZIF-8/DW-CA and ZIF/TW-CA were 1.37 and 1.24, respectively, indicating that ZIF-8/TW-CA exhibited larger defects than ZIF-8/DW-CA, which can be attributed to the high degree of N-doping [40,66]. Next, thermogravimetric analysis (TGA) was conducted to investigate the thermal properties and compare the ZIF-8 loading amounts of the ZIF-8/wood aerogels (Figure 4b). The decomposition of the ZIF-8/wood aerogels was observed to begin at approximately 240 °C, which can be attributed to the thermal degradation of the cellulose content. The residual content of ZIF-8/TW-A (20%) was higher than that of ZIF-8/DW-A (12%), indicating the higher ZIF-8 loadings of ZIF-8/TW-A. This difference between their ZIF-8 loadings can be attributed to the different amounts of carboxyl groups from the TEMPO oxidation. To examine the specific surface area and pore-size distribution, we conducted argon adsorption–desorption measurements for the pristine wood, ZIF-8/DW-CA, and ZIF-8/TW-CA. The pristine wood hardly adsorbed argon and only slightly adsorbed argon towards the end of the measurement, indicating that the wood was only composed of macropores (Figure 4c and Table 1).

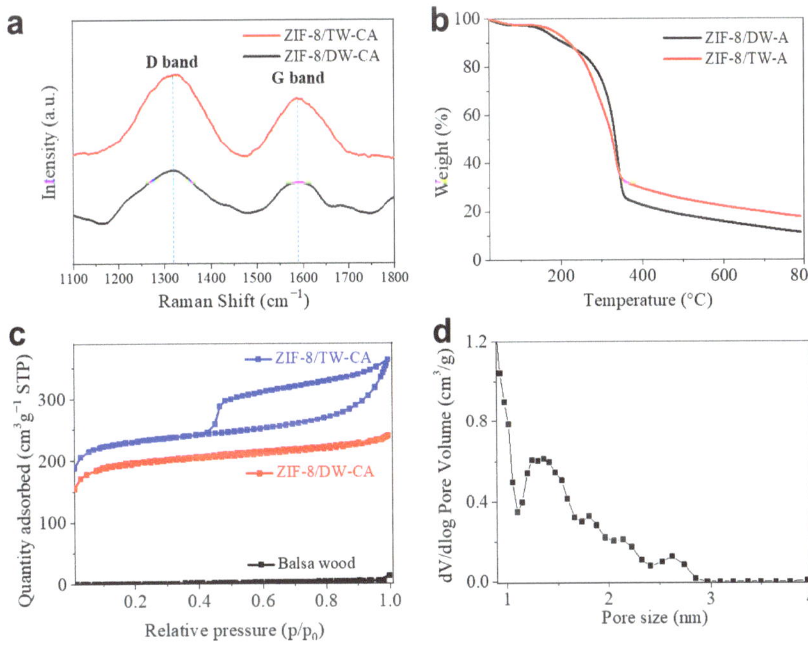

Figure 4. (**a**) Raman spectra of ZIF-8/TW-CA and ZIF-8/DW-CA. (**b**) Thermal stability of the ZIF-8/wood composite aerogels. Thermogravimetric analysis (TGA) curve of ZIF-8/DW-A and ZIF-8/TW-A under nitrogen atmosphere. (**c**) Argon adsorption/desorption isotherms of ZIF-8/wood carbon aerogel composites. (**d**) The corresponding pore-size distributions of ZIF-8/TW-CA.

Table 1. BET surface area (S_{BET}), average pore volume (V_m), and pore-size distribution of the samples.

Samples	S_{BET} (m^2 g^{-1})	V_m (cm^3 g^{-1})	Pore Size (nm)
Balsa wood	1.5	0.02	46.0
ZIF-8/DW-CA	662.1	0.30	1.9
ZIF-8/TW-CA	785.8	0.46	2.4

Compared to those of ZIF-8/DW-CA (662.1 m^2 g^{-1}, 0.30 cm^3 g^{-1}), ZIF-8/TW-CA exhibited a large specific surface area (785.8 m^2/g) and pore volume (0.46 cm^3 g^{-1}). These results confirm that the high ZIF-8 loading mass owing to the TEMPO oxidation promoted the nanoporous structure of the wood-based CA. The pore-size distribution range of ZIF-8/DW-CA and ZIF-8/TW-CA was 0.8–2.8 nm, indicating that carbonization and activation processes led to the presence of micro- and mesopores in both wood-based CAs because of ZIF-8 and chemical activation process. (Figure 4d and Figure S5). Next, the adsorption ability of ZIF-8/TW-CA was investigated using RhB, which is a model dye.

We investigated the effect of the adsorption temperature and pH on the dye solution. The effect of temperature on RhB adsorption by the samples was investigated using a kinetic study at three different temperatures (Figure 5a and Table 2). The results reveal that there was no significant difference in the adsorption capacities (9.5, 10.1, and 10.0 mg g^{-1} at 293, 308, and 318 K) of ZIF-8/TW-CA at the three temperatures, implying that the adsorption temperature had no significant effect on the dye adsorption in the temperature range of 293–318 K. RhB is a zwitterion dye in which the ionic state depends on pH; therefore, RhB changes from a cation to zwitterion form when the pH exceeds 3.7 (pKa of RhB). Consequently, this can affect the electrostatic interaction with the adsorbent. As shown in Figure 5b, the removal efficiency of ZIF-8/TW-CA was maintained within the pH range of 3–11, suggesting that RhB molecules can interact with other bonds through various mechanisms, such as hydrophobic interactions (π–π bonds), hydrogen bonds, and cation–π bonds of ZIF-8/TW-CA, except electrostatic interactions [71].

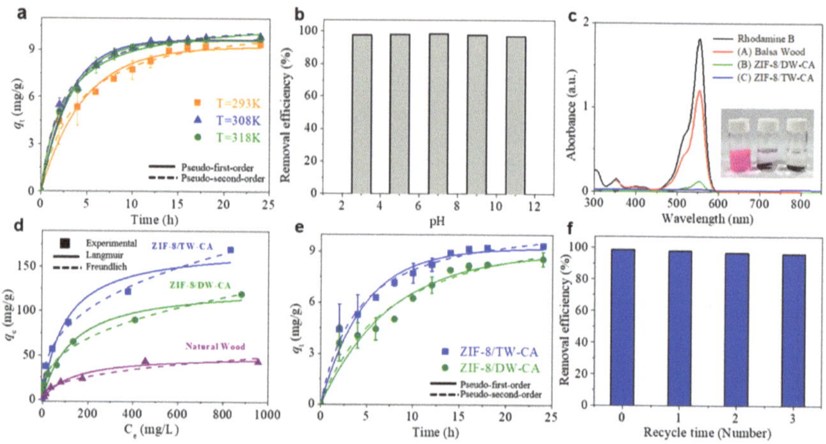

Figure 5. (**a**) Adsorption kinetic of RhB on ZIF-8/TW-CA at 293, 308, and 318 K fitted using pseudo-first-order and pseudo-second-order kinetic models. (**b**) Removal efficiency of RhB (10 mg/L) for ZIF-8/TW-CA in the pH range of 3–11. (**c**) Absorbance spectra of pristine wood, ZIF-8/DW-CA, and ZIF-8/TW-CA for RhB solution (10 mg L^{-1}). (**d**) Adsorption isotherms of pristine wood, ZIF-8/DW-CA, and ZIF-8/TW-CA at various RhB concentrations in water (Ce) fitted using Langmuir and Freundlich models. (**e**) Adsorption kinetic of RhB (10 mg L^{-1}) exposed to ZIF-8/DW-CA and ZIF-8/TW-CA fitted using pseudo-first-order and pseudo-second-order kinetic models. (**f**) Recycling ability of ZIF-8/TW-CA for RhB (10 mg L^{-1}).

Figure 5c shows the UV–VIS spectra of the RhB solution (10 mg L^{-1}) after 24 h adsorption by ZIF-8/DW-CA and ZIF-8/DW-CA. The color of the RhB solution disappeared when ZIF-8/TW-CA was immersed, whereas the color remained in the case of ZIF-8/DW-CA, indicating the enhanced porosity of ZIF-8/TW-CA for dye adsorption. Compared to that of DW, the loading of ZIF-8, with a high surface area, on TW was higher, owing to the presence of carboxyl groups (Figure 4a). In addition, the nanoporous structure of the cell wall (Figure 2g) caused by TEMPO treatment could facilitate the efficient capture of dye molecules in the solution. These factors can increase the specific area and micro/mesopores of ZIF-8/TW-CA (Figure 4b and Table 1). Figure 5d describes the effect of the initial RhB concentration on the adsorption capacity of ZIF-8/TW-CA. In this study, we employed two well-known adsorption isotherm models (Langmuir and Freundlich) to evaluate the experimental results. The parameters of the Langmuir and Freundlich isotherms are summarized in Table S1.

Table 2. Adsorption kinetic parameters for RhB on ZIF-8/TW-CA at three different temperatures (293, 308, and 318 K).

Samples	Pseudo-First-Order			Pseudo-Second-Order		
	q_t (mg/g)	K_1	R^2	q_t (mg/g)	K_1	R^2
ZIF-8/DW-CA-293 K	8.6	0.13	0.94	8.7	0.01	0.95
ZIF-8/TW-CA-293 K	9.2	0.20	0.96	9.5	0.02	0.98
ZIF-8/TW-CA-308 K	9.7	0.32	0.98	10.1	0.04	0.99
ZIF-8/TW-CA-318 K	9.6	0.30	0.99	10.0	0.03	0.99

The Freundlich model was more suitable for the analysis of the adsorption isotherm, and it exhibited a high correlation coefficient (R^2 = 0.99). The maximum adsorption capacities (q_{max}) of balsa wood, ZIF-8/DW-CA, and ZIF-8/TW-CA were 43.15, 119.4, and 169.4 mg g^{-1}, respectively. Particularly, the adsorption capacity of ZIF-8/TW-CA for RhB was approximately 1.4 times higher than that of ZIF-8/DW-CA. Importantly, the ZIF-8/TW-CA exhibited a good adsorption capacity (169.4 mg/g) for RhB comparable to previously reported values of other adsorbents, including biomass-based activated carbons (35.7 to 161.8 mg/g) (Table 3) [71–82]. The kinetic adsorption of ZIF-8/DW-CA and ZIF-8/TW-CA was observed for 24 h (Figure 5e and Table 2), and the result revealed that the pseudo-second-order kinetic model (R^2 = 0.99) for chemical adsorption was more suitable for explaining the experimental data. A recycling test was conducted to investigate the reusability of ZIF-8/TW-CA. The used adsorbent was washed several times with a mixed solution of DIW and ethanol (1:1, 60 °C) until no dye was released, and then immersed in a fresh RhB solution (10 mg L^{-1}). As shown in Figure 5f, even after three adsorption rounds, there was only a slight difference in the removal efficiency (96%) of ZIF-8/TW-CA. This result demonstrates the good reusability of ZIF-8/TW-CA via a simple desorption method. Porous materials with high porosity and a good affinity for hydrophobic solvents can also be used for the removal of organic solvents from polluted water. Lastly, we tested the ability of ZIF-8/TW-CA for the removal of organic solvents (Figure S6), and the results revealed that it took just 3 and 15 s to remove n-hexane and chloroform from the surface of water and underwater, respectively. These results demonstrate the good affinity of ZIF-8/TW-CA for hydrophobic organic solvents and its usefulness for the removal of organic solvents from polluted water. Considering the cost-effectiveness, renewability, and eco-friendliness of the ZIF-8/wood-based CA, these functional wood-based nanocomposites can serve as a promising adsorbent platform for various practical applications.

Table 3. Comparison of the maximum adsorption capacity of various adsorbents for RhB.

Adsorbent	Adsorbent (mg)	Adsorbate (mL)	q_e (mg/g)	References
Sugar-based carbon	100	1000	123.5	[71]
RGO-Ni nanocomposite	0.5	50	65.3	[72]
Cal-ZIF-67/AC	20	100	46.2	[73]
Iron-pillared bentonite	10	20	98.6	[74]
MWCNT-COOH	50	25	42.7	[75]
Fe_3O_4/Humic acid	50	100	161.8	[76]
ZIF-8@ZIF-67	0.4	20	143.3	[77]
Carbon xerogel	20	50	132.0	[78]
Coffee-activated carbon	200	200	83.4	[79]
Sawdust-activated carbon	10	10	35.7	[80]
Lignocellulosic-activated carbon	2000	1000	39.98	[81]
$ZnCl_2$-activated carbon	200	100	46.7	[82]
ZIF-8/TW-CA	10	10	169.4	This work

4. Conclusions

In this study, we developed a highly porous, heteroatom (O, N)-doped CA based on ZIF-8 and wood via TEMPO oxidation, in situ synthesis, and carbonization/activation processes. The results demonstrate that the introduction of carboxyl groups through TEMPO oxidation enabled a high ZIF-8 loading content. The proposed ZIF-8/TW-CA exhibited a 3D hierarchical porous structure, a high specific area, and an excellent adsorption capacity toward organic pollutants. These results can be attributed to the successful combination of the natural wood template with an ordered open-channel structure and the microporous ZIF-8 nanocrystals. The ZIF-8/TW-CA exhibited a high specific surface area of 785.8 $m^2\ g^{-1}$ and a high adsorption capacity of 169.4 mg g^{-1} for RhB. During the adsorption–desorption cycle experiments, ZIF-8/TW-CA retained approximately 96% of its initial adsorption capacity after three cycles. The findings of this study will provide insight into the development of functional wood-based nanocomposites.

Supplementary Materials: The following supporting information can be downloaded at: https://www.mdpi.com/article/10.3390/nano13192695/s1, Table S1. Adsorption isotherm model parameters for RhB on balsa wood, TW-CA, ZIF-8/DW-CA, and ZIF-8/TW-CA; Figure S1. Photographs of (a) pristine wood, (b) delignified wood (DW), (c) TEMPO-oxidized wood (TW), (d) ZIF-8/TW-A, and (e) ZIF-8/TW-CA. Scale bar = 0.5 mm; Figure S2. (a) Lignin contents of balsa wood, delignified wood, and TEMPO wood after Klason lignin analysis. (b) Fourier transform infrared (FT-IR) spectra of balsa wood and delignified wood; Figure S3. X-ray diffraction (XRD) pattern of the (a) synthesized ZIF-8 and (b) ZIF-8/TW-CA; Figure S4. (a) Adsorption spectra of RhB dye at various concentrations (0.1–10 mg/L). (b) Calibration curve for the adsorption of RhB dye; Figure S5. The corresponding pore-size distributions of ZIF-8/DW-CA. Figure S6. Photograph of the removal of (a) *n*-hexane (stained with Sudan III) and (b) chloroform (stained with Sudan III) by ZIF-8/TW-CA.

Author Contributions: Conceptualization, Y.J. and D.K.; methodology, Y.J., D.K., S.L., K.L., Y.K., G.K. and U.-J.K.; software, Y.J.; validation, Y.J.; formal analysis, Y.J.; investigation, Y.J.; writing—original draft preparation, Y.J.; writing—review and editing, J.K. and J.Y.; supervision, J.K. and J.Y.; project administration, J.Y.; data curation and investigation, J.P.; funding acquisition, S.Y.H., J.K. and J.Y. All authors have read and agreed to the published version of the manuscript.

Funding: This research was funded by the R&D Program for Forest Science Technology (Project No. "2023489B10-2325-AA01") provided by Korea Forest Service (Korea Forestry Promotion Institute), the National Research Foundation of Korea (NRF) Grant funded by the Ministry of Science, ICT & Future Planning (NRF-2022M3J4A1091450), the NRF grant (2022R1A4A1020543) funded by the Ministry of Science, ICT & Future Planning (MSIP), the Korea Institute of Energy Technology Evaluation and Planning (KETEP) and the Ministry of Trade, Industry & Energy (MOTIE) of the Republic of Korea (No. 20213030010430), and the Technology Development Program (RS-2022-00143349) funded by the Ministry of SMEs and Startups (MSS, Korea).

Data Availability Statement: Data can be made available on request.

Conflicts of Interest: The authors declare no conflict of interest.

References

1. Lellis, B.; Fávaro-Polonio, C.Z.; Pamphile, J.A.; Polonio, J.C. Effects of textile dyes on health and the environment and bioremediation potential of living organisms. *Biotechnol. Res. Innov.* **2019**, *3*, 275–290. [CrossRef]
2. Cordes, E.E.; Jones, D.O.; Schlacher, T.A.; Amon, D.J.; Bernardino, A.F.; Brooke, S.; Carney, R.; DeLeo, D.M.; Dunlop, K.M.; Escobar-Briones, E.G. Environmental impacts of the deep-water oil and gas industry: A review to guide management strategies. *Front. Environ. Sci.* **2016**, *4*, 58. [CrossRef]
3. Uzma, N.; Khaja Mohinuddin Salar, B.; Kumar, B.S.; Aziz, N.; David, M.A.; Reddy, V.D. Impact of organic solvents and environmental pollutants on the physiological function in petrol filling workers. *Int. J. Environ. Res. Public Health* **2008**, *5*, 139–146. [CrossRef]
4. Velho, S.R.; Brum, L.F.; Petter, C.O.; dos Santos, J.H.Z.; Šimunić, Š.; Kappa, W.H. Development of structured natural dyes for use into plastics. *Dyes Pigments* **2017**, *136*, 248–254. [CrossRef]
5. Benkhaya, S.; M'rabet, S.; El Harfi, A. A review on classifications, recent synthesis and applications of textile dyes. *Inorg. Chem. Commun.* **2020**, *115*, 107891. [CrossRef]

6. Pérez-Ibarbia, L.; Majdanski, T.; Schubert, S.; Windhab, N.; Schubert, U.S. Safety and regulatory review of dyes commonly used as excipients in pharmaceutical and nutraceutical applications. *Eur. J. Pharm. Sci.* **2016**, *93*, 264–273. [CrossRef]
7. Wainwright, M. Dyes in the development of drugs and pharmaceuticals. *Dyes Pigments* **2008**, *76*, 582–589. [CrossRef]
8. Levitan, H. Food, drug, and cosmetic dyes: Biological effects related to lipid solubility. *Proc. Natl. Acad. Sci. USA* **1977**, *74*, 2914–2918. [CrossRef]
9. Dey, S.; Nagababu, B.H. Applications of food color and bio-preservatives in the food and its effect on the human health. *Food Chem. Adv.* **2022**, *1*, 100019. [CrossRef]
10. Orts, F.; Del Río, A.; Molina, J.; Bonastre, J.; Cases, F. Electrochemical treatment of real textile wastewater: Trichromy Pro-cion HEXL®. *J. Electroanal. Chem.* **2018**, *808*, 387–394. [CrossRef]
11. Emam, M. Thermal stability of some textile dyes. *J. Therm. Anal. Calorim.* **2001**, *66*, 583–591. [CrossRef]
12. Tkaczyk, A.; Mitrowska, K.; Posyniak, A. Synthetic organic dyes as contaminants of the aquatic environment and their implications for ecosystems: A review. *Sci. Total Environ.* **2021**, *717*, 137222. [CrossRef]
13. Zhao, C.; Zhou, J.; Yan, Y.; Yang, L.; Xing, G.; Li, H.; Wu, P.; Wang, M.; Zheng, H. Application of coagulation/flocculation in oily wastewater treatment: A review. *Sci. Total Environ.* **2021**, *765*, 142795. [CrossRef] [PubMed]
14. Ramos, S.; Homem, V.; Alves, A.; Santos, L. A review of organic UV-filters in wastewater treatment plants. *Environ. Int.* **2016**, *86*, 24–44. [CrossRef] [PubMed]
15. Sena, M.; Hicks, A. Life cycle assessment review of struvite precipitation in wastewater treatment. *Resour. Conserv. Recycl.* **2018**, *139*, 194–204. [CrossRef]
16. Rashid, R.; Shafiq, I.; Akhter, P.; Iqbal, M.J.; Hussain, M. A state-of-the-art review on wastewater treatment techniques: The effectiveness of adsorption method. *Environ. Sci. Pollut. Res.* **2021**, *28*, 9050–9066. [CrossRef]
17. Huang, X.; Guida, S.; Jefferson, B.; Soares, A. Economic evaluation of ion-exchange processes for nutrient removal and recovery from municipal wastewater. *NPJ Clean Water* **2020**, *3*, 7. [CrossRef]
18. Ma, D.; Yi, H.; Lai, C.; Liu, X.; Huo, X.; An, Z.; Li, L.; Fu, Y.; Li, B.; Zhang, M. Critical review of advanced oxidation processes in organic wastewater treatment. *Chemosphere* **2021**, *275*, 130104. [CrossRef]
19. Lu, T.; Cao, W.; Liang, H.; Deng, Y.; Zhang, Y.; Zhu, M.; Ma, W.; Xiong, R.; Huang, C. Blow-Spun nanofibrous membrane for simultaneous treatment of emulsified oil/water mixtures, dyes, and bacteria. *Langmuir* **2022**, *38*, 15729–15739. [CrossRef]
20. Banerjee, P.; Dinda, P.; Kar, M.; Uchman, M.; Mandal, T.K. Ionic liquid cross-linked high-absorbent polymer hydrogels: Kinetics of swelling and dye adsorption. *Langmuir* **2023**, *39*, 9757–9772. [CrossRef]
21. Ugwu, E.; Othmani, A.; Nnaji, C. A review on zeolites as cost-effective adsorbents for removal of heavy metals from aqueous environment. *Int. J. Environ. Sci. Technol.* **2022**, *19*, 8061–8084. [CrossRef]
22. Qiang, Z.; Li, R.; Yang, Z.; Guo, M.; Cheng, F.; Zhang, M. Zeolite X adsorbent with high stability synthesized from bauxite tailings for cyclic adsorption of CO2. *Energy Fuels* **2019**, *33*, 6641–6649. [CrossRef]
23. Awasthi, A.; Jadhao, P.; Kumari, K. Clay nano-adsorbent: Structures, applications and mechanism for water treatment. *SN Appl. Sci.* **2019**, *1*, 1–21. [CrossRef]
24. Wang, Z.K.; Li, T.T.; Peng, H.K.; Ren, H.T.; Lin, J.H.; Lou, C.W. Natural-clay-reinforced hydrogel adsorbent: Rapid ad-sorption of heavy-metal ions and dyes from textile wastewater. *Water Env. Res.* **2022**, *94*, e10698. [CrossRef]
25. Raj, S.I.; Jaiswal, A.; Uddin, I. Tunable porous silica nanoparticles as a universal dye adsorbent. *RSC Adv.* **2019**, *9*, 11212–11219. [CrossRef]
26. Huang, C.H.; Chang, K.P.; Ou, H.D.; Chiang, Y.C.; Wang, C.F. Adsorption of cationic dyes onto mesoporous silica Microporous. *Mesoporous Mater.* **2011**, *141*, 102–109. [CrossRef]
27. Gan, G.; Li, X.; Fan, S.; Wang, L.; Qin, M.; Yin, Z.; Chen, G. Carbon aerogels for environmental clean-up. *Eur. J. Inorg. Chem.* **2019**, *27*, 3126–3141. [CrossRef]
28. Geng, S.; Wei, J.; Jonasson, S.; Hedlund, J.; Oksman, K. Multifunctional carbon aerogels with hierarchical anisotropic structure derived from lignin and cellulose nanofibers for CO_2 capture and energy storage. *ACS Appl. Mater. Interfaces* **2020**, *12*, 7432–7441. [CrossRef]
29. Moreno-Castilla, C.; Maldonado-Hódar, F. Carbon aerogels for catalysis applications: An overview. *Carbon* **2005**, *43*, 455–465. [CrossRef]
30. Zhuo, H.; Hu, Y.; Chen, Z.; Peng, X.; Liu, L.; Luo, Q.; Yi, J.; Liu, C.; Zhong, L. A carbon aerogel with super mechanical and sensing performances for wearable piezoresistive sensors. *J. Mater. Chem. A* **2019**, *7*, 8092–8100. [CrossRef]
31. Luo, Q.; Zheng, H.; Hu, Y.; Zhuo, H.; Chen, Z.; Peng, X.; Zhong, L. Carbon nanotube/chitosan-based elastic carbon aerogel for pressure sensing. *Indus. Eng. Chem. Res.* **2019**, *58*, 17768–17775. [CrossRef]
32. Ma, Y.; Liu, Q.; Li, W.; Zheng, Y.; Shi, Q.; Zhou, Z.; Shao, G.; Yang, W.; Chen, D.; Fang, X. Ultralight and robust carbon nano-fiber aerogels for advanced energy storage. *J. Mater. Chem. A* **2021**, *9*, 900–907. [CrossRef]
33. Fan, W.; Shi, Y.; Gao, W.; Sun, Z.; Liu, T. Graphene–carbon nanotube aerogel with a scroll-interconnected-sheet structure as an advanced framework for a high-performance asymmetric supercapacitor electrode. *ACS Appl. Nano Mater.* **2018**, *1*, 4435–4441. [CrossRef]
34. Lee, Y.J.; Jung, J.C.; Yi, J.; Baeck, S.-H.; Yoon, J.R.; Song, I.K. Preparation of carbon aerogel in ambient conditions for electrical double-layer capacitor. *Curr. Appl. Phys.* **2010**, *10*, 682–686. [CrossRef]

35. Pekala, R.; Farmer, J.; Alviso, C.; Tran, T.; Mayer, S.; Miller, J.; Dunn, B. Carbon aerogels for electrochemical applications. *J. Non-Cryst. Solids* **1998**, *225*, 74–80. [CrossRef]
36. Chen, G.; He, S.; Shi, G.; Ma, Y.; Ruan, C.; Jin, X.; Chen, Q.; Liu, X.; Dai, H.; Chen, X. In-situ immobilization of ZIF-67 on wood aerogel for effective removal of tetracycline from water. *Chem. Eng. J.* **2021**, *423*, 130184. [CrossRef]
37. Zhang, X.-F.; Wang, Z.; Song, L.; Yao, J. In situ growth of ZIF-8 within wood channels for water pollutants removal. *Sep. Purif. Technol.* **2021**, *266*, 118527. [CrossRef]
38. Kaushik, J.; Kumar, V.; Garg, A.K.; Dubey, P.; Tripathi, K.M.; Sonkar, S.K. Bio-mass derived functionalized graphene aerogel: A sustainable approach for the removal of multiple organic dyes and their mixtures. *New J. Chem.* **2021**, *45*, 9073–9083. [CrossRef]
39. Wang, M.-L.; Zhang, S.; Zhou, Z.-H.; Zhu, J.-L.; Gao, J.-F.; Dai, K.; Huang, H.-D.; Li, Z.-M. Facile heteroatom doping of bio-mass-derived carbon aerogels with hierarchically porous architecture and hybrid conductive network: Towards high electro-magnetic interference shielding effectiveness and high absorption coefficient. *Compos. B Eng.* **2021**, *224*, 109175. [CrossRef]
40. Wang, C.; Kim, J.; Tang, J.; Na, J.; Kang, Y.M.; Kim, M.; Lim, H.; Bando, Y.; Li, J.; Yamauchi, Y. Large-scale synthesis of MOF-derived superporous carbon aerogels with extraordinary adsorption capacity for organic solvents. *Angew. Chem.* **2020**, *132*, 2082–2086. [CrossRef]
41. Grossman, P.; Wold, M. Compression fracture of wood parallel to the grain. *Wood Sci. Technol.* **1971**, *5*, 147–156. [CrossRef]
42. Kelley, S.S.; Rials, T.G.; Glasser, W.G. Relaxation behaviour of the amorphous components of wood. *J. Mater. Sci.* **1987**, *22*, 617–624. [CrossRef]
43. Zhu, W.; Zhang, Y.; Wang, X.; Wu, Y.; Han, M.; You, J.; Jia, C.; Kim, J. Aerogel nanoarchitectonics based on cellulose nanocrystals and nanofibers from eucalyptus pulp: Preparation and comparative study. *Cellulose* **2022**, *29*, 817–833. [CrossRef]
44. Gibson, L.J. The hierarchical structure and mechanics of plant materials. *J. R. Soc. Interface* **2012**, *9*, 2749–2766. [CrossRef]
45. Xia, L.; Li, X.; Wu, Y.; Rong, M. Wood-derived carbons with hierarchical porous structures and monolithic shapes prepared by biological-template and self-assembly strategies. *ACS Sustain. Chem. Eng.* **2015**, *3*, 1724–1731. [CrossRef]
46. Zhang, W.; Taheri-Ledari, R.; Saeidirad, M.; Qazi, F.S.; Kashtiaray, A.; Ganjali, F.; Tian, Y.; Maleki, A. Regulation of Porosity in MOFs: A Review on Tunable Scaffolds and Related Effects and Advances in Different Applications. *J. Environ. Chem. Eng.* **2022**, *10*, 108836. [CrossRef]
47. Baumann, A.E.; Burns, D.A.; Liu, B.; Thoi, V.S. Metal-organic framework functionalization and design strategies for advanced electrochemical energy storage devices. *Commun. Chem.* **2019**, *2*, 86. [CrossRef]
48. Qian, Q.; Asinger, P.A.; Lee, M.J.; Han, G.; Mizrahi Rodriguez, K.; Lin, S.; Benedetti, F.M.; Wu, A.X.; Chi, W.S.; Smith, Z.P. MOF-based membranes for gas separations. *Chem. Rev.* **2020**, *120*, 8161–8266. [CrossRef]
49. Han, M.; Zhu, W.; Hossain, M.S.A.; You, J.; Kim, J. Recent progress of functional metal–organic framework materials for water treatment using sulfate radicals. *Environ. Res.* **2022**, *211*, 112956. [CrossRef]
50. Shet, S.P.; Priya, S.S.; Sudhakar, K.; Tahir, M. A review on current trends in potential use of metal-organic framework for hydrogen storage. *Int. J. Hydro. Energy* **2021**, *46*, 11782–11803. [CrossRef]
51. Zhu, W.; Han, M.; Kim, D.; Zhang, Y.; Kwon, G.; You, J.; Jia, C.; Kim, J. Facile preparation of nanocellulose/Zn-MOF-based catalytic filter for water purification by oxidation process. *Environ. Res.* **2022**, *205*, 112417. [CrossRef] [PubMed]
52. Wang, C.; Kim, J.; Tang, J.; Kim, M.; Lim, H.; Malgras, V.; You, J.; Xu, Q.; Li, J.; Yamauchi, Y. New strategies for novel MOF-derived carbon materials based on nanoarchitectures. *Chem* **2020**, *6*, 19–40. [CrossRef]
53. Kim, J.; Kim, J.; Kim, J.H.; Park, H.S. Hierarchically open-porous nitrogen-incorporated carbon polyhedrons derived from metal-organic frameworks for improved CDI performance. *Chem. Eng. J.* **2020**, *382*, 122996. [CrossRef]
54. Kim, M.; Xu, X.; Xin, R.; Earnshaw, J.; Ashok, A.; Kim, J.; Park, T.; Nanjundan, A.K.; El-Said, W.A.; Yi, J.W. KOH-activated hollow ZIF-8 derived porous carbon: Nanoarchitectured control for upgraded capacitive deionization and supercapacitor. *ACS Appl. Mater. Interfaces* **2021**, *13*, 52034–52043. [CrossRef] [PubMed]
55. Kim, M.; Xin, R.; Earnshaw, J.; Tang, J.; Hill, J.P.; Ashok, A.; Nanjundan, A.K.; Kim, J.; Young, C.; Sugahara, Y. MOF-derived nanoporous carbons with diverse tunable nanoarchitectures. *Nat. Protoc.* **2022**, *17*, 2990–3027. [CrossRef]
56. Kajal, N.; Singh, V.; Gupta, R.; Gautam, S. Metal organic frameworks for electrochemical sensor applications: A review. *Environ. Res.* **2022**, *204*, 112320. [CrossRef]
57. Lu, T.; Liang, H.; Cao, W.; Deng, Y.; Qu, Q.; Ma, W.; Xiong, R.; Huang, C. Blow-spun nanofibrous composite Self-cleaning membrane for enhanced purification of oily wastewater. *J. Colloid Interface Sci.* **2022**, *608*, 2860–2869. [CrossRef]
58. Ma, X.; Zhao, S.; Tian, Z.; Duan, G.; Pan, H.; Yue, Y.; Li, S.; Jian, S.; Yang, W.; Liu, K. MOFs meet wood: Reusable magnetic hydrophilic composites toward efficient water treatment with super-high dye adsorption capacity at high dye concentration. *Chem. Eng. J.* **2022**, *446*, 136851. [CrossRef]
59. Zhu, X.; Fan, Z.; Zhang, X.F.; Yao, J. Metal-organic frameworks decorated wood aerogels for efficient particulate matter removal. *J. Colloid Interface Sci.* **2023**, *629*, 182–188. [CrossRef]
60. Wang, S.; Wang, C.; Zhou, Q. Strong foam-like composites from highly mesoporous wood and metal-organic frameworks for efficient CO2 capture. *ACS Appl. Mater. Interfaces* **2021**, *13*, 29949–29959. [CrossRef]
61. Tu, K.; Puértolas, B.; Adobes-Vidal, M.; Wang, Y.; Sun, J.; Traber, J.; Burgert, I.; Pérez-Ramírez, J.; Keplinger, T. Green synthesis of hierarchical metal–organic framework/wood functional composites with superior mechanical properties. *Adv. Sci.* **2022**, *7*, 1902897. [CrossRef] [PubMed]

62. Santoso, E.; Ediati, R.; Istiqomah, Z.; Sulistiono, D.O.; Nugraha, R.E.; Kusumawati, Y.; Bahruji, H.; Prasetyoko, D. Facile syn-thesis of ZIF-8 nanoparticles using polar acetic acid solvent for enhanced adsorption of methylene blue. *Microporous Mesoporous Mater.* **2021**, *310*, 110620. [CrossRef]
63. Fu, Q.; Ansari, F.; Zhou, Q.; Berglund, L.A. Wood nanotechnology for strong, mesoporous, and hydrophobic biocomposites for selective separation of oil/water mixtures. *ACS Nano* **2018**, *12*, 2222–2230. [CrossRef] [PubMed]
64. Frey, M.; Widner, D.; Segmehl, J.S.; Casdorff, K.; Keplinger, T.; Burgert, I. Delignified and densified cellulose bulk materials with excellent tensile properties for sustainable engineering. *ACS Appl. Mater. Interfaces* **2018**, *10*, 5030–5037. [CrossRef]
65. Khakalo, A.; Tanaka, A.; Korpela, A.; Orelma, H. Delignification and ionic liquid treatment of wood toward multifunctional high-performance structural materials. *ACS Appl. Mater. Interfaces* **2020**, *12*, 23532–23542. [CrossRef] [PubMed]
66. Zhang, W.; Li, M.; Zhong, L.; Huang, J.; Liu, M. A family of MOFs@ Wood-Derived hierarchical porous composites as free-standing thick electrodes of solid supercapacitors with enhanced areal capacitances and energy densities. *Mater. Today Energy* **2022**, *24*, 100951. [CrossRef]
67. Abbasi, Z.; Shamsaei, E.; Leong, S.K.; Ladewig, B.; Zhang, X.; Wang, H. Effect of carbonization temperature on adsorption property of ZIF-8 derived nanoporous carbon for water treatment. *Microporous Mesoporous Mater.* **2016**, *236*, 28–37. [CrossRef]
68. Wang, Y.; Zhu, M.; Li, Y.; Zhang, M.; Xue, X.; Shi, Y.; Dai, B.; Guo, X.; Yu, F. Heteroatom-doped porous carbon from methyl orange dye wastewater for oxygen reduction. *Green Energy Environ.* **2018**, *3*, 172–178. [CrossRef]
69. Zhang, L.; Tu, L.-Y.; Liang, Y.; Chen, Q.; Li, Z.-S.; Li, C.-H.; Wang, Z.-H.; Li, W. Coconut-based activated carbon fibers for efficient adsorption of various organic dyes. *RSC Adv.* **2018**, *8*, 42280–42291. [CrossRef]
70. Treeweranuwat, P.; Boonyoung, P.; Chareonpanich, M.; Nueangnoraj, K. Role of nitrogen on the porosity, surface, and electro-chemical characteristics of activated carbon. *ACS Omega* **2020**, *5*, 1911–1918. [CrossRef]
71. Xiao, W.; Garba, Z.N.; Sun, S.; Lawan, I.; Wang, L.; Lin, M.; Yuan, Z. Preparation and evaluation of an effective activated carbon from white sugar for the adsorption of rhodamine B dye. *J. Clean. Prod.* **2020**, *253*, 119989. [CrossRef]
72. Üner, O.; Geçgel, Ü.; Kolancilar, H.; Bayrak, Y. Adsorptive removal of rhodamine B with activated carbon obtained from okra wastes. *Chem. Eng. Commun.* **2017**, *204*, 772–783. [CrossRef]
73. Li, Y.; Yan, X.; Hu, X.; Feng, R.; Zhou, M. Trace pyrolyzed ZIF-67 loaded activated carbon pellets for enhanced adsorption and catalytic degradation of Rhodamine B in water. *Chem. Eng. J.* **2019**, *375*, 122003. [CrossRef]
74. Hou, M.-F.; Ma, C.-X.; Zhang, W.-D.; Tang, X.-Y.; Fan, Y.-N.; Wan, H.-F. Removal of rhodamine B using iron-pillared bentonite. *J. Hazard. Mater.* **2011**, *186*, 1118–1123. [CrossRef]
75. Oyetade, O.A.; Nyamori, V.O.; Martincigh, B.S.; Jonnalagadda, S.B. Effectiveness of carbon nanotube–cobalt ferrite nanocomposites for the adsorption of rhodamine B from aqueous solutions. *RSC Adv.* **2015**, *5*, 22724–22739. [CrossRef]
76. Peng, L.; Qin, P.; Lei, M.; Zeng, Q.; Song, H.; Yang, J.; Shao, J.; Liao, B.; Gu, J. Modifying Fe_3O_4 nanoparticles with humic acid for removal of Rhodamine B in water. *J. Hazard. Mater.* **2012**, *209*, 193–198. [CrossRef]
77. Nazir, M.A.; Najam, T.; Shahzad, K.; Wattoo, M.A.; Hussain, T.; Tufail, M.K.; Shah, S.S.A.; Rehman, A.-U. Heterointerface engineering of water stable ZIF-8@ZIF-67: Adsorption of rhodamine B from water. *Surf. Interfaces* **2022**, *34*, 102324. [CrossRef]
78. Koniarz, M.P.; Goscianska, J.R.; Pietrzak, R. Removal of rhodamine B from water by modified carbon xerogels. *Colloid Surf. A Physicochem. Eng. Asp.* **2018**, *543*, 109–117. [CrossRef]
79. Chen, H.M.; Lau, W.M.; Zhou, D. Waste-Coffee-Derived activated carbon as efficient adsorbent for water treatment. *Materials* **2022**, *15*, 8684. [CrossRef] [PubMed]
80. Nguyen, D.T.; Nguyen, T.T.; Nguyen, H.P.T.; Khuat, H.B.; Nguyen, T.H.; Tran, V.K.; Chang, S.W.; Nguyen-Tri, P.; Nguyen, D.D.; La, D.D. Activated carbon with ultrahigh surface area derived from sawdust biowaste for the removal of rhodamine B in water. *Environ. Technol. Innov.* **2021**, *24*, 101811. [CrossRef]
81. Da Silva Lacerda, V.; López-Sotelo, J.B.; Correa-Guimarães, A.; Hernández-Navarro, S.; Sánchez-Báscones, M.; Navas-Gracia, L.M.; Martín-Ramos, P.; Martín-Gil, J. Rhodamine B removal with activated carbons obtained from lignocellulosic waste. *J. Environ. Manag.* **2015**, *155*, 67–76. [CrossRef] [PubMed]
82. Bhadusha, N.; Ananthabaskaran, T. Kinetic, thermodynamic and equilibrium studies on uptake of rhodamine B onto $ZnCl_2$ activated low cost carbon. *J. Chem.* **2012**, *9*, 137–144.

materials

MDPI

Article

A PEGylated PVDF Antifouling Membrane Prepared by Grafting of Methoxypolyethylene Glycol Acrylate in Gama-Irradiated Homogeneous Solution

Ting Wang [1,2], Zhengchi Hou [3,*], Haijun Yang [3] and Jun Hu [1,3]

1 Shanghai Institute of Applied Physics, Chinese Academy of Sciences, Shanghai 201800, China; wangting@sinap.ac.cn (T.W.); hujun@sinap.ac.cn (J.H.)
2 University of Chinese Academy of Sciences, Beijing 100049, China
3 Shanghai Synchrotron Radiation Facility (SSRF), Shanghai Advanced Research Institute, Chinese Academy of Sciences, Shanghai 201204, China; yanghj@sari.ac.cn
* Correspondence: houzc@sari.ac.cn; Tel.: +86-13818279326

Abstract: In this study, methoxypolyethylene glycol acrylate (mPEGA) served as a PEGylated monomer and was grafted onto polyvinylidene fluoride (PVDF) through homogeneous solution gamma irradiation. The grafting process was confirmed using several techniques, including infrared spectroscopy (FTIR), thermodynamic stability assessments, and rotational viscosity measurements. The degree of grafting (DG) was determined via the gravimetric method. By varying the monomer concentration, a range of DGs was achieved in the PVDF-g-mPEGA copolymers. Investigations into water contact angles and scanning electron microscopy (SEM) images indicated a direct correlation between increased hydrophilicity, membrane porosity, and higher DG levels in the PVDF-g-mPEGA membrane. Filtration tests demonstrated that enhanced DGs resulted in more permeable PVDF-g-mPEGA membranes, eliminating the need for pore-forming agents. Antifouling tests revealed that membranes with a lower DG maintained a high flux recovery rate, indicating that the innate properties of PVDF could be largely preserved.

Keywords: polyvinylidene fluoride; methoxypolyethylene glycol acrylate; antifouling; simultaneous irradiation; ultrafiltration membrane

Citation: Wang, T.; Hou, Z.; Yang, H.; Hu, J. A PEGylated PVDF Antifouling Membrane Prepared by Grafting of Methoxypolyethylene Glycol Acrylate in Gama-Irradiated Homogeneous Solution. *Materials* **2024**, *17*, 873. https://doi.org/10.3390/ma17040873

Academic Editors: Guangxu Lan and Yi-Nan Wu

Received: 18 January 2024
Revised: 7 February 2024
Accepted: 10 February 2024
Published: 14 February 2024

1. Introduction

The issue of water scarcity has garnered growing attention in recent years [1]. Membrane technology, due to its energy efficiency, scalability, and consistent effluent quality, has emerged as a promising solution for ensuring a dependable water supply [2,3]. Specifically, ultrafiltration membranes, characterized by pore sizes ranging between 5 and 100 nm, demonstrate notable efficacy in eliminating viruses, proteins, and colloidal contaminants at relatively low applied pressures, setting them apart from alternative membrane technologies [4]. Consequently, ultrafiltration membranes have found extensive application in drinking water treatment facilities and urban wastewater treatment plants, capitalizing on their inherent advantages [5–8]. Polyvinylidene fluoride (PVDF), renowned for its high mechanical strength, exceptional chemical resistance, and robust thermal stability, is a popular choice in separation membrane applications [9–12]. However, its inherent hydrophobicity often leads to membrane fouling [13,14], underscoring the necessity of modifying PVDF membranes to enhance their performance.

Numerous studies have demonstrated the effectiveness of various hydrophilic materials in improving the antifouling properties of PVDF membranes [10,15–23]. In this context, our research group has explored grafting different monomers onto membrane materials, including N-vinylpyrrolidone (NVP) [9], methacrylic acid (MAA) [24–26], acrylic acid (AA) [27], and polyethylene glycol methacrylate (PEGMA) [28]. Among these, PEGylated monomers have shown superior antifouling capabilities. Previous research has established PEG materials as effective in reducing nonspecific protein adsorption [16,17]. In aqueous environments, PEG's ability to form a hydration layer on its surface, coupled with the rapid mobility of its hydrated chains, can alter the microscopic thermodynamics at the protein–surface interface. This alteration potentially hinders the adsorption or adhesion of proteins [18].

Methoxypolyethylene glycol acrylate (mPEGA), a notable PEGylation reagent, effectively enhances the hydrophilicity and antifouling properties of materials [19,20]. Wang et al. [29] utilized UV-induced free radical polymerization to graft mPEGA onto silicone hydrogels, noting a significant decrease in the static water contact angle as the grafting degree increased. This modification reduced the adsorption of a single protein to the modified silicone hydrogel by 70–80%. Bozukova et al. [30] implemented surface modification of poly (HEMA-co-MMA) hydrogels with oligoethylene glycol methacrylate (OEGA) using atom transfer radical polymerization (ATRP) at 25 °C. This process enabled the grafting of poly(ethylene glycol) (PEG) chains onto the hydrogel surface, with findings indicating that even short poly(OEGA) brushes significantly enhanced the hydrogels' antifouling characteristics, as verified by in vitro biotests. Similarly, Asatekin et al. [31] and Barroso et al. [32] used acrylonitrile and mPEGA as raw materials, producing PAN-g-PEO through free radical polymerization. They then prepared a PAN-based UF membrane with added graft polymer, exhibiting substantial resistance to irreversible fouling. Further, studies involving grafting or blending methoxy polyethylene glycol (mPEG) into PSf [33,34], PES [35], and PVDF [36] membranes have shown remarkable improvements in their antifouling characteristics, thereby validating the effectiveness of mPEG chain segments in enhancing these properties.

Numerous methods have been explored for PVDF PEGylation, including ozone-activated surface grafting [37] and bulk grafting [38], atom-transfer radical-polymerization (ATRP) surface grafting [39] and bulk grafting [40,41], plasma surface grafting [42], and radiation grafting [43]. While these techniques have unique characteristics, they also face specific limitations. Surface grafting can lead to irregular membrane surface structures, potentially altering pore size and distribution, which may subsequently reduce filtration performance [44,45]. In ozone-activated bulk grafting of PVDF, significant material degradation can occur [46], potentially impairing membrane properties. Moreover, the high bond energy of C-F bonds in PVDF affects both the degree and uniformity of ATRP grafting [47], and the monomer-to-PVDF concentration ratio is substantial (10:1) [41].

Gamma radiation-induced grafting, known for its superior penetrating power and mild reaction conditions, is an effective method for polymer modification. This grafting reaction is initiated by high-energy radiation, eliminating the need for an additional initiator [48,49]. Our research group's findings indicate that gamma radiation-induced grafting yields more uniform grafting compared to heterogeneous grafting methods. However, it is noteworthy that, unlike NVP, common small molecular monomers generally cannot undergo homogeneous radiation grafting at high concentrations, leading to challenges such as insolubility and significant homopolymerization. In contrast, PEGylated monomers do not encounter these issues. Despite numerous studies demonstrating mPEGA as an effective antifouling material, reports of mPEGA-grafted modified PVDF membranes are scarce.

In this study, mPEGA was successfully grafted onto PVDF through homogeneous solution radiation grafting. The resultant polymer, PVDF-g-mPEGA, was then directly formed into a membrane using non-solvent-induced phase separation (NIPS). The degree of grafting (DG) was determined using the gravimetric method, and a kinetic study was conducted to investigate factors influencing DG. Additionally, the thermodynamic stability of the grafted product was analyzed by constructing a phase diagram via cloud point titration. The viscosity of the copolymer was measured using a rotational viscometer, and its molecular structure was characterized by Fourier transform infrared spectroscopy (FTIR). The hydrophilicity, surface morphology, and filtration performance of the membrane were examined using scanning electron microscopy (SEM) and water contact angle measurements. The antifouling performance of PVDF-g-mPEGA was assessed by evaluating the recovery rate of membrane filtration of bovine serum albumin (BSA) solution and determining water flux post-water wash, with comparisons drawn to the antifouling performance of PVDF-g-NVP membranes. In this research, BSA served as the model pollutant. The schematic diagram of the study is shown in Figure 1.

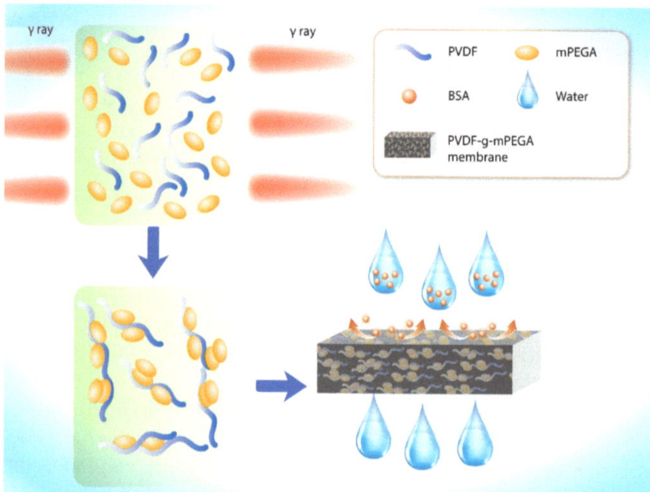

Figure 1. Schematic diagram of the synthesis/fabrication process and antifouling performance of membrane.

2. Materials and Methods

2.1. Materials

PVDF (Solef 6020, Mn = 670,000) was procured from Solvay Co. Ltd. and subjected to pre-use drying at 80 °C in a vacuum oven for 24 h. NMP, NaCl, Na_2HPO_4, KCl, KH_2PO_4, HCl, polyethylene glycols (PEG2000), and BSA were acquired from Sinopharm Chemical Reagent Co. Ltd., Beijing, China, while mPEGA (Mn = 480) was obtained from Sigma-Aldrich (Shanghai) Trading Co. Ltd., Shanghai, China. Additionally, water purified through a Milli-Q system from Millipore was employed.

2.2. Synthesis of Graft Copolymer

PVDF powder and N-methyl-2-pyrrolidone (NMP) were combined in an Erlenmeyer flask as per the ratios specified in Table 1. This mixture was stirred continuously at 70 °C for 24 h to achieve complete dissolution of PVDF in NMP. Once stirring was completed, the solution was allowed to cool to room temperature. Subsequently, varying amounts of mPEGA monomer were added to the flask and stirred until the solution became uniform and transparent. The mixture was then transferred to a glass tube, which was purged with nitrogen gas for 30 min before being sealed. The prepared samples were subjected to gamma-ray irradiation from a source at room temperature for a predetermined duration. Post-irradiation, the solution was carefully poured into an ethanol solution and subsequently into hot water to remove any homopolymer and unreacted monomer. Finally, the resulting precipitates were meticulously dried at 80 °C in a vacuum oven until a constant weight was obtained, yielding the PVDF-g-mPEGA product.

Table 1. Recipe for synthetic PVDF-g-mPEGA.

No.	PVDF/g	mPEGA/g	NMP/g	Total/g
1	10	1	89	100
2	10	4	86	100
3	10	7	83	100
4	10	10	80	100
5	10	13	77	100

The degree of grafting (DG) of PVDF-g-mPEGA is defined as Equation (1):

$$DG = \frac{W_1 - W_0}{W_0} \times 100\% \quad (1)$$

where W_1 denotes the mass of PVDF-g-mPEGA, and W_0 is the mass of pristine PVDF.

2.3. FTIR Characterization

The attenuated total reflectance infrared (ATR-IR) spectrum of the pristine PVDF and the grafted polymer was analyzed using a Nicolet Avatar 370 infrared spectrometer (Thermo Nicolet Instrument Corporation, Madison, WI, USA). Spectral scans were conducted over a range of 400 cm^{-1} to 4000 cm^{-1} with a resolution of 4 cm^{-1}, and each scan was repeated 32 times to ensure accuracy.

Given the challenges in pulverizing the grafted polymer PVDF-g-mPEGA, the solution membrane method was selected for sample preparation. The procedure involved dissolving the grafted polymer in NMP solvent at a 1:9 ratio, stirring the mixture at 70 °C for 24 h, and then allowing it to stand for defoaming. The homogenized solution was then poured onto a glass plate in a controlled environment with air humidity maintained at 45% ± 5% and a temperature of 24 °C ± 1 °C. The solution was evenly spread using a 50 μm spatula and left to stand in air for approximately 15 s. Subsequently, the glass plate was placed horizontally in water at 20 °C, allowing the membrane to detach automatically. The membrane was then marked and immersed in deionized water, with water changes conducted several times. After 24 h of soaking, the membrane was transferred to a vacuum drying oven for drying, making it ready for subsequent use.

2.4. Phase Diagram Drawing

The phase diagram was constructed using cloud point titration. Initially, a series of polymer solutions were prepared with concentrations of 0.2%, 0.4%, 0.6%, and 0.8%, using NMP as the solvent. These solutions were then stirred continuously at a stable temperature of 25 °C. Water was gradually added dropwise until the initially transparent polymer solution turned cloudy. The composition of the solution at the moment of cloudiness, marked by the addition of the last water drop, is defined as the cloud point. Based on the data obtained from cloud point titration, phase diagrams for different solution systems were then plotted.

2.5. Rotational Viscosity

The viscosity of polymer solutions with varying compositions was analyzed using an NDJ-79A rotary viscometer (Tryte Technology (H.K.) Limited, Hong Kong, China) as the rotational speed was altered. Initially, the samples were heated and stirred in a solute-to-solvent ratio of 1:9 to achieve a homogeneous solution. After cooling the solution to room temperature, the viscosity measurements were initiated at a rotational speed of 80 r/min. The speed was then incrementally increased by 20 r/min steps, reaching a maximum of 400 r/min.

2.6. Membrane Preparation

The membranes were fabricated using the non-solvent-induced phase separation (NIPS) method, as described in our group's previous research [9]. Initially, PVDF (5.0 g), graft polymer (5.0 g), and PEG20000-enhanced PVDF (4.5 g PVDF with 0.5 g PEG20000) were dissolved in NMP (35.0 g). This mixture was heated and stirred at 70 °C for one week, followed by a 12-h stabilization period to allow bubble release. Subsequently, the prepared solution was spread onto a glass plate using a casting knife to achieve a uniform thickness of 200 µm. The spread solution was then left to evaporate in ambient air for 15 s before being immersed in deionized water. The membranes were carefully stored in water for two days to ensure complete removal of any residual solvent before testing. The water was changed several times during this period.

2.7. Water Contact Angle Measurements

Membrane contact angle measurements were performed utilizing an Attension Theta system (KSV Instruments Ltd., Helsinki, Finland). A droplet of 5 µL was deposited from a needle tip onto the membrane surface, and its magnified image was recorded using a digital camera. Static contact angles were then determined from these images using computational software. To guarantee accuracy and consistency, the contact angle values were averaged from three separate locations on each membrane.

2.8. SEM Analysis

Membrane morphology was examined using a LEO1530vp scanning electron microscope (Zeiss, Jena, Germany). To obtain the cross-section images, membranes were immersed in liquid nitrogen to freeze them, and then they were cracked in a brittle state. Samples were then mounted on the stage, and a thin gold coating was applied before examination. The scanning was conducted at a voltage of 10 kV and a current of 10 mA. The collected images were processed with the software ImageJ to determine the number of pores per unit area, pore sizes, and pore size distribution.

2.9. Filtration Experiment

The performance of various membranes was assessed using a custom-built cross-flow filtration apparatus (Figure 2). Circular samples, each with a surface area of 15.9 cm^2, were prepared and installed in flat cross-flow cells. Membranes were first subjected to a 20-min pre-compression at 0.1 MPa. Three independent trials were performed for each membrane. Flux was calculated by measuring the volume of solution that permeated over time and normalizing it to the membrane area.

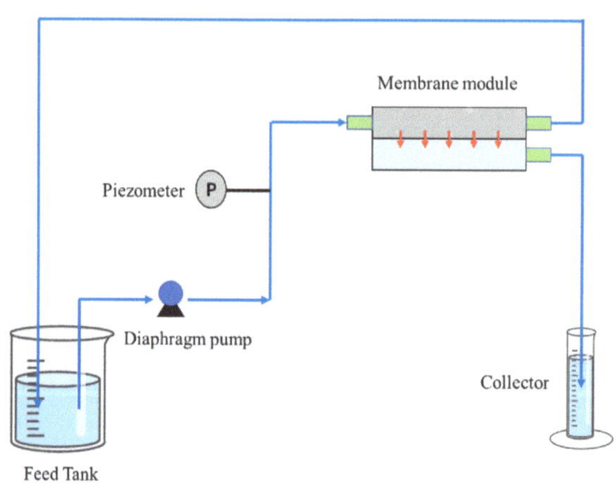

Figure 2. Schematic diagram of the cross-flow device used in the filtration experiments (The control of trans-membrane pressure is regulated through the utilization of a diaphragm pump).

The water flux (F_0) was calculated according to Equation (2):

$$F_0 = \frac{Q}{A\Delta T} \quad (2)$$

where Q refers to the volume of the filtered water; A, the area of the membrane; and ΔT, the time of the filtration.

The preparation of the BSA solution entailed several steps. Initially, a mixture comprising 8 g of NaCl, 1.42 g of Na_2HPO_4, 0.2 g of KCl, 0.27 g of KH_2PO_4, and 800 mL of deionized water was thoroughly mixed in a volumetric flask. The pH of the resulting solution was adjusted to 7.4 using HCl. The final volume was then brought up to 1 L. Following this, 1 g of BSA was dissolved in the prepared buffer solution, resulting in a feed solution with a BSA concentration of 1 g/L. Samples before and after filtration were carefully collected for subsequent analysis. The determination of BSA concentration in the solution was conducted using a UV-1100 visible-ultraviolet spectrophotometer (Ruili Instruments Ltd., Beijing, China), as outlined in the referenced literature [50].

2.10. Anti-Fouling Evaluation

The antifouling properties of the membranes were assessed from two perspectives. The first involved analyzing the recovery ratio of pure water flux post-BSA solution filtration. The second entailed evaluating membrane performance under repeated fouling and cleaning cycles. After a 30-min filtration of the BSA solution, the solution in the cross-flow system was replaced with deionized water. This was followed by a 30-min dedicated cleaning process focused on the membrane surface, culminating in the measurement of water flux recovery.

The flux recovery ratio (FRR) is calculated according to Equation (3):

$$FRR = \frac{F_1}{F_0} \times 100\% \quad (3)$$

where F_0 and F_1 denote the water flux prior to and following fouling, respectively.

3. Results and Discussions

3.1. Kinetics of the Grafted Polymer

The influence of grafting was investigated, considering three variables: Monomer concentration, absorbed radiation dose, and radiation duration. Figure 3 depicts how monomer concentration affects grafting. With a constant absorbed dose of 20 kGy and a radiation time of 18 h, the degree of grafting noticeably increased as the monomer concentration rose from 1% to 7%. Beyond 8% concentration, the grafting level plateaued and then decreased. This pattern is likely due to the enhanced potential for free radical-initiated graft polymerization at higher monomer concentrations, which initially increases grafting. However, at certain absorbed doses, the number of active sites becomes saturated. Consequently, homopolymerization competes with graft copolymerization for these sites. An increase in monomer concentration leads to a higher viscosity in the graft copolymerization solution, affecting monomer solution diffusion. Furthermore, the homopolymerization reaction between monomers can reduce the degree of grafting (DG). Therefore, an increase in monomer concentration does not continuously correlate with an increase in DG.

Figure 4 illustrates the effect of absorbed dose on grafting degree under specified conditions, with a constant monomer concentration of 7% and a radiation dose rate of 0.83 kGy/h. The analysis indicated a moderate impact of the absorbed dose on grafting degree. For absorbed doses ranging from 5 kGy to 40 kGy, the grafting degree of the polymer graft mostly ranged between 8.5% and 10%. It was also noted that up to an absorbed dose of 25 kGy, there was a marginal increase in grafting degree with a rising absorbed dose. Beyond 30 kGy, however, this trend reversed. This shift is attributed to the degradation or cross-linking of PVDF at higher radiation doses [51], resulting in a decrease in free radical active sites available for polymerization, thereby reducing the DG.

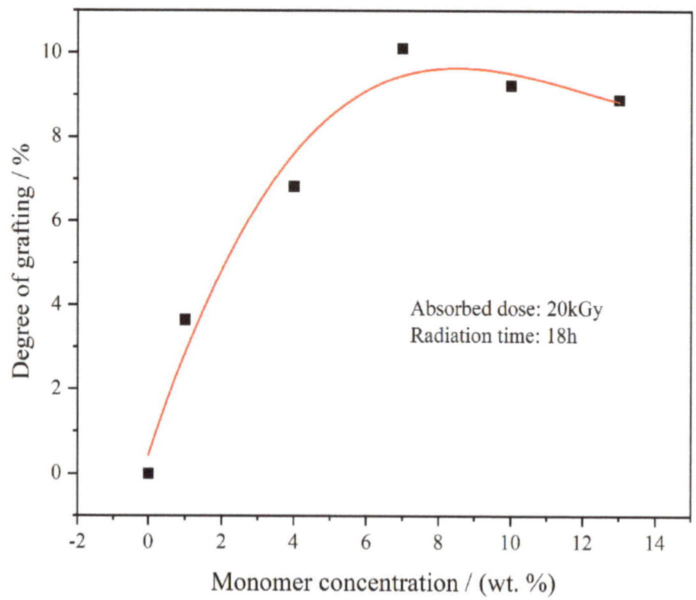

Figure 3. DGs of PVDF-g-mPEGA at different monomer concentration (absorbed dose = 20 kGy; radiation time = 18 h).

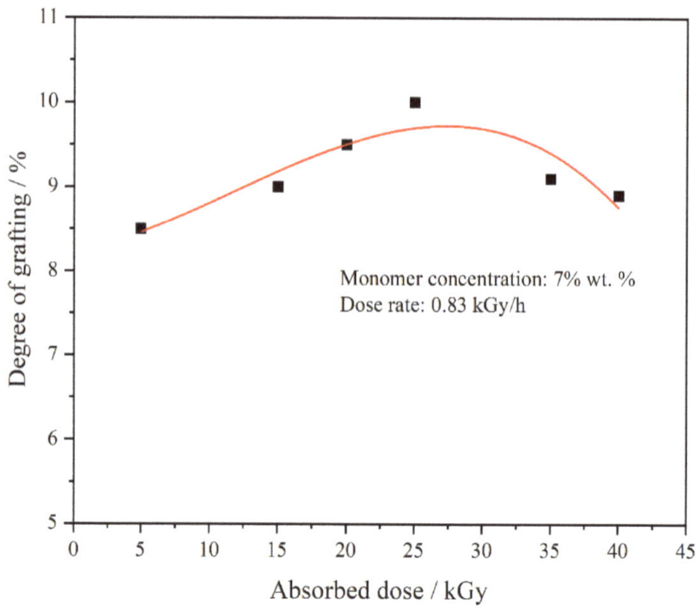

Figure 4. DGs of PVDF-g-mPEGA at different dose rate (monomer concentration = 7 wt.%, dose rate = 0.83 kGy/h).

Figure 5 illustrates the effect of irradiation time on the degree of grafting. Under fixed monomer concentration and absorbed dose conditions, variations in irradiation time had minimal impact on grafting degree. This outcome significantly deviates from the established behavior of homogeneous radiation grafting in small-molecule monomers [9]. The larger molecular weight of mPEGA, compared to smaller molecular monomers, results in a reduced molar concentration of carbon-carbon double bonds available for free radical combination during irradiation, given an equal mass of monomers. Additionally, homogeneous solution radiation grafting causes molecular chains to extend within the solution, thereby accelerating the diffusion process of the graft reaction. Consequently, the grafting reaction tends to reach saturation rapidly.

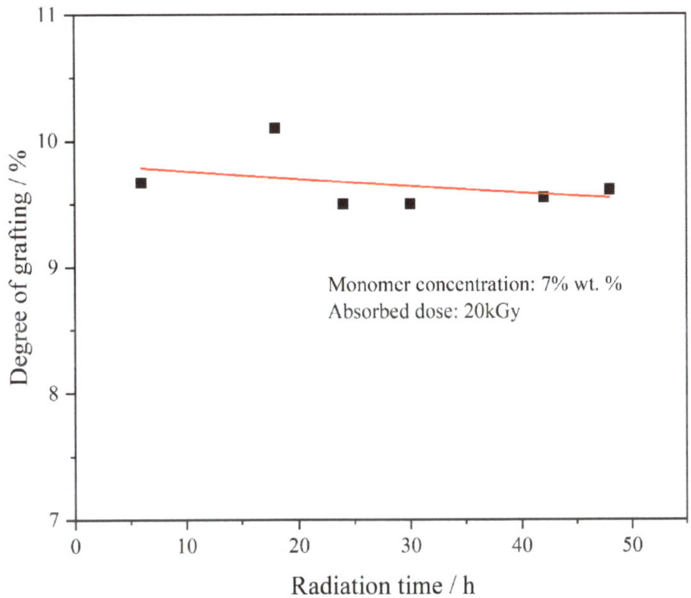

Figure 5. DGs of PVDF-g-mPEGA at different radiation time (monomer concentration = 7 wt.%, absorbed dose = 20 kGy).

3.2. FT-IR Spectroscopy of Grafted Copolymers

Figure 6 displays the infrared spectra of pristine PVDF and PVDF-g-mPEGA with different degrees of grafting (DG) after vector normalization. The characteristic C-F absorption peak, around 1072 cm^{-1}, was noticeable in both the grafted polymer and the original PVDF, while the CF_2 vibrational absorption peak appeared at 1182 cm^{-1}. Additionally, the deformation vibration absorption peak of CH_2 was present at 1404 cm^{-1}. Notably, the grafted polymer exhibited a new characteristic peak at 1726 cm^{-1}, associated with the stretching vibration of the carbonyl group in the mPEGA molecule [52,53]. This peak confirms the successful irradiation grafting of the monomer onto PVDF. Furthermore, a trend was observed where the intensity of the carbonyl peak increased with the degree of grafting.

3.3. Ternary Phase Diagram

This study examined pristine PVDF, irradiated PVDF, PVDF/PEG20000 blends with varying ratios, and PVDF-g-mPEGA with different DG, all titrated with ultra-pure water using NMP as the solvent. Figure 7 demonstrates that phase separation in the PVDF ternary system was more challenging post-irradiation, necessitating greater amounts of water. The propensity for phase separation increased with higher DG and PEG20000 content. However, it was noted that blending's effect on the system was considerably less than that of grafting. The enhanced thermodynamic stability post-irradiation could be attributed to PVDF degradation and the presence of smaller molecules. Conversely, increased grafting and blending ratios led to a decrease in thermodynamic stability, likely due to the reduced PVDF content in the mixture. A more significant decline in thermodynamic stability was observed in graft copolymers, possibly due to an increase in graft copolymer molecules.

Consequently, during phase separation, the speed of this process in the casting solution of PVDF-g-mPEGA was accelerated with increasing DG. This finding implies that membranes manufactured using the same concentration of the polymer-solvent system, but with higher grafting degrees in PVDF-g-mPEGA, are more prone to exhibit pronounced surface loosening.

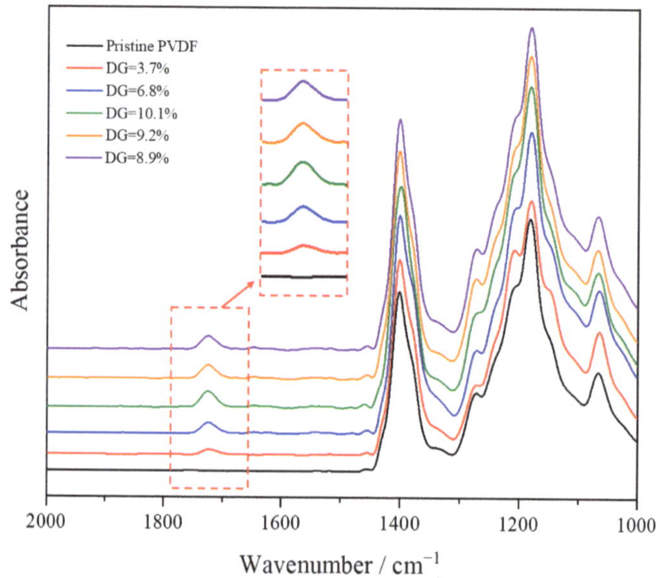

Figure 6. FTIR spectra of the pristine PVDF and PVDF-g-mPEGA copolymers with different DGs.

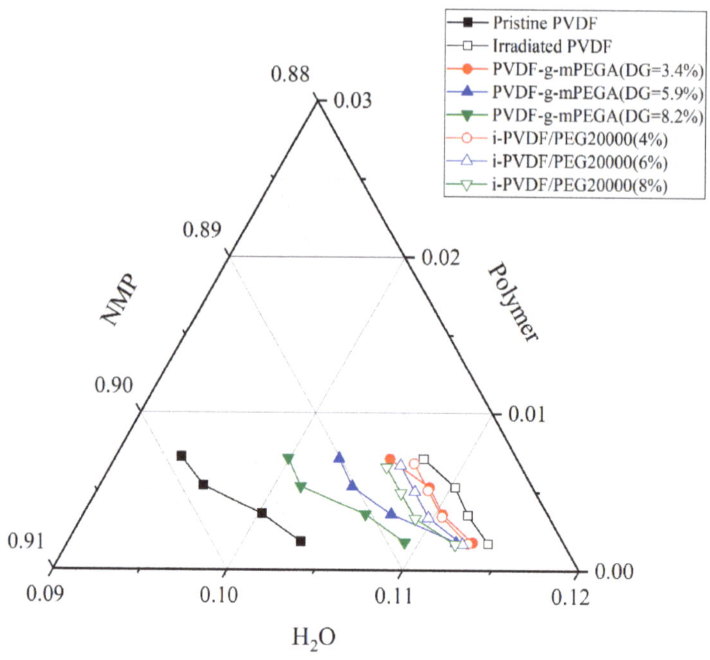

Figure 7. Polymer-NMP-H_2O ternary phase diagram.

3.4. Rotational Viscosity Test

Figure 8 presents a graph illustrating the variation in rotational viscosity of various samples with changes in rotational speed. As depicted, the rotational viscosities of all samples demonstrated a decline to different extents as the rotational speed increased, categorizing PVDF as a pseudoplastic non-Newtonian fluid [54,55]. Concurrently, it was observed that irradiation reduced the rotational viscosity of PVDF, indicating its degradation. However, the impact of alterations in grafting ratio and blending ratio on rotational viscosity was contrary: Both grafting and blending individually led to a decrease in rotational viscosity. Notably, an increase in the DG resulted in higher rotational viscosity, while an increased blending ratio lowered it. This viscosity reduction can be attributed to the decreased PVDF content in the total composition. In contrast, compared to the mixed system, a rise in DG led to increased viscosity, possibly due to the augmentation of graft content and molecule size. This distinction further confirms that mPEGA was grafted onto PVDF, rather than merely blended.

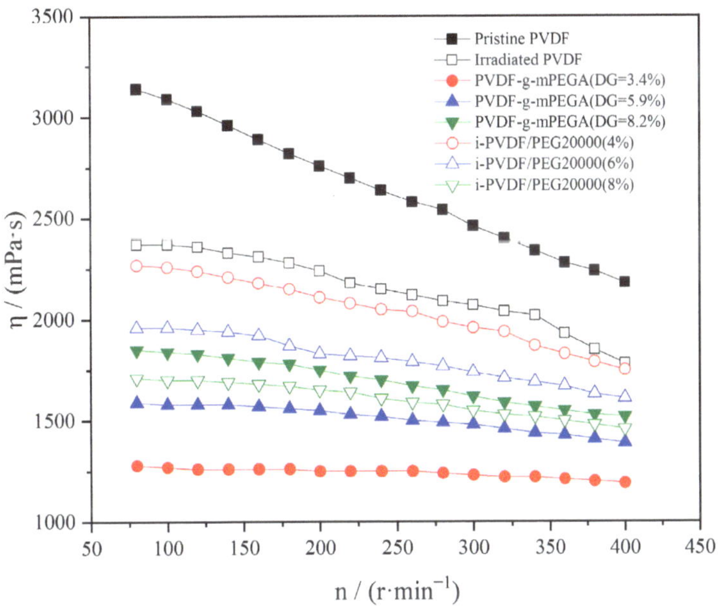

Figure 8. Diagram of the rotating viscosity of different samples as a function of rotational speed.

3.5. Water Contact Angle Test

It is widely recognized that enhancing the hydrophilicity of a membrane typically improves both its flux and antifouling properties [56,57]. Figure 9 demonstrates this principle, showing the contact angle of pristine PVDF prior to grafting as 93.8°, indicative of significant hydrophobicity, consistent with the inherent characteristics of PVDF materials. At a grafting degree of 3.4%, the contact angle decreased to 77.4°. Further, there was a continuous decrease in water contact angle with increasing degrees of grafting, reaching 66.4° at a grafting degree of 8.2%. Thus, it is evident that the grafted PVDF exhibited a substantial improvement in hydrophilicity.

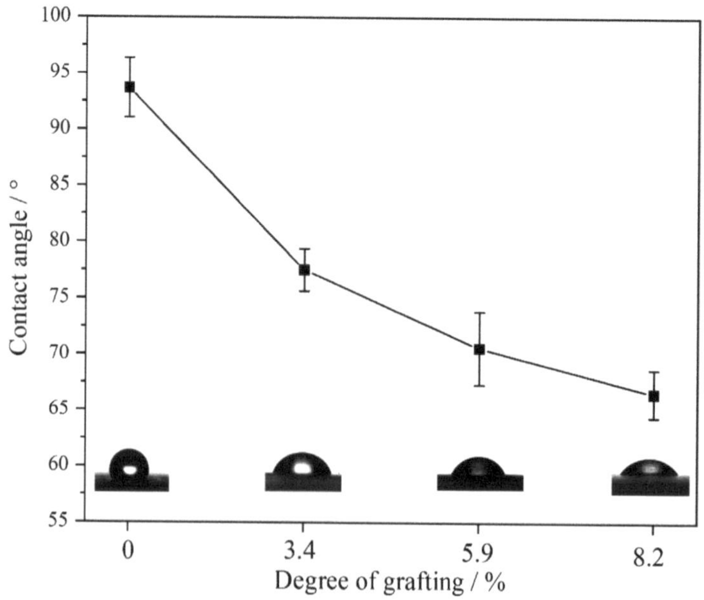

Figure 9. Water contact angle of pristine PVDF and PVDF-g-mPEGA with different DGs. The static water contact angle images corresponding to individual membranes are presented below the line graph.

3.6. Membrane Morphology Characterization

Figure 10 illustrates the surface and cross-sectional structures of the membrane, while Figure 11 displays the pore size distribution histogram derived from surface SEM images calculated by using ImageJ. The findings reveal that all membrane surfaces exhibit improved nanoscale pore structures. The pore size shows an increasing trend with the increase in DG. The cross-sectional SEM image of the PVDF membrane confirmed its asymmetric structure, characteristic of such membranes. The uppermost layer of the PVDF membrane consists of a dense skin layer, whereas the lower surface features a supportive structure. As depicted in Figure 10, grafting did not modify the skin-finger structure of the membrane. However, a comparative analysis of the support layer structures under different membranes revealed that, unlike the pristine PVDF membrane, the PVDF-g-mPEGA graft-modified membrane exhibited large macrovoids structures. These structures were interconnected with the pores, extending throughout the entire membrane. The following explanations are proposed, integrating this study with previous experimental results: On the one hand, the increased thermodynamic stability of the grafted polymer will lead to slow initial demixing; that is, liquid–liquid demixing occurs before the gelation process. This creates a polymer-rich phase and a polymer-poor phase, corresponding to the membrane matrix and pores, respectively, and the tiny macrovoids formed in the skin layer. Subsequently, some of the advancing polymer-lean phases may coalesce and advance in a disordered manner, with tiny macrovoids in the skin layer transforming into larger and irregularly shaped macrovoids at the base of the sublayer [58]. On the other hand, as the grafting rate of the graft polymer increases, the content of the hydrophilic graft segment increases, and the diffusion rate of the solvent becomes faster, resulting in a greater tendency to form macrovoids during the membrane formation process [59].

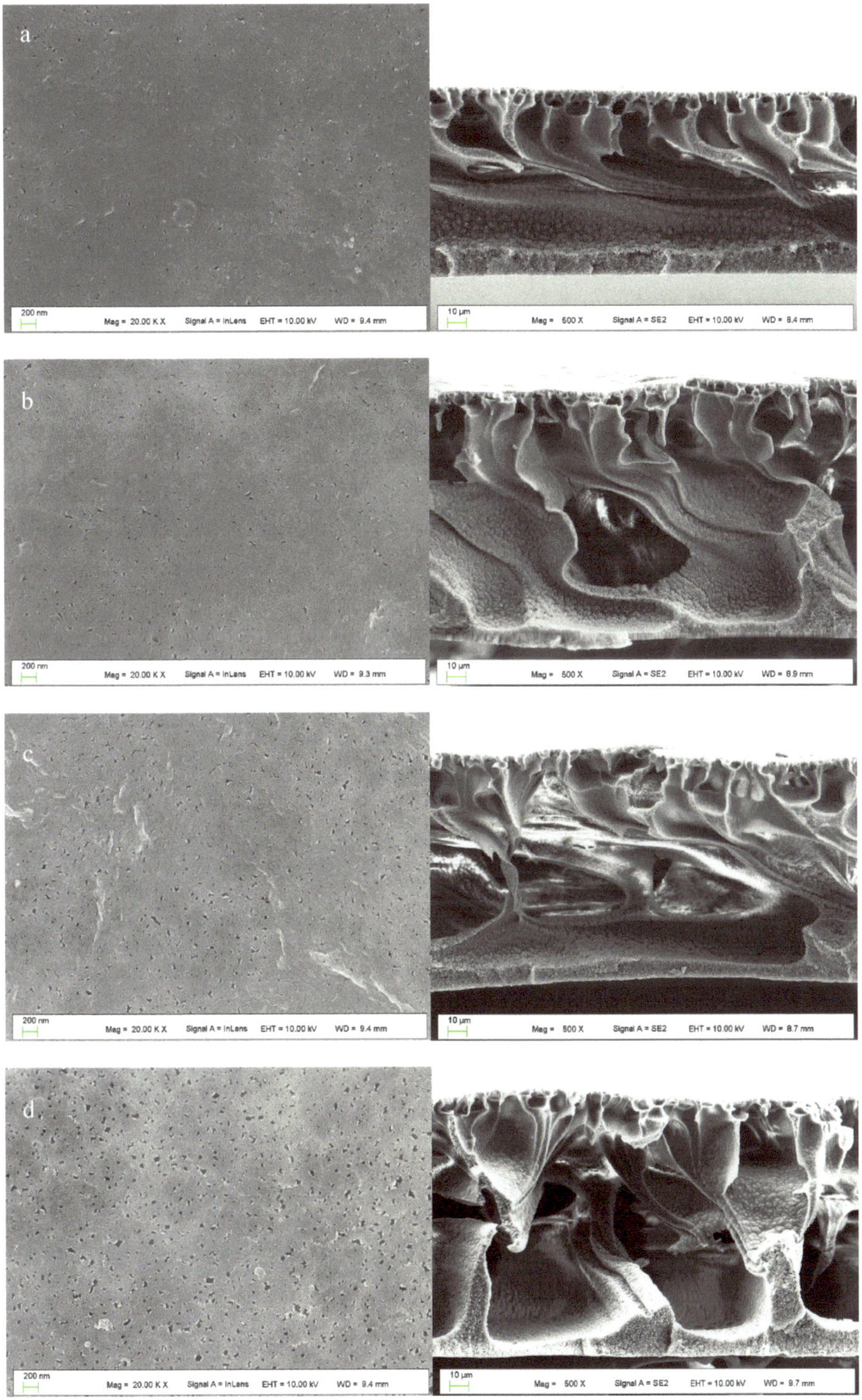

Figure 10. SEM images of the membrane surface (left) and cross section (right): (**a**) pristine PVDF, (**b**) PVDF-g-mPEGA (DG = 3.4%), (**c**) PVDF-g-mPEGA (DG = 5.9%), (**d**) PVDF-g-mPEGA (DG = 8.2%).

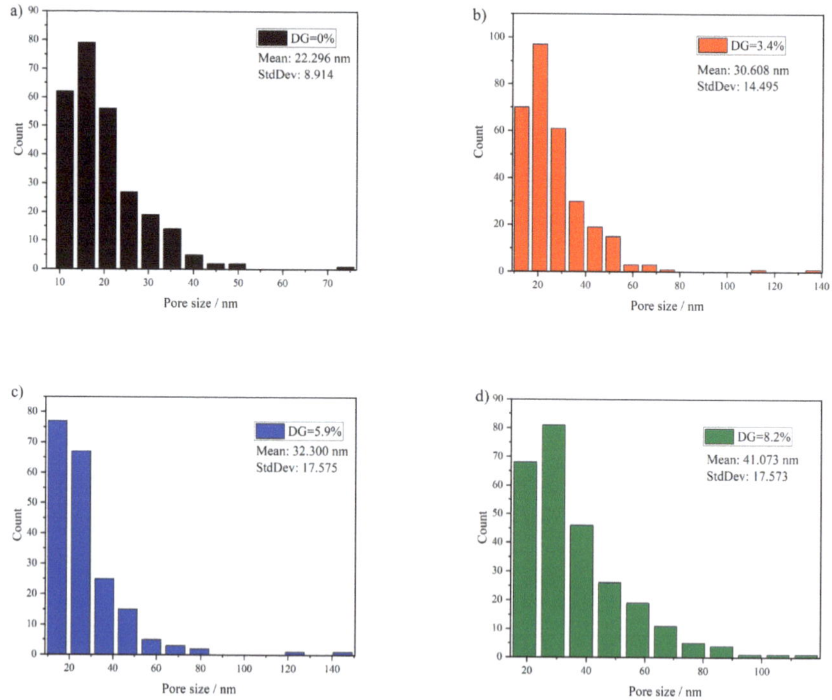

Figure 11. Histogram of the pore size distribution calculated using the surface SEM images. (**a**) Pristine PVDF, (**b**) PVDF-g-mPEGA (DG = 3.4%), (**c**) PVDF-g-mPEGA (DG = 5.9%), (**d**) PVDF-g-mPEGA (DG = 8.2%).

3.7. Membrane Filtration Performance

Figure 12 presents the water flux and BSA rejection rates of pristine PVDF, irradiated PVDF, and PVDF-g-mPEGA membranes with varying DG. The water flux and BSA rejection rates for the PVDF membrane, both pre- and post-irradiation, showed negligible differences, indicating that the irradiation dose employed had minimal impact on the membrane's properties. As DG increased from 0 to 3.4%, 5.9%, and 8.2%, the water flux correspondingly rose from 5.2 LMH to 18.6 LMH, 23.9 LMH, and 43.5 LMH. This trend indicates that graft modification of PVDF significantly enhanced its ultrafiltration membrane flux, with a clear positive correlation between higher DG and improved flux. The reasons for this are twofold: Firstly, pure PVDF is known for its strong hydrophobicity, which results in substantial filtration resistance in the PVDF membrane during the water filtration process, necessitating a high driving force for achieving high flux [60,61]. The contact angle test results earlier demonstrated that grafting with the hydrophilic monomer mPEGA significantly increased the hydrophilicity of the modified membrane. This increase in hydrophilicity reduced filtration resistance, leading to a higher flux. The greater the DG, the more pronounced the improvement in hydrophilicity, thereby enhancing the flux. Secondly, the membrane morphology characterization revealed an increase in surface porosity after modification. Additionally, SEM analysis showed that large cavity structures within the support layer of the modified membrane were interconnected with the pores, extending throughout the membrane. This structural arrangement facilitated increased membrane flux. In BSA rejection tests, all types of PVDF membranes consistently achieved BSA rejection ratios over 80%.

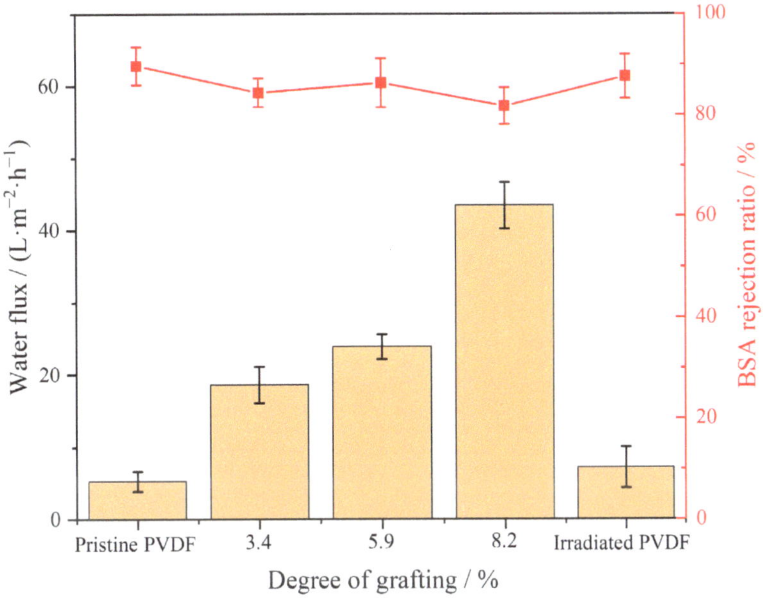

Figure 12. Water flux and BSA rejection of the pristine PVDF, irradiated PVDF, and PVDF-g-mPEGA membranes with different DGs.

3.8. Anti-Fouling Evaluation

The assessment of antifouling performance for each membrane typically includes calculating the pure water FRR [62,63]. In this experiment, an in situ measurement technique was employed, integrating online cleaning of the membrane via a cross-flow filtration method. This process entailed alternating the fluid composition, specifically switching between water and a BSA solution.

Figure 13a displays the fouling and cleaning flux profiles over multiple cycles for different types of PVDF membranes, whereas Figure 13b presents the corresponding flux recovery rates (FRR) calculated using the data from Figure 13a. PEG20000 was incorporated to enhance the flux of pure PVDF for antifouling evaluation. Notably, membranes grafted with mPEGA demonstrated a significantly higher flux recovery rate compared to PVDF/PEG20000 membranes. The FRR of PVDF-g-mPEGA membranes with various degrees of grafting (DG) remained above 90% and was stable across multiple cycles. These findings highlight the effectiveness of PVDF-g-mPEGA in achieving optimal antifouling performance with minimal grafting, thereby preserving the bulk material properties of the membrane. The flux and FRR of irradiated PVDF/PEG20000 (i-PVDF/PEG20000) membranes were similar to those of PVDF/PEG20000, indicating that irradiation had a negligible impact on the membrane's formative properties. PVDF-g-NVP, a modified PVDF membrane created using NVP (a commonly used small-molecule monomer suitable for homogeneous radiation) as the graft monomer through the aforementioned methods, showed improved FRR compared to the PVDF/PEG20000 membrane after repeated fouling and cleaning tests. However, its FRR was still considerably lower than that of the PVDF-g-mPEGA membrane. This difference accentuates the unique benefits of mPEGA as a graft monomer in enhancing the antifouling properties of PVDF membranes, especially when compared to NVP.

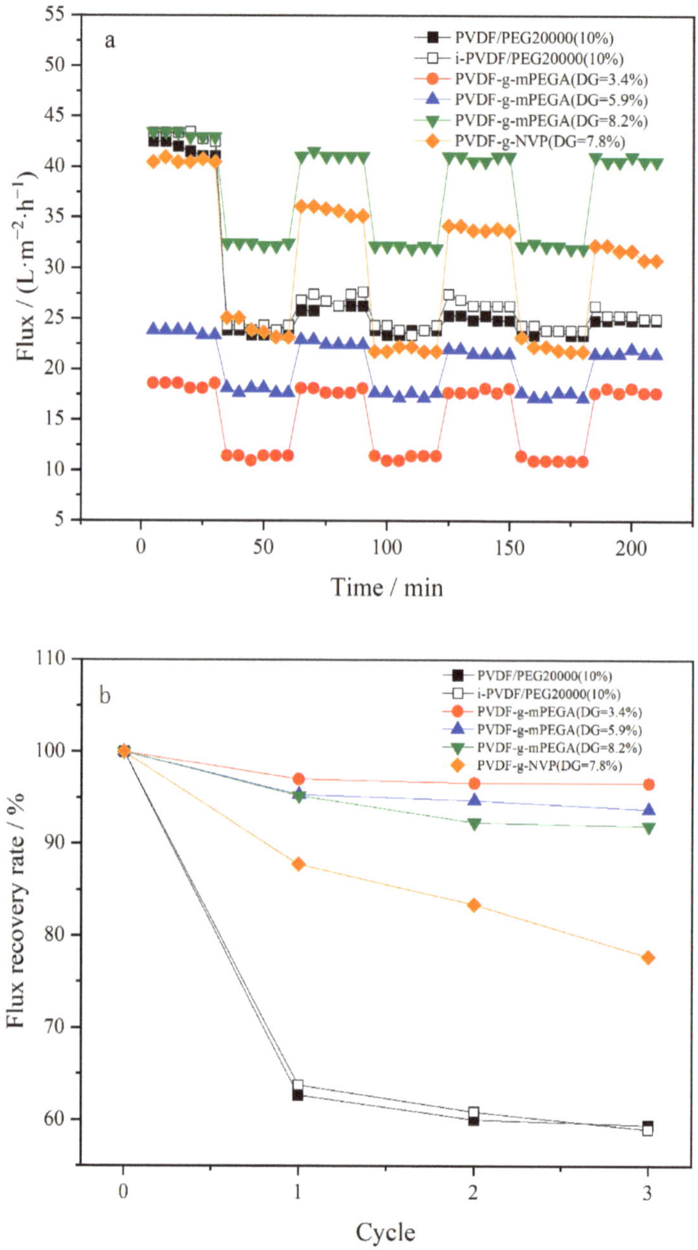

Figure 13. Anti-fouling test of the membranes: (**a**) multiple cycle fouling and cleaning fluxes of different types of PVDF membranes; (**b**) FRRS of different types of PVDF membranes under multiple cycles.

4. Conclusions

In this study, mPEGA, a novel PEGylated monomer, was grafted onto PVDF using homogeneous γ-ray irradiation. Analysis of grafting kinetics revealed that the DG was primarily influenced by monomer concentration, with an increase in DG corresponding to higher concentrations. Thermodynamic analysis of the grafted products indicated that during membrane formation, the phase separation rate of the PVDF-g-mPEGA casting solution accelerated with an increase in DG, thereby enhancing membrane hydrophilicity.

Although irradiation grafting induced some degradation of PVDF, as evidenced by the rotational viscosity test and phase diagram, both the flux and anti-fouling tests demonstrated minimal impact on the membrane-forming properties of PVDF. Moreover, a range of membranes was fabricated using the NIPS method, and SEM analyses confirmed an expansion in membrane pore size commensurate with increasing DGs. The flux and anti-fouling assessments consistently highlighted the superior anti-fouling properties of the PVDF-g-mPEGA membrane compared to the PVDF-g-NVP membrane. Significantly, PVDF-g-mPEGA achieved the highest anti-fouling effect at a very low DG, thus minimizing alterations to the membrane's bulk material.

From an overall perspective, in addition to better performance, the scalability, processability, and cost of the manufacturing method, as well as the long-term stability of the membrane, all need to be considered [64]. In this study, the anti-fouling cycle test was only conducted for more than 200 min, so the long-term sustainability of the membrane's performance requires further study. However, the fabrication method of the membrane in this work is simple, making it more scalable and having a low manufacturing cost from an industrial perspective.

Author Contributions: Conceptualization, T.W. and Z.H.; methodology, Z.H.; validation, T.W. and Z.H.; data curation, T.W., Z.H. and H.Y.; writing—original draft preparation, T.W.; writing—review and editing, Z.H., H.Y. and J.H.; project administration, Z.H.; funding acquisition, Z.H., H.Y. and J.H. All authors have read and agreed to the published version of the manuscript.

Funding: This work was financially supported by the National Natural Foundation of China (NSFC, No. 11375252 and No. U1632135) and the National Key Research and Development Program (2022YFA1603600).

Institutional Review Board Statement: Not applicable.

Informed Consent Statement: Not applicable.

Data Availability Statement: Data are contained within the article.

Conflicts of Interest: The authors declare no conflicts of interest.

References

1. Liao, Z.; Zhu, J.; Li, X.; Van Der Bruggen, B. Regulating Composition and Structure of Nanofillers in Thin Film Nanocomposite (TFN) Membranes for Enhanced Separation Performance: A Critical Review. *Sep. Purif. Technol.* **2021**, *266*, 118567. [CrossRef]
2. Zhao, S.; Liao, Z.; Fane, A.; Li, J.; Tang, C.; Zheng, C.; Lin, J.; Kong, L. Engineering Antifouling Reverse Osmosis Membranes: A Review. *Desalination* **2021**, *499*, 114857. [CrossRef]
3. Li, X.; Sotto, A.; Li, J.; Van Der Bruggen, B. Progress and Perspectives for Synthesis of Sustainable Antifouling Composite Membranes Containing in Situ Generated Nanoparticles. *J. Membr. Sci.* **2017**, *524*, 502–528. [CrossRef]
4. Xue, W.; Jian, M.; Lin, T.; Ma, B.; Wu, R.; Li, X. A Novel Strategy to Alleviate Ultrafiltration Membrane Fouling by Rotating Membrane Module. *Chemosphere* **2020**, *260*, 127535. [CrossRef] [PubMed]
5. Yang, J.; Monnot, M.; Eljaddi, T.; Ercolei, L.; Simonian, L.; Moulin, P. Ultrafiltration as Tertiary Treatment for Municipal Wastewater Reuse. *Sep. Purif. Technol.* **2021**, *272*, 118921. [CrossRef]
6. Dmitrenko, M.; Kuzminova, A.; Zolotarev, A.; Markelov, D.; Komolkin, A.; Loginova, E.; Plisko, T.; Burts, K.; Bildyukevich, A.; Penkova, A. Modification Strategies of Polyacrylonitrile Ultrafiltration Membrane Using TiO2 for Enhanced Antifouling Performance in Water Treatment. *Sep. Purif. Technol.* **2022**, *286*, 120500. [CrossRef]
7. Ahmad, T.; Guria, C.; Mandal, A. A Review of Oily Wastewater Treatment Using Ultrafiltration Membrane: A Parametric Study to Enhance the Membrane Performance. *J. Water Process Eng.* **2020**, *36*, 101289. [CrossRef]
8. Desiriani, R.; Susanto, H.; Istirokhatun, T.; Lin, Y.; Aryanti, N.; Abriyanto, H.; Saputra, H.; Matsuyama, H. Preparation of Polyethersulfone Ultrafiltration Membrane Coated Natural Additives toward Antifouling and Antimicrobial Agents for Surface Water Filtration. *J. Environ. Chem. Eng.* **2024**, *12*, 111797. [CrossRef]
9. Qin, Q.; Hou, Z.; Lu, X.; Bian, X.; Chen, L.; Shen, L.; Wang, S. Microfiltration Membranes Prepared from Poly(N-Vinyl-2-Pyrrolidone) Grafted Poly(Vinylidene Fluoride) Synthesized by Simultaneous Irradiation. *J. Membr. Sci.* **2013**, *427*, 303–310. [CrossRef]

10. Chen, L.; Hou, Z.; Lu, X.; Chen, P.; Liu, Z.; Shen, L.; Bian, X.; Qin, Q. Antifouling Microfiltration Membranes Prepared from Poly(Vinylidene Fluoride)-*Graft*-Poly(*N*-Vinyl Pyrrolidone) Powders Synthesized via Pre-Irradiation Induced Graft Polymerization. *J. Appl. Polym. Sci.* **2013**, *128*, 3949–3956. [CrossRef]
11. Valizadeh, K.; Heydarinasab, A.; Hosseini, S.S.; Bazgir, S. Preparation of Modified Membrane of Polyvinylidene Fluoride (PVDF) and Evaluation of Anti-Fouling Features and High Capability in Water/Oil Emulsion Separation. *J. Taiwan Inst. Chem. Eng.* **2021**, *126*, 36–49. [CrossRef]
12. Salahshoori, I.; Mohseni, A.; Namayandeh Jorabchi, M.; Ghasemi, S.; Afshar, M.; Wohlrab, S. Study of Modified PVDF Membranes with High-Capacity Adsorption Features Using Quantum Mechanics, Monte Carlo, and Molecular Dynamics Simulations. *J. Mol. Liq.* **2023**, *375*, 121286. [CrossRef]
13. Liu, F.; Hashim, N.A.; Liu, Y.; Abed, M.R.M.; Li, K. Progress in the Production and Modification of PVDF Membranes. *J. Membr. Sci.* **2011**, *375*, 1–27. [CrossRef]
14. Kang, G.; Cao, Y. Application and Modification of Poly(Vinylidene Fluoride) (PVDF) Membranes—A Review. *J. Membr. Sci.* **2014**, *463*, 145–165. [CrossRef]
15. Xu, C.; Huang, W.; Lu, X.; Yan, D.; Chen, S.; Huang, H. Preparation of PVDF Porous Membranes by Using PVDF-g-PVP Powder as an Additive and Their Antifouling Property. *Radiat. Phys. Chem.* **2012**, *81*, 1763–1769. [CrossRef]
16. Lowe, S.; O'Brien-Simpson, N.M.; Connal, L.A. Antibiofouling Polymer Interfaces: Poly(Ethylene Glycol) and Other Promising Candidates. *Polym. Chem.* **2015**, *6*, 198–212. [CrossRef]
17. Knop, K.; Hoogenboom, R.; Fischer, D.; Schubert, U.S. Poly(Ethylene Glycol) in Drug Delivery: Pros and Cons as Well as Potential Alternatives. *Angew. Chem. Int. Ed.* **2010**, *49*, 6288–6308. [CrossRef]
18. Lee, J.H.; Kopecek, J.; Andrade, J.D. Protein-Resistant Surfaces Prepared by PEO-Containing Block Copolymer Surfactants. *J. Biomed. Mater. Res.* **1989**, *23*, 351–368. [CrossRef]
19. Cai, L.; Wang, K.; Wang, S. Poly(Ethylene Glycol)-Grafted Poly(Propylene Fumarate) Networks and Parabolic Dependence of MC3T3 Cell Behavior on the Network Composition. *Biomaterials* **2010**, *31*, 4457–4466. [CrossRef]
20. Ma, N.; Cao, J.; Li, H.; Zhang, Y.; Wang, H.; Meng, J. Surface Grafting of Zwitterionic and PEGylated Cross-Linked Polymers toward PVDF Membranes with Ultralow Protein Adsorption. *Polymer* **2019**, *167*, 1–12. [CrossRef]
21. Valizadeh, K.; Heydarinasab, A.; Hosseini, S.S.; Bazgir, S. Fabrication of Modified PVDF Membrane in the Presence of PVI Polymer and Evaluation of Its Performance in the Filtration Process. *J. Ind. Eng. Chem.* **2022**, *106*, 411–428. [CrossRef]
22. Ahmad, T.; Guria, C. Progress in the Modification of Polyvinyl Chloride (PVC) Membranes: A Performance Review for Wastewater Treatment. *J. Water Process Eng.* **2022**, *45*, 102466. [CrossRef]
23. Alipoor, Z.; Mousavi, S.M.; Saljoughi, E.; Karkhanechi, H. Toward Fabrication of Fouling Resistant Pervaporation Membrane for Desalination: Surface Modification of TFC Membrane via Grafting of mPEG-NH2. *Desalination* **2023**, *567*, 116970. [CrossRef]
24. Fan, K.; Huang, J.; Yang, H.; Lu, R.; Sun, X.; Hu, J.; Hou, Z. pH and Thermal-Dependent Ultrafiltration Membranes Prepared from Poly (Methacrylic Acid) Grafted onto Polyethersulfone Synthesized by Simultaneous Irradiation in Homogenous Phase. *J. Membr. Sci.* **2017**, *543*, 335–341. [CrossRef]
25. Deng, B.; Yang, X.; Xie, L.; Li, J.; Hou, Z.; Yao, S.; Liang, G.; Sheng, K.; Huang, Q. Microfiltration Membranes with pH Dependent Property Prepared from Poly(Methacrylic Acid) Grafted Polyethersulfone Powder. *J. Membr. Sci.* **2009**, *330*, 363–368. [CrossRef]
26. Fan, K.; Su, J.; Zeng, Z.; Hu, J.; Yang, H.; Hou, Z. Anti-Fouling and Protein Separation of PVDF-g-PMAA@MnO2 Filtration Membrane with in-Situ Grown MnO2 Nanorods. *Chemosphere* **2022**, *286*, 131756. [CrossRef]
27. Deng, B.; Li, J.; Hou, Z.; Yao, S.; Shi, L.; Liang, G.; Sheng, K. Microfiltration Membranes Prepared from Polyethersulfone Powder Grafted with Acrylic Acid by Simultaneous Irradiation and Their pH Dependence. *Radiat. Phys. Chem.* **2008**, *77*, 898–906. [CrossRef]
28. Lou, D.; Hou, Z.; Yang, H.; Liu, Y.; Wang, T. Antifouling Membranes Prepared from Polyethersulfone Grafted with Poly(Ethylene Glycol) Methacrylate by Radiation-Induced Copolymerization in Homogeneous Solution. *ACS Omega* **2020**, *5*, 27094–27102. [CrossRef]
29. Wang, J.J.; Liu, F. Imparting Antifouling Properties of Silicone Hydrogels by Grafting Poly(Ethylene Glycol) Methyl Ether Acrylate Initiated by UV Light. *J. Appl. Polym. Sci.* **2012**, *125*, 548–554. [CrossRef]
30. Bozukova, D.; Pagnoulle, C.; De Pauw-Gillet, M.-C.; Ruth, N.; Jérôme, R.; Jérôme, C. Imparting Antifouling Properties of Poly(2-Hydroxyethyl Methacrylate) Hydrogels by Grafting Poly(Oligoethylene Glycol Methyl Ether Acrylate). *Langmuir* **2008**, *24*, 6649–6658. [CrossRef]
31. Asatekin, A.; Kang, S.; Elimelech, M.; Mayes, A.M. Anti-Fouling Ultrafiltration Membranes Containing Polyacrylonitrile-Graft-Poly(Ethylene Oxide) Comb Copolymer Additives. *J. Membr. Sci.* **2007**, *298*, 136–146. [CrossRef]
32. Barroso, T.; Temtem, M.; Casimiro, T.; Aguiar-Ricardo, A. Antifouling Performance of Poly(Acrylonitrile)-Based Membranes: From Green Synthesis to Application. *J. Supercrit. Fluids* **2011**, *56*, 312–321. [CrossRef]
33. Sinha, M.K.; Purkait, M.K. Increase in Hydrophilicity of Polysulfone Membrane Using Polyethylene Glycol Methyl Ether. *J. Membr. Sci.* **2013**, *437*, 7–16. [CrossRef]
34. Saini, B.; Sinha, M.K. Effect of Hydrophilic Poly(Ethylene Glycol) Methyl Ether Additive on the Structure, Morphology, and Performance of Polysulfone Flat Sheet Ultrafiltration Membrane. *J. Appl. Polym. Sci.* **2019**, *136*, 47163. [CrossRef]

35. Huang, J.; Xue, J.; Xiang, K.; Zhang, X.; Cheng, C.; Sun, S.; Zhao, C. Surface Modification of Polyethersulfone Membranes by Blending Triblock Copolymers of Methoxyl Poly(Ethylene Glycol)–Polyurethane–Methoxyl Poly(Ethylene Glycol). *Colloids Surf. B Biointerfaces* **2011**, *88*, 315–324. [CrossRef]
36. Singh, R.; Purkait, M.K. Evaluation of mPEG Effect on the Hydrophilicity and Antifouling Nature of the PVDF-Co-HFP Flat Sheet Polymeric Membranes for Humic Acid Removal. *J. Water Process Eng.* **2016**, *14*, 9–18. [CrossRef]
37. Chang, Y.; Shih, Y.-J.; Ruaan, R.-C.; Higuchi, A.; Chen, W.-Y.; Lai, J.-Y. Preparation of Poly(Vinylidene Fluoride) Microfiltration Membrane with Uniform Surface-Copolymerized Poly(Ethylene Glycol) Methacrylate and Improvement of Blood Compatibility. *J. Membr. Sci.* **2008**, *309*, 165–174. [CrossRef]
38. Wang, P.; Tan, K.L.; Kang, E.T.; Neoh, K.G. Synthesis, Characterization and Anti-Fouling Properties of Poly(Ethylene Glycol) Grafted Poly(Vinylidene Fluoride) Copolymer Membranes. *J. Mater. Chem.* **2001**, *11*, 783–789. [CrossRef]
39. Chang, Y.; Ko, C.-Y.; Shih, Y.-J.; Quémener, D.; Deratani, A.; Wei, T.-C.; Wang, D.-M.; Lai, J.-Y. Surface Grafting Control of PEGylated Poly(Vinylidene Fluoride) Antifouling Membrane via Surface-Initiated Radical Graft Copolymerization. *J. Membr. Sci.* **2009**, *345*, 160–169. [CrossRef]
40. Hester, J.F.; Banerjee, P.; Won, Y.-Y.; Akthakul, A.; Acar, M.H.; Mayes, A.M. ATRP of Amphiphilic Graft Copolymers Based on PVDF and Their Use as Membrane Additives. *Macromolecules* **2002**, *35*, 7652–7661. [CrossRef]
41. Liu, B.; Chen, C.; Li, T.; Crittenden, J.; Chen, Y. High Performance Ultrafiltration Membrane Composed of PVDF Blended with Its Derivative Copolymer PVDF-g-PEGMA. *J. Membr. Sci.* **2013**, *445*, 66–75. [CrossRef]
42. Wang, P.; Tan, K.L.; Kang, E.T.; Neoh, K.G. Plasma-Induced Immobilization of Poly(Ethylene Glycol) onto Poly(Vinylidene Fluoride) Microporous Membrane. *J. Membr. Sci.* **2002**, *195*, 103–114. [CrossRef]
43. Liu, F.; Du, C.-H.; Zhu, B.-K.; Xu, Y.-Y. Surface Immobilization of Polymer Brushes onto Porous Poly(Vinylidene Fluoride) Membrane by Electron Beam to Improve the Hydrophilicity and Fouling Resistance. *Polymer* **2007**, *48*, 2910–2918. [CrossRef]
44. Susanto, H.; Balakrishnan, M.; Ulbricht, M. Via Surface Functionalization by Photograft Copolymerization to Low-Fouling Polyethersulfone-Based Ultrafiltration Membranes. *J. Membr. Sci.* **2007**, *288*, 157–167. [CrossRef]
45. Susanto, H.; Arafat, H.; Janssen, E.M.L.; Ulbricht, M. Ultrafiltration of Polysaccharide–Protein Mixtures: Elucidation of Fouling Mechanisms and Fouling Control by Membrane Surface Modification. *Sep. Purif. Technol.* **2008**, *63*, 558–565. [CrossRef]
46. Zhai, G.; Kang, E.T.; Neoh, K.G. Poly(2-Vinylpyridine)- and Poly(4-Vinylpyridine)-Graft-Poly(Vinylidene Fluoride) Copolymers and Their pH-Sensitive Microfiltration Membranes. *J. Membr. Sci.* **2003**, *217*, 243–259. [CrossRef]
47. Lanzalaco, S.; Fantin, M.; Scialdone, O.; Galia, A.; Isse, A.A.; Gennaro, A.; Matyjaszewski, K. Atom Transfer Radical Polymerization with Different Halides (F, Cl, Br, and I): Is the Process "Living" in the Presence of Fluorinated Initiators? *Macromolecules* **2017**, *50*, 192–202. [CrossRef]
48. Xu, Z.; Chen, L.; Zhou, B.; Li, Y.; Li, B.; Niu, J.; Shan, M.; Guo, Q.; Wang, Z.; Qian, X. Nano-Structure and Property Transformations of Carbon Systems under γ-Ray Irradiation: A Review. *RSC Adv.* **2013**, *3*, 10579–10597. [CrossRef]
49. Dargaville, T.R.; George, G.A.; Hill, D.J.T.; Whittaker, A.K. High Energy Radiation Grafting of Fluoropolymers. *Prog. Polym. Sci.* **2003**, *28*, 1355–1376. [CrossRef]
50. Noble, J.E.; Bailey, M.J.A. Chapter 8 Quantitation of Protein. In *Methods in Enzymology*, Burgess, R.R., Deutscher, M.P., Eds.; Guide to Protein Purification, 2nd ed.; Academic Press: Cambridge, MA, USA, 2009; Volume 463, pp. 73–95.
51. Zhudi, Z.; Wenxue, Y.; Xinfang, C. Study on Increase in Crystallinity in γ-Irradiated Poly(Vinylidene Fluoride). *Radiat. Phys. Chem.* **2002**, *65*, 173–176. [CrossRef]
52. Lanceros-Méndez, S.; Mano, J.F.; Costa, A.M.; Schmidt, V.H. FTIR and DSC Studies of Mechanically Deformed Β-PVDF Films. *J. Macromol. Sci. Part B* **2001**, *40*, 517–527. [CrossRef]
53. Ross, G.J.; Watts, J.F.; Hill, M.P.; Morrissey, P. Surface Modification of Poly(Vinylidene Fluoride) by Alkaline Treatment1. The Degradation Mechanism. *Polymer* **2000**, *41*, 1685–1696. [CrossRef]
54. Hussain, A.; Akbar, S.; Sarwar, L.; Nadeem, S.; Iqbal, Z. Effect of Time Dependent Viscosity and Radiation Efficacy on a Non-Newtonian Fluid Flow. *Heliyon* **2019**, *5*, e01203. [CrossRef] [PubMed]
55. Chhabra, R.P. Non-Newtonian Fluids: An Introduction. In *Rheology of Complex Fluids*; Krishnan, J.M., Deshpande, A.P., Kumar, P.B.S., Eds.; Springer: New York, NY, USA, 2010; pp. 3–34. ISBN 978-1-4419-6494-6.
56. Zhang, M.; Nguyen, Q.T.; Ping, Z. Hydrophilic Modification of Poly (Vinylidene Fluoride) Microporous Membrane. *J. Membr. Sci.* **2009**, *327*, 78–86. [CrossRef]
57. Subasi, Y.; Cicek, B. Recent Advances in Hydrophilic Modification of PVDF Ultrafiltration Membranes—A Review: Part I. *Membr. Technol.* **2017**, *2017*, 7–12. [CrossRef]
58. Lim, Y.J.; Goh, K.; Lai, G.S.; Zhao, Y.; Torres, J.; Wang, R. Unraveling the Role of Support Membrane Chemistry and Pore Properties on the Formation of Thin-Film Composite Polyamide Membranes. *J. Membr. Sci.* **2021**, *640*, 119805. [CrossRef]
59. Ying, L.; Kang, E.T.; Neoh, K.G. Covalent Immobilization of Glucose Oxidase on Microporous Membranes Prepared from Poly(Vinylidene Fluoride) with Grafted Poly(Acrylic Acid) Side Chains. *J. Membr. Sci.* **2002**, *208*, 361–374. [CrossRef]
60. Wang, Z.; Yu, H.; Xia, J.; Zhang, F.; Li, F.; Xia, Y.; Li, Y. Novel GO-Blended PVDF Ultrafiltration Membranes. *Desalination* **2012**, *299*, 50–54. [CrossRef]
61. Otitoju, T.A.; Ahmad, A.L.; Ooi, B.S. Polyvinylidene Fluoride (PVDF) Membrane for Oil Rejection from Oily Wastewater: A Performance Review. *J. Water Process Eng.* **2016**, *14*, 41–59. [CrossRef]

62. Shao, X.-S.; Li, J.-H.; Zhou, Q.; Miao, J.; Zhang, Q.-Q. Amphiphilic Poly(Vinyl Chloride)-g-Poly[Poly(Ethylene Glycol) Methylether Methacrylate] Copolymer for the Surface Hydrophilicity Modification of Poly(Vinylidene Fluoride) Membrane. *J. Appl. Polym. Sci.* **2013**, *129*, 2472–2478. [CrossRef]
63. Chen, C.; Tang, L.; Liu, B.; Zhang, X.; Crittenden, J.; Chen, K.L.; Chen, Y. Forming Mechanism Study of Unique Pillar-like and Defect-Free PVDF Ultrafiltration Membranes with High Flux. *J. Membr. Sci.* **2015**, *487*, 1–11. [CrossRef]
64. Lim, Y.J.; Goh, K.; Wang, R. The Coming of Age of Water Channels for Separation Membranes: From Biological to Biomimetic to Synthetic. *Chem. Soc. Rev.* **2022**, *51*, 4537–4582. [CrossRef] [PubMed]

separations

MDPI

Article

The Effect of Rotation on Gas Storage in Nanoporous Materials

Athanasios Ch. Mitropoulos [1,*], Ramonna I. Kosheleva [1,2], Margaritis Kostoglou [2] and Thodoris D. Karapantsios [2]

[1] Hephaestus Laboratory, Department of Chemistry, School of Sciences, International Hellenic University, St. Lucas, 654 04 Kavala, Greece
[2] Faculty of Chemistry, Division of Chemical Technology, Aristotle University of Thessaloniki, University Box 116, 541 24 Thessaloniki, Greece
* Correspondence: amitrop@chem.ihu.gr

Abstract: Nanoporous materials offer a promising solution for gas storage applications in various scientific and engineering domains. However, several crucial challenges need to be addressed, including adsorptive capacity, rapid loading, and controlled gas delivery. A potential approach to tackle these issues is through rotation-based methods. In this study, we investigate the impact of rotation on CO_2 adsorption using activated carbon, both at the early and late stages of the adsorption process. Towards this direction, three sets of experiments were conducted: (i) adsorption isotherm with rotation at each gas loading, (ii) adsorption kinetics with multiple rotations performed in sequence 15 min after CO_2 introduction, and (iii) adsorption kinetics with a single rotation after 40 h of adsorption and repetition after another 20 h. For the first two cases, the comparison was performed by respective measurements without rotation, while for the last case, results were compared to a theoretical pseudo-first-order kinetic curve. Our findings demonstrate that rotation enhances the adsorptive capacity by an impressive 54%, accelerates kinetics by a factor of 3.25, and enables controllable gas delivery by adjusting the angular velocity. These results highlight rotation as a promising technique to optimize gas storage in nanoporous materials, facilitating advancements in numerous scientific and engineering applications.

Keywords: gas storage; physisorption; gas rotation; adsorption kinetics

Citation: Mitropoulos, A.C.; Kosheleva, R.I.; Kostoglou, M.; Karapantsios, T.D. The Effect of Rotation on Gas Storage in Nanoporous Materials. *Separations* **2024**, *11*, 72. https://doi.org/10.3390/separations11030072

Academic Editors: Guangxu Lan and Yi-Nan Wu

Received: 31 January 2024
Revised: 21 February 2024
Accepted: 22 February 2024
Published: 24 February 2024

1. Introduction

Gas storage plays a pivotal role in ensuring energy security, grid stability, and the transition to a more sustainable energy future. Challenges that impact gas storage technologies [1,2] include effectiveness, scalability, and overall implementation, especially for mobile devices and vehicles.

The storage of gases in nanoporous materials and their subsequent transport and use is a modern challenge in chemical technology. There are various gases whose storage in porous media is useful and desirable. The storage of gases in porous media is useful because porous materials offer the possibility of storing larger quantities in proportion to the volume available compared to cylinders and tanks. It is also desirable because they require lower pressures and therefore provide higher safety in the workplace.

Historically, gas storage technology in porous media referred to the use of depleted reservoirs for the storage of transported natural gas. It was later realized that other gases also needed storage with functional transport and release capabilities. Today, storage technology is applied (a) for the environmental remediation of greenhouse gases (e.g., CO_2, NOx, SOx, etc.) and (b) for the efficient and safer energy recovery of gases such as CH_4 and H_2 [3,4]. The first case aims at the permanent removal of gases that, according to the European Union as well as the US Environmental Protection Agency, are considered pollutants, while the second case concerns their temporary storage until they are used [5]. Among these gases, this paper focuses on CO_2 as the most typical greenhouse gas [6].

Various methods have been developed for CO_2 capture and storage [7,8]. Among them are (a) carbon capture and sequestration (CCS) and (b) carbon capture and storage (CCS). Classical CO_2 capture techniques are carried out at the point of production and include (i) pre-combustion or stream purification [3] and (ii) post-combustion CO_2 capture [9]. However, due to the severe problem of climate change, direct capture of CO_2 from the atmosphere (direct air capture—DAC) is also highly desirable [10]. The above-mentioned techniques make use of either chemical compounds (e.g., ionic liquids) or porous materials (adsorption) [11].

Although adsorption of gases into porous materials is one promising method, optimizations are still needed. The main solids in use are activated carbon and metal–organic frameworks [12,13]. Again, the effort to produce new porous materials goes on very intensively [14,15], but there are still challenges to overcome.

There are three major requirements that a storage system must fulfill [16,17]: (a) sufficient storage capacity, (b) fast loading, and (c) controllable delivery. However, the methods to satisfy these requirements vary widely depending on the gas and the solid to be used. A popular method is the physisorption of the gas on activated carbon (AC) at ambient temperature [18,19]. One reason is that AC is an economical material; another reason is that the sample holder is simply a vessel filled with a closed random pack of AC granules that may easily accept modifications, e.g., for the proper control of input and output of the gas [20,21]. Opposite to the case of H_2 storage materials, studies on CO_2 adsorbents are scattered, and there is a lack of a systematic categorization based on criteria like adsorption capacity and working conditions. In this framework, a rigorous literature review has been conducted on carbon-based materials for CO_2 adsorption; the results are presented in Table 1.

Table 1. Carbon-based adsorbents with their CO_2 adsorption capacity under the related conditions.

Adsorbents	q_{max} mmol/g	T °C	P Bar	Ref.
Cellulosic AC doped with Cu	48	25	15	[22]
Carbon foam	15.2	0	5	[23]
Chemically activated AC from olive stone	10	50	1	[9]
Modified AC from fir bark	71	0	1	[24]
AC Norit RX *	2.5	30	5	[25]
AC from black locust activated with KOH	5.05	25	1	[26]
AC doped with N,S and activated with potassium salts	3.04–3.99	25	1	[27]

The asterisk (*) denotes the material used in this study.

It is well-known that under rotation, a packed bed promotes process intensification [28,29]. A porous packed bed (PPB) is characterized by two porosities: intra- and inter-granular [30,31]. Again, the behavior of the spinning gas is important for a rotating device. Geyko and Fisch [32] have outlined a theoretical study on ideal spinning gas. Liu et al. [33] have extended this study on van der Waals gases, and Zhang et al. [34] have generalized the theory for a non-ideal gas by using the virial equation.

To the best of our knowledge, although there are a number of studies regarding the effect of a rotational field in systems including adsorption such as rotating packed beds (RPBs), they are limited in investigating only phenomena like mass and heat transfer for continuous flow which most of the times refers to a three-phase system. However, such conditions are inadequate for the examination of valuable parameters in regard to the impact of rotation on inherited properties (i.e., adsorption capacity) required for the optimization of a system with no steady-state flow, like gas storage applications based on adsorption. To this end, this study attempts to shed light on the mechanism that is enabled when adsorption occurs under rotation in a closed system.

Kosheleva et al. [35] have conducted an experimental study on the effect of spinning gas on the adsorption process. They have suggested that rotation increases the amount adsorbed. Guo et al. [36] studied the mechanism of gas–solid adsorption under rotation based on the particle diffusion model involved. Their finding suggests that in a rotating PPB, intraparticle diffusion is dominant and is the driving force of the so-called deep adsorption.

For an ideally behaved gas, rotating a disc-shaped container causes the concentration to increase at the perimeter and decrease at the center due to centrifugation. In this study, we have examined the effect of rotation on the adsorption process in the aforementioned frame of gas storage. Experimental measurements on adsorption and adsorption kinetics with and without rotation have been recorded. We have organized this work in three sections regarding the experimental procedure. The first section presents the adsorption isotherm experiment. Two adsorption isotherms were produced and compared, with and without rotation. Next, CO_2 kinetics were examined after multiple rotations at an early stage of the adsorption, while a comparison with kinetics for the same period of time but without rotation was conducted as well. Finally, rotation at a late stage was performed after ~40 h, and a second one after a total of ~60 h CO_2 adsorption, and the results were fitted to a pseudo-first-order (PFO) kinetic curve in order to showcase the effect of the rotation. The results are analyzed accordingly, and the mechanism constituting each case is discussed.

2. Experimental Details

The effect of rotation on the adsorption and adsorption kinetics of CO_2 on activated carbon was examined. A specially designed sample cell that allows adsorption in situ with rotation was constructed (Figure 1). The device consisted of a low-vibration rotating motor and a rotating sample chamber of radius r_{cell} = 4.5 cm and height h = 1 mm. All rotations were conducted for 1 min at 5000 rpm. Isothermal temperature was maintained by air-conditioning. During rotation, a negligible temperature increase of about +0.5 °C was observed. To monitor this, a digital thermometer was used to record °C/s. A reservoir tank, connected to the sample cell on one end and to the gas container on the other, was employed as a vessel of known volume that served as a gas dosage regulator.

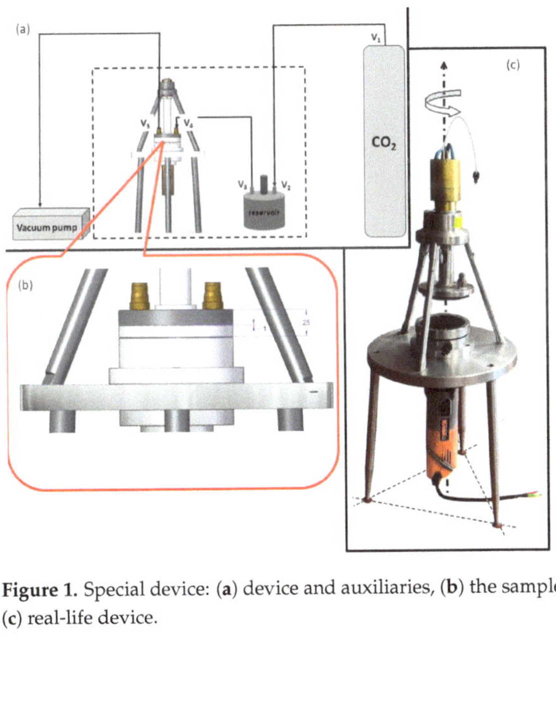

Figure 1. Special device: (**a**) device and auxiliaries, (**b**) the sample cell with dimensions in mm, and (**c**) real-life device.

2.1. Adsorption Isotherms

For conducting the adsorption isotherms experiments, 3.6 g AC was placed into the sample cell. The cell was tightly closed, and the system was set for evacuation overnight. After that, the sample cell was isolated (by closing the valves) from the reservoir tank. A desired amount of CO_2 was introduced into the reservoir tank. For the first pressure step, the valve to the sample cell was opened and closed immediately after achieving pressure equilibration between the reservoir tank and the sample cell (<2 s). Finally, the pressure was recorded for 60 min at 1 s intervals. Rotation was enabled at each pressure step after 15 min of CO_2 introduction.

2.2. Adsorption Kinetics

For the performance of kinetics, the steps until system evacuation were the same. Next, CO_2 was introduced at a desired pressure while recording the pressure drop over time started. Rotation took place according to each experiment; see text under results. More experimental details are given elsewhere [37].

Commercially available activated carbon (AC), with grains of size 100 μm, was used in this study. A N_2-adsorption isotherm of AC at 77 K was measured with a Nova 4200e porosimeter (Quantachrome, Boynton Beach, FL, USA). The BET area was estimated to be equal to about 1000 m^2/g, and the average pore size, based on the BJH method, was equal to 16 Å. The mass of the AC bed was m_{AC} = 3.6 g with solid and bulk densities of $\rho_s = 2\ g/cm^3$ and $\rho_b = 0.9\ g/cm^3$, respectively. The intra-, inter-, and total porosities ε of the system were then calculated as follows:

$$\varepsilon_{intra} = (\rho_s - \rho_b)/\rho_s,\ \varepsilon_{inter} = (V_G - V_b)/V_G,\ \text{and}\ \varepsilon_{tot} = (\varepsilon_{intra} V_b + \varepsilon_{inter} V_G)/V_G, \quad (1)$$

where $V_G = 6.36\ cm^3$ is the geometric volume of the cell, and $V_b = m_{AC}/\rho_b = 4\ cm^3$ is the bulk volume of the solid; $\varepsilon_{intra} = 55\%$, $\varepsilon_{inter} = 37\%$, and $\varepsilon_{tot} = 72\%$.

2.3. Adsorption Isotherms under Rotation

Adsorption isotherms at 20 °C with and without rotation were measured by the aforementioned device. The isotherm measurements were conducted in a sequence of pressure steps, where the sample cell started from vacuum conditions, and gas was introduced in controllable dosages up to 10 bar. The control of the step was provided by the reservoir tank described above. For the former case and at each pressure step, rotation was conducted for 1 min after 15 min of CO_2 introduction. When rotation ceased, the system was left for 45 min before receiving the measuring point.

Figure 2 shows the results. Subsequently, the adsorption isotherms were fitted to the Langmuir model:

$$q_t = \frac{q_e P K_{eq}}{1 + P K_{eq}} \quad (2)$$

where q_t is the amount adsorbed, q_e is the maximum amount adsorbed, and K_{eq} is the Langmuir constant. For the case without rotation, $K_{eq} = 0.213$ and $q_e = 0.016$ kg/kg, whereas for the case with rotation, $K'_{eq} = 0.082$ and $q'_e = 0.035$ kg/kg (the prime denotes rotation). Rotation changes the Langmuir constant in a way that $K_{eq} > K'_{eq}$ while $q_e < q'_e$. We have verified this trend by repeating the experiment with a single rotation as well as with a different type of AC. Although the decrease in the Langmuir constant indicates weaker adsorption, the increase in q_e indicates a higher concentration of gas molecules over the solid surface [38,39]. Rotation increases the strikes of the adsorbed molecules onto the surface and also pushes some of them to previously inaccessible sites following deep adsorption, as described by Guo et al. [40].

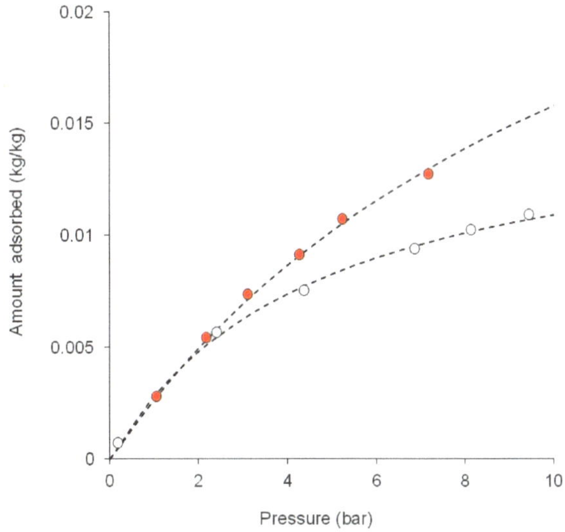

Figure 2. Adsorption isotherm of CO_2 on AC bed at 20 °C: open points correspond to no rotation; solid red points correspond to rotation at every pressure step; broken lines correspond to Langmuir model fit.

Figure 3 shows this mechanism where an element of the solid surface is depicted. The system at a given pressure P has 2 occupied sites and 1 unoccupied site; therefore, $k_aPS = k_dA$ where S are the unoccupied sites and A are the adsorbed molecules. Rotation makes accessible previously inaccessible sites, leading to 3 adsorbed molecules and 2 unoccupied sites; therefore, at the same given pressure, P, $k_a'PS' = k_d'A'$. Since $K_{eq} = k_a/k_d > k'_a/k'_d = K'_{eq}$, by comparing these two situations, it becomes clear that A/S must be greater than A′/S′ while A < A′ and S < S′. The above inequality is compatible with the case described in Figure 3; for example, the ratio for the no-rotation case is A/S = 2/1, while the ratio for the rotational one is A′/S′ = 3/2.

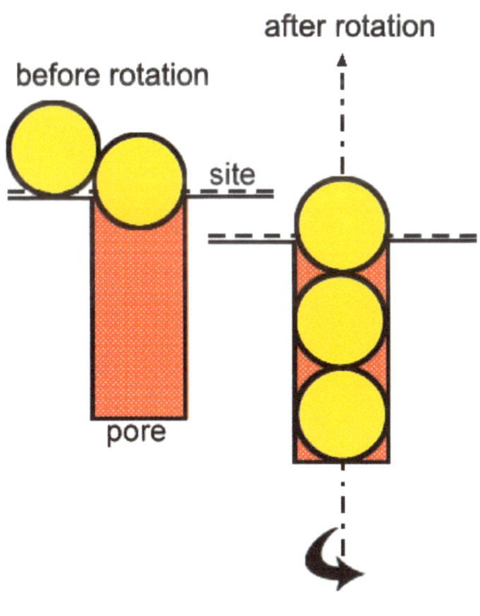

Figure 3. The effect of rotation on the adsorption process. Yellow cycles represents gas adsorbed molecules and red area represents a pore.

3. Kinetics

We have conducted two adsorption kinetic experiments with CO_2 on AC PPB. In the first experiment, multiple rotations at an early stage of the adsorption process were conducted at 10 °C. In the second one, a single rotation at a late stage of the adsorption process was conducted at 20 °C, followed by another rotation in a similar manner after some time.

Figure 4 shows the results of the first experiment. The reference curve (i.e., without rotation) was fitted to a non-linear pseudo-first-order (PFO) curve by least squares:

$$P_{th} = P_{eq} + (P_{max} - P_{eq})e^{-k_1 t} \tag{3}$$

where P_{th} is the theoretical pressure to be compared with the experimental one, P_{max} = 9190 mbar is the initial pressure, P_{eq} = 2555 mbar is the equilibrium pressure, k_1 = 0.206 is the PFO constant, and t is the time. Multiple rotations have started 20 min after gas entry into the cell and continue for 60 min with 1 min rotation every 15 min, i.e., four rotations in total. Without rotation, the pressure drops to 2270 mbar after 26 h. With multiple rotations, the system reaches the same pressure within 8 h, while after 26 h, the pressure drops to 830 mbar. Rotation makes the process 3.25 times faster with 63% more amount to be adsorbed.

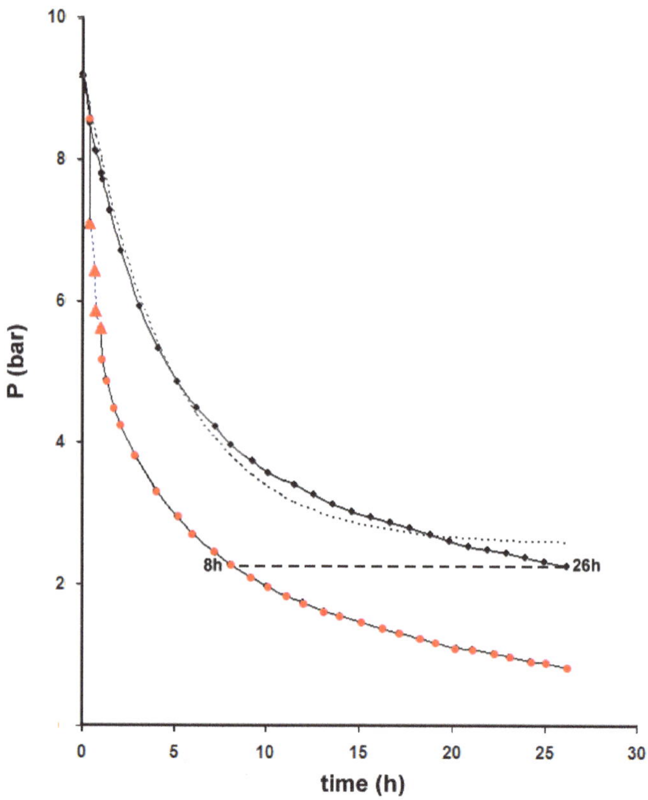

Figure 4. Early-stage adsorption kinetics at 10 °C: Curve with black points correspond to no rotation, where the dashed line indicates the PFO fit, and curve with red points correspond to multiple rotations. Solid red triangles indicate the rotation period. The broken line shows the time required to reach the same adsorption stage with and without rotation.

Rotation increases the number of molecules that strike on the adsorbent surface. At an early adsorption stage, the bare surface is eager to accommodate the striking molecules, while rotation forces some of them to gain access to commonly inaccessible sites. Multiple rotations intensify this process. As a result, faster kinetics and higher adsorptive capacity are observed.

Figure 5 shows the results of the second experiment. Long time kinetics of about 90 h were attained. More than 300,000 data points are collected, and for practical reasons, they are reduced to 10,000. The curve may be divided into five sections from A to J. Additionally, the sections after rotation may be divided into three subsections each (see Figure 5 for details). Section AB corresponds to ordinary adsorption kinetics. The data were fitted to a non-linear pseudo-first-order (PFO) curve with an initial pressure of 6190 mbar, an equilibrium pressure of 5023 mbar, and $k_1 = 0.0934$ (1/h).

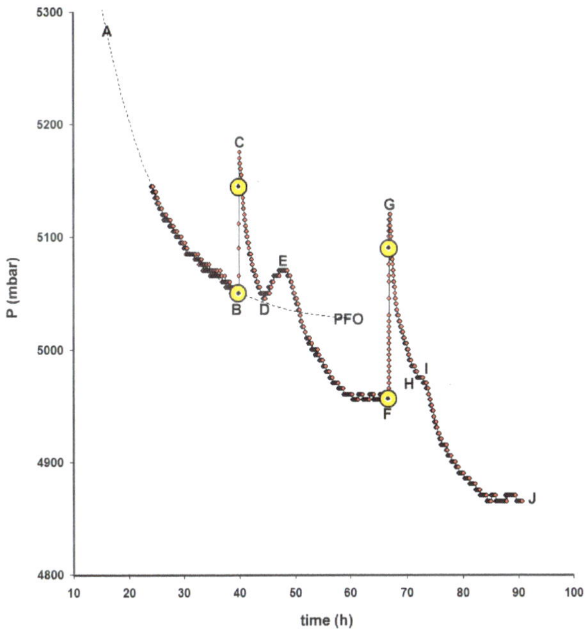

Figure 5. Late-stage adsorption kinetics at 20 °C. The resulting curve is divided into the following sections: (1) AB adsorption with a PFO fit (dashed line); (2) BC 1st rotation; (3) CF after 1st rotation; (4) FG 2nd rotation; and (5) GJ after 2nd rotation. Sections CF and GJ are further divided into the following subsections: CD and GI, gas flow from the edge of the rotating cell to the center; DE and HI, secondary desorption; EF and IK, enhanced adsorption. Open circles show the start/end of 1 min rotation.

In section BC, rotation causes a redistribution of gas molecules in the inter-granular space and, consequently, a pressure redistribution between the center and the edge of the rotating cell. Just before rotation starts, the pressure is $P_B = 5050$ mbar (first yellow circle). After 1 min of rotation, the pressure increases to 5145 mbar (second yellow circle) and keeps increasing to $P_C = 5175$ mbar.

For an ideal gas spinning in an empty cylinder, Geyko [32] introduced a ratio φ:

$$\varphi = \frac{M\omega^2 r^2}{2RT} \tag{4}$$

where M is the molecular weight of the gas, ω is the angular velocity, r is the distance from the axis of rotation, T is the constant temperature, and R is the gas constant. For this experiment, φ = 0.005 and $P_{(rmax)}$ = 5063 mbar, where $P_{(rmax)}$ is the pressure at the edge of the rotating cell ($r_{cell} = r_{max}$). However, the theory fails to predict $P_{(rmax)}$. Apparently, the presence of the PPB plays a role in this discrepancy. Since the temperature increment due to rotation was not more than +0.5 °C, it was concluded that this abnormal pressure increase was due to desorption. By assuming static load, before rotation starts, the mass m of free gas molecules (i.e., not adsorbed) will be equal to:

$$m = P\frac{V_p M}{RT} \tag{5}$$

where *P* is the applied pressure (global), and V_p is the inter-granular pore volume of the PPB in the rotating cell. During rotation, the radial force balance is given as:

$$dP = \frac{m(r)\omega^2}{V_p} rdr, \tag{6}$$

where *dP* is the pressure differential, and *m*(*r*) is the mass distributed within the different zones at a distance *r* from the axis of rotation. Rotation promotes desorption from intra-pores of the inner zone, resulting in the addition of more free gas into the inter-granular space. By substituting Equations (5) and (6):

$$\frac{dP}{P} = \frac{M\omega^2 r}{RT}dr \Rightarrow ln\frac{P(r)}{P} = \frac{M\omega^2 r^2}{2RT} + ln\kappa \tag{7}$$

where $ln\kappa$ is the integration constant that takes care of the mass (or pressure) differences; $\kappa \geq 1$ depending on the stage at which the adsorption process has proceeded. The distribution of pressure along the rotating PPB will be:

$$P(r) = \kappa P\frac{\varphi e^{\varphi}}{e^{\varphi} - 1} \tag{8}$$

Notice that $\frac{\varphi e^{\varphi}}{e^{\varphi}-1} \rightarrow 1$ as $\varphi \rightarrow 0$. Since, in our experiment, φ = 0.005, by neglecting this factor for simplicity reasons, κ = 1.019, indicating that desorption occurs during rotation. After rotation ceases, kicking effects, due to abrupt stopping, lead to further desorption until point C.

In section CF, the system undergoes a settling-down procedure. However, this is a complicated phase comprising different movements and processes. The mean free path of CO_2 molecules is about 200 Å, which is much less than the inter-granular space (~10 μm) but bigger than the intra-pore space (pore size ~16 Å). Due to pressure difference, gas will flow from the edge to the center of the rotating cell within the inter-granular space, while in the intra-pore space, Knudsen diffusion will occur.

We have assumed two zones within the rotating PPB cell (Figure 6). A lower concentration zone forms around the axis of rotation toward a critical radius r_c, where a transition from negative to positive pressure differences occur, $\Delta P = P(r) - P$, and a hallow circular zone of higher concentration from r_c to the edge of the cell (see also Figure 7). For the present experiment, r_c = 3.2 cm, depending on the geometry of the sample cell.

In the CD subsection, the gas flows from the outer to the inner zone. At point D, P_D = 5045 mbar. Since the driving force $\Delta P \leq 125$ mbar and continues to decrease over time, this subsection is completed in about 5 h. However, inside the pores, there is an unequal distribution of gas molecules, with the pores in the outer zone having a pressure higher than P_D and the pores in the inner zone having a pressure less than P_D. As a result, secondary desorption occurs from inside the pores in the over-pressure zone, and the bulk pressure reaches after 3 h the E point where $P_D < P_E$ = 5070 mbar. Since the two zones are of equal volume, but the bulk pressure is higher than the pressure inside the pores in the under-pressure zone, additional adsorption takes place from E to F. It is obvious that the whole process is not discrete but continuous as the kinetics of the molecules in the bulk with the kinetics of adsorption inside the pores is different, resulting in the presented profile.

At point F, the pressure has dropped to P_F = 4955 mbar, whereas according to the PFO model, it was expected to be P_{eq} = 5023 mbar, i.e., a small amount of gas has further been adsorbed; ΔP = 68 mbar. However, at a late adsorption stage, most of the surface is occupied by adsorbed molecules. Therefore, rotation primarily triggers desorption and only marginally pushes them into previously inaccessible sites, as it happens in an early adsorption stage instead.

A similar behavior is also observed for the second rotation event. Just before the rotation, the global pressure was P_F = 4955 mbar (yellow circle). During rotation, the pressure jumps to 5090 mbar (second yellow circle) and keeps increasing to P_G = 5120 mbar. According to Equation (8), κ = 1.027. Since the system is now at a later stage, the amount desorbed is higher than the previous one, as it is expected. After rotation, the pressure keeps increasing for a while due to the kick-stop effect and then starts decaying towards $P_H = P_I$ = 4975 mbar, where less pronounced secondary desorption occurs. Finally, the system reaches point J at P_J = 4865 mbar. Another small amount of gas has further been adsorbed, ΔP = 4955 − 4865 = 90 mbar, indicating that multiple rotations keep pushing molecules deeper into the pore matrix even when the whole process has become marginal.

The magnitude of the effect depends on various factors as the whole process is quite complex, and many different events take place either simultaneously or in sequence. In order to find the criterion with the highest importance, a number of scenarios were examined by the theoretical model. The results provide a guide for device optimization as well as know-how on configuration parameters according to any potential application.

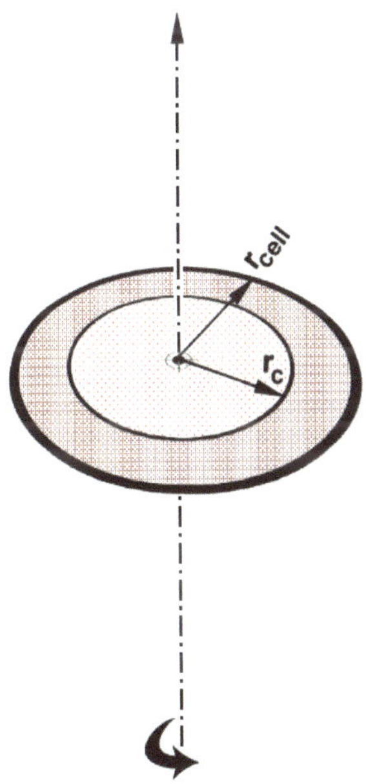

Figure 6. Concentration zones. During rotation, at the edge of the cell, the concentration is higher than at the center. The critical radius, where the pressure difference changes from negative to positive, is equal to r_c = 3.2 cm.

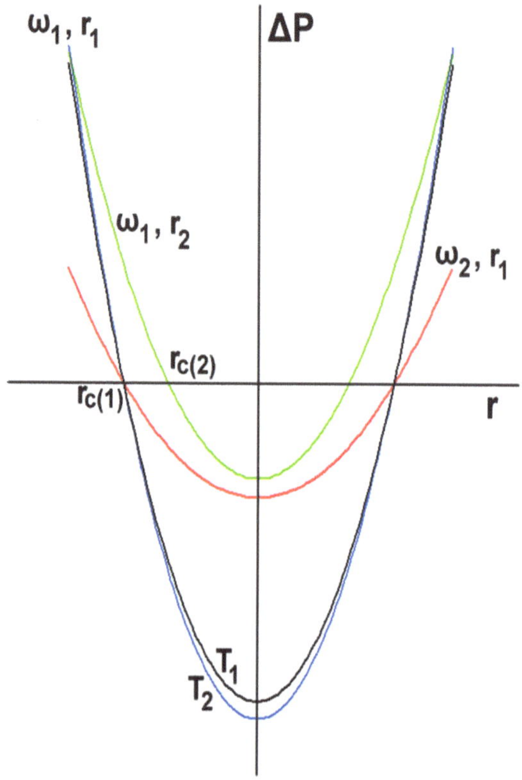

Figure 7. The effect of rotating parameters on spinning gas. Broken lines indicate the limits of the rotating cell, r_{cell}. Line 1—ω_1, T_1, $r_1 = r_{cell}$; line 2—$\omega_2 < \omega_1$, $T_2 = T_1$, $r_2 = r_{cell}$; line 3—$\omega_3 = \omega_1$, $T_3 < T_1$, $r_3 = r_{cell}$; and line 4—$\omega_4 > \omega_1$, $T_4 = T_1$, $r_4 > r_{cell}$. Notice that r_c and r'_c are the radii of the zones where the transitions from negative to positive pressures occur; $\Delta P = P(r) - P$.

Figure 7 shows the effect of ω, T, and r on the delivery process. As the angular velocity decreases, ΔP decreases while the critical radius remains constant. As the temperature decreases, ΔP slightly increases, and r_c remains constant. Finally, as the radius of the cell increases, ΔP and r_c increase. Apparently, for a given rotating cell, by controlling the angular velocity, a controllable delivery is allowed.

4. Conclusions

In this study, we have examined the effect of rotation as to whether or not it can assist gas storage in nanoporous materials in a way that (a) increases the storage capacity of the adsorbent, (b) promotes a fast loading, and (c) allows a controllable delivery. To this end, we have conducted adsorption and adsorption kinetic experiments with and without rotation on a PPB.

Multiple rotations cause an increase in the amount adsorbed by allowing previously inaccessible sites of the solid surface to become accessible. It was noticed, however, that although the adsorptive capacity of the solid increases, the Langmuir constant decreases compared to that of the isotherm without rotation. Additionally, a possible mechanism was presented. It was concluded that rotation provides an in situ method for increasing the storage capacity of the adsorbent.

In order to gain a better insight into the involved mechanisms, we have conducted the kinetic experiments with (a) multiple rotations at an early adsorption stage and (b) a single rotation at a late stage of the adsorption process followed by another single rotation after some time for repeatability reasons. All rotations were performed for 1 min at 5000 rpm. Multiple rotations accelerate adsorption kinetics, bestowing 3.25 times faster loading and 63% more amount adsorbed.

Rotation at a late adsorption stage is characterized by two main phases: (a) during rotation, fast desorption of the already adsorbed molecules, and (b) after rotation, slow and complicated procedures to settle the system down. At equilibrium, some marginal increase in the amount adsorbed is recorded too. We have given possible explanations for all observed sections of the kinetic curve, and we have also outlined a correction for the spinning gas when a PPB is present in the rotating cylinder. It was concluded that a single and short rotation triggers a fast delivery of the adsorbed gas.

Author Contributions: R.I.K.: conceptualization, data analysis, and writing—original draft preparation. T.D.K.: writing—review and editing. M.K.: writing—review and editing. A.C.M.: writing—review and editing. All authors have read and agreed to the published version of the manuscript.

Funding: This research received no external funding.

Institutional Review Board Statement: Not applicable. No human and/or animal studies have been conducted.

Data Availability Statement: The data that support the findings of this study are available from the corresponding author upon reasonable request.

Conflicts of Interest: The authors declare no conflicts of interest.

References

1. Yue, M.; Lambert, H.; Pahon, E.; Roche, R.; Jemei, S.; Hissel, D. Hydrogen energy systems: A critical review of technologies, applications, trends and challenges. *Renew. Sustain. Energy Rev.* **2021**, *146*, 111180. [CrossRef]
2. Bahman, N.; Al-Khalifa, M.; Al Baharna, S.; Abdulmohsen, Z.; Khan, E. Review of carbon capture and storage technologies in selected industries: Potentials and challenges. *Rev. Environ. Sci. Bio/Technol.* **2023**, *22*, 451–470. [CrossRef]
3. Dixit, F.; Zimmermann, K.; Alamoudi, M.; Abkar, L.; Barbeau, B.; Mohseni, M.; Kandasubramanian, B.; Smith, K. Application of MXenes for air purification, gas separation and storage: A review. *Renew. Sustain. Energy Rev.* **2022**, *164*, 112527. [CrossRef]
4. Ghazvini, M.F.; Vahedi, M.; Nobar, S.N.; Sabouri, F. Investigation of the MOF adsorbents and the gas adsorptive separation mechanisms. *J. Environ. Chem. Eng.* **2021**, *9*, 104790. [CrossRef]
5. Tarkowski, R.; Uliasz-Misiak, B.; Tarkowski, P. Storage of hydrogen; natural gas, and carbon dioxide—Geological and legal conditions. *Int. J. Hydrogen Energy* **2021**, *46*, 20010–20022. [CrossRef]
6. Yuan, B.; Wu, X.; Chen, Y.; Huang, J.; Luo, H.; Deng, S. Adsorption of CO_2, CH_4, and N_2 on Ordered Mesoporous Carbon: Approach for Greenhouse Gases Capture and Biogas Upgrading. *Environ. Sci. Technol.* **2013**, *47*, 5474–5480. [CrossRef]
7. Diwan, A.D.; Harke, S.N.; Panche, A.N. Application of proteomics in shrimp and shrimp aquaculture. *Comp. Biochem. Physiol. Part D Genom. Proteom.* **2022**, *43*, 101015. [CrossRef] [PubMed]
8. Madejski, P.; Chmiel, K.; Subramanian, N.; Kuś, T. Methods and techniques for CO_2 capture: Review of potential solutions and applications in modern energy technologies. *Energies* **2022**, *15*, 887. [CrossRef]
9. Moussa, M.; Bader, N.; Querejeta, N.; Durán, I.; Pevida, C.; Ouederni, A. Toward sustainable hydrogen storage and carbon dioxide capture in post-combustion conditions. *J. Environ. Chem. Eng.* **2017**, *5*, 1628–1637. [CrossRef]
10. Zhu, X.; Xie, W.; Wu, J.; Miao, Y.; Xiang, C.; Chen, C.; Ge, B.; Gan, Z.; Yang, F.; Zhang, M.; et al. Recent advances in direct air capture by adsorption. *Chem. Soc. Rev.* **2022**, *51*, 6574–6651. [CrossRef] [PubMed]
11. de Queiroz Fernandes Araújo, O.; de Medeiros, J.L. Carbon capture and storage technologies: Present scenario and drivers of innovation. *Curr. Opin. Chem. Eng.* **2017**, *17*, 22–34. [CrossRef]
12. Ma, S. Gas storage in porous metal–organic frameworks for clean energy applications. *Chem. Commun.* **2010**, *46*, 44–53. [CrossRef]
13. Deegan, M.M.; Lorzing, G.R.; Korman, K.J.; Rowland, C.A.; Dworzak, M.R.; Antonio, A.M.; Bloch, E.D. Hydrogen Storage in Porous Cages. *ACS Mater. Au* **2023**, *3*, 66–74. [CrossRef]
14. Lee, S.-Y.; Lee, J.-H.; Kim, Y.-H.; Kim, J.-W.; Lee, K.-J.; Park, S.-J. Recent Progress Using Solid-State Materials for Hydrogen Storage: A Short Review. *Processes* **2022**, *10*, 304. [CrossRef]
15. Ding, M.; Liu, X.; Ma, P.; Yao, J. Porous materials for capture and catalytic conversion of CO_2 at low concentration. *Coord. Chem. Rev.* **2022**, *465*, 214576. [CrossRef]
16. Berenguer-Murcia, Á.; Marco-Lozar, J.P.; Cazorla-Amorós, D. Hydrogen storage in porous materials: Status, milestones, and challenges. *Chem. Rec.* **2018**, *18*, 900–912. [CrossRef] [PubMed]
17. Heinemann, N.; Alcalde, J.; Miocic, J.M.; Hangx, S.J.T.; Kallmeyer, J.; Ostertag-Henning, C.; Hassanpouryouzband, A.; Thaysen, E.M.; Strobel, G.J.; Schmidt-Hattenberger, C. Enabling large-scale hydrogen storage in porous media–the scientific challenges. *Energy Environ. Sci.* **2021**, *14*, 853–864. [CrossRef]

18. Marco-Lozar, J.P.; Kunowsky, M.; Suarez-Garcia, F.; Carruthers, J.D.; Linares-Solano, A. Activated carbon monoliths for gas storage at room temperature. *Energy Environ. Sci.* **2012**, *5*, 9833–9842. [CrossRef]
19. Zhang, W.J.; Rabiei, S.; Bagreev, A.; Zhuang, M.S.; Rasouli, F. Study of NO adsorption on activated carbons. *Appl. Catal. B Environ.* **2008**, *83*, 63–71. [CrossRef]
20. Feroldi, M.; Neves, A.C.; Borba, C.E.; Alves, H.J. Methane storage in activated carbon at low pressure under different temperatures and flow rates of charge. *J. Clean. Prod.* **2018**, *172*, 921–926. [CrossRef]
21. Strizhenov, E.M.; Shkolin, A.V.; Chugaev, S.S.; Men'shchikov, I.E.; Solovtsova, O.V.; Shiryaev, A.A.; Nickolsky, M.S. Adsorbed natural gas storage facility based on activated carbon of wood waste origin. *Adsorption* **2022**, *29*, 291–307. [CrossRef]
22. Conte, G.; Policicchio, A.; De Luca, O.; Rudolf, P.; Desiderio, G.; Agostino, R.G. Copper-doped activated carbon from amorphous cellulose for hydrogen, methane and carbon dioxide storage. *Int. J. Hydrogen Energy* **2022**, *47*, 18384–18395. [CrossRef]
23. Kusdhany, M.I.M.; Ma, Z.; Mufundirwa, A.; Li, H.-W.; Sasaki, K.; Hayashi, A.; Lyth, S.M. Hydrogen and carbon dioxide uptake on scalable and inexpensive microporous carbon foams. *Microporous Mesoporous Mater.* **2022**, *343*, 112141. [CrossRef]
24. Luo, L.; Chen, T.; Li, Z.; Zhang, Z.; Zhao, W.; Fan, M. Heteroatom self-doped activated biocarbons from fir bark and their excellent performance for carbon dioxide adsorption. *J. CO2 Util.* **2018**, *25*, 89–98. [CrossRef]
25. Peredo-Mancilla, D.; Hort, C.; Jeguirim, M.; Ghimbeu, C.M.; Limousy, L.; Bessieres, D. Experimental Determination of the CH_4 and CO_2 Pure Gas Adsorption Isotherms on Different Activated Carbons. *J. Chem. Eng. Data* **2018**, *63*, 3027–3034. [CrossRef]
26. Zhang, C.; Song, W.; Ma, Q.; Xie, L.; Zhang, X.; Guo, H. Enhancement of CO_2 Capture on Biomass-Based Carbon from Black Locust by KOH Activation and Ammonia Modification. *Energy Fuels* **2016**, *30*, 4181–4190. [CrossRef]
27. Cui, H.; Shi, J.; Xu, J.; Yan, N.; Liu, Y. Direct synthesis of N, S co-doped porous carbons using novel organic potassium salts as activators for efficient CO_2 adsorption. *Fuel* **2023**, *342*, 127824. [CrossRef]
28. Neumann, K.; Gladyszewski, K.; Groß, K.; Qammar, H.; Wenzel, D.; Górak, A.; Skiborowski, M. A guide on the industrial application of rotating packed beds. *Chem. Eng. Res. Des.* **2018**, *134*, 443–462. [CrossRef]
29. Sang, L.; Luo, Y.; Chu, G.-W.; Liu, Y.-Z.; Liu, X.-Z.; Chen, J.-F. Modeling and experimental studies of mass transfer in the cavity zone of a rotating packed bed. *Chem. Eng. Sci.* **2017**, *170*, 355–364. [CrossRef]
30. Guo, Z.; Sun, Z.; Zhang, N.; Cao, X.; Ding, M. Mean porosity variations in packed bed of monosized spheres with small tube-to-particle diameter ratios. *Powder Technol.* **2019**, *354*, 842–853. [CrossRef]
31. Dolamore, F.; Fee, C.; Dimartino, S. Modelling ordered packed beds of spheres: The importance of bed orientation and the influence of tortuosity on dispersion. *J. Chromatogr. A* **2018**, *1532*, 150–160. [CrossRef]
32. Geyko, V.I. Reduced Compressibility and an Inverse Problem for a Spinning Gas. *Phys. Rev. Lett.* **2013**, *110*, 150604. [CrossRef]
33. Liu, H.-S.; Lin, C.-C.; Wu, S.-C.; Hsu, H.-W. Characteristics of a rotating packed bed. *Ind. Eng. Chem. Res.* **1996**, *35*, 3590–3596. [CrossRef]
34. Zhang, J.; Burke, N.; Zhang, S.; Liu, K.; Pervukhina, M. Thermodynamic analysis of molecular simulations of CO_2 and CH_4 adsorption in FAU zeolites. *Chem. Eng. Sci.* **2014**, *113*, 54–61. [CrossRef]
35. Kosheleva, R.I.; Karapantsios, T.D.; Kostoglou, M.; Mitropoulos, A.C. Thermodynamic analysis of the effect of rotation on gas adsorption. *J. Non-Equilib. Thermodyn.* **2023**, *48*, 403–416. [CrossRef]
36. Guo, Q.; Zhao, Y.; Qi, G.; Liu, Y. Performance and Mechanism of Gas-Solid Adsorption in a Rotating Adsorption Bed. *Chem. Eng. Technol.* **2022**, *45*, 844–852. [CrossRef]
37. Kosheleva, R.I.; Karapantsios, T.D.; Kostoglou, M.; Mitropoulos, A.C. A novel device for in situ study of gas adsorption under rotation. *Rev. Sci. Instrum.* **2021**, *92*, 45106. [CrossRef] [PubMed]
38. Azizian, S. Chapter 6—Adsorption isotherms and kinetics. In *Adsorption: Fundamental Processes and Applications*; Ghaedi, M., Ed.; Elsevier: Amsterdam, The Netherlands, 2021; pp. 445–509. [CrossRef]
39. Bénard, P. 10—Carbon nanostructures for hydrogen storage. In *Solid-State Hydrogen Storage: Materials and Chemistry*; Woodhead Publishing Series in Electronic and Optical Materials; Woodhead Publishing: Sawston, UK, 2008; pp. 261–287. [CrossRef]
40. Guo, Q.; Liu, Y.; Qi, G. Application of high-gravity technology NaOH-modified activated carbon in rotating packed bed (RPB) to adsorb toluene. *J. Nanoparticle Res.* **2019**, *21*, 175. [CrossRef]

nanomaterials

MDPI

Article

Insight into the Storage Mechanism of Sandwich-Like Molybdenum Disulphide/Carbon Nanofibers Composite in Aluminum-Ion Batteries

Xiaobing Wang [1], Ruiyuan Zhuang [2,*], Xinyi Liu [1], Mingxuan Hu [1], Panfeng Shen [1], Jintao Luo [1], Jianhong Yang [3] and Jianchun Wu [3,4,*]

1 School of Advanced Materials Engineering, Jiaxing Nanhu University, Jiaxing 314000, China
2 School of Mechanical and Electrical Engineering, Jiaxing Nanhu University, Jiaxing 314000, China
3 School of Materials Science and Engineering, Jiangsu University, Zhenjiang 212013, China
4 Institute of Nuclear Science and Technology, Sichuan University, Chengdu 610064, China
* Correspondence: zhuangryuan@jxnhu.edu.cn (R.Z.); jcwu@ujs.edu.cn (J.W.)

Abstract: Aluminum-ion batteries (AIBs) have become a research hotspot in the field of energy storage due to their high energy density, safety, environmental friendliness, and low cost. However, the actual capacity of AIBs is much lower than the theoretical specific capacity, and their cycling stability is poor. The exploration of energy storage mechanisms may help in the design of stable electrode materials, thereby contributing to improving performance. In this work, molybdenum disulfide (MoS_2) was selected as the host material for AIBs, and carbon nanofibers (CNFs) were used as the substrate to prepare a molybdenum disulfide/carbon nanofibers (MoS_2/CNFs) electrode, exhibiting a residual reversible capacity of 53 mAh g^{-1} at 100 mA g^{-1} after 260 cycles. The energy storage mechanism was understood through a combination of electrochemical characterization and first-principles calculations. The purpose of this study is to investigate the diffusion behavior of ions in different channels in the host material and its potential energy storage mechanism. The computational analysis and experimental results indicate that the electrochemical behavior of the battery is determined by the ion transport mechanism between MoS_2 layers. The insertion of ions leads to lattice distortion in the host material, significantly impacting its initial stability. CNFs, serving as a support material, not only reduce the agglomeration of MoS_2 grown on its surface, but also effectively alleviate the volume expansion caused by the host material during charging and discharging cycles.

Keywords: aluminum-ion batteries; electrochemical behavior; energy storage mechanism; MoS_2/CNFs; composite materials

Citation: Wang, X.; Zhuang, R.; Liu, X.; Hu, M.; Shen, P.; Luo, J.; Yang, J.; Wu, J. Insight into the Storage Mechanism of Sandwich-Like Molybdenum Disulphide/Carbon Nanofibers Composite in Aluminum-Ion Batteries. *Nanomaterials* **2024**, *14*, 442. https://doi.org/10.3390/nano14050442

Academic Editor: Guangxu Lan

Received: 25 January 2024
Revised: 15 February 2024
Accepted: 22 February 2024
Published: 28 February 2024

1. Introduction

In response to global climate change, the construction of a green, low-carbon, and clean energy system with renewable energy as the mainstay is the energy strategy of the world. However, the intermittency and randomness of renewable energy have brought great challenges to maintaining the balance of power and electricity in power systems. Addressing the imbalance between renewable energy generation and electricity load demands the utilization of energy storage technology [1,2]. Therefore, large-scale energy storage technology is perceived as a pivotal strategy to facilitate the widespread adoption of renewable energy, serving as a strategic underpinning for the future transformation of energy structures and shifts in power production and consumption patterns. At present, lithium-ion batteries (LIBs) are facing challenges, including the scarcity of lithium resources and safety concerns, hindering their widespread adoption in the energy storage field [3,4]. Consequently, the development of cost-effective and high-safety secondary batteries has garnered widespread attention from society.

Aluminum-ion batteries (AIBs) have the advantages of abundant aluminum resources, high safety, and high theoretical volumetric capacity (8040 mAh cm^{-3}). Compared with traditional LIBs, AIBs have multiple electrons participating in the electrode reaction during the charging and discharging processes, resulting in a high theoretical specific capacity, making AIBs a growing research hotspot in the field of energy storage [5,6]. Since Holleck used aluminum metal as the anode; metal chloride as the cathode; and molten salt electrolyte consisting of $AlCl_3$, KCl, and NaCl to form aluminum batteries, research on non-aqueous AIBs has begun to receive attention from researchers [7]. However, due to the high electrode potential of aluminum and the large ionic radius involved in the electrode reaction, AIBs have lower energy density and insufficient electrochemical performance to meet practical demands. In 2015, Lin introduced a model of AIBs utilizing a novel ionic liquid ($AlCl_3$/[EMIm]Cl) as the electrolyte, marking a significant breakthrough in the field of AIBs [8]. Subsequently, various materials have been integrated into the cathode materials of non-aqueous AIBs, including metal oxides (such as MnO_2 and V_2O_5) [9,10], transition metal chalcogenides (such as CuS_2, VS_2, and Co_9S_8) [11–13], carbon-based materials (such as graphene, graphitic carbon, and graphitic foam) [14–16], and other materials (such as MXenes and conducting polymers) [17,18]. Among them, carbon-based materials are one of the most used in AIBs, with advantages such as high discharge voltage and good cycle stability. However, in the storage mechanism of carbon-based materials, the guest ion is $AlCl_4^-$, whose theoretical specific capacity is only one-third of Al^{3+} under the same amount of ion insertion, resulting in insufficient discharge capacity (approximately 70 mAh g^{-1}). Transition metal oxides have gained significant attention for their higher discharge specific capacity compared to carbon materials. Nevertheless, current research on transition metal oxides mainly focuses on vanadium-based oxides, mainly due to the high surface charge density of Al^{3+}, coupled with the high transmission barrier of multivalent ions in oxide crystal structures, which leads to strong coulombic interactions between Al^{3+} and the material lattice, hindering its interlattice diffusion [19,20]. Compared to other cathode materials (such as carbon-based materials, transition metal chalcogenides, et al.), transition metal chalcogenides can facilitate the exchange of three electrons during aluminum storage, thereby exhibiting a high theoretical specific capacity. Additionally, transition metal chalcogenides possess advantages such as wide interlayer spacing, low electronegativity, and abundant resources. Moreover, the coulombic interaction between Al^{3+} and transition metal chalcogenides is comparatively weaker than that observed with metal oxides. Consequently, they are considered highly promising cathode materials for AIBs [21].

Molybdenum disulfide (MoS_2) is a sulfide with a sandwich-like layered structure, similar to graphene, but with a distinct stacking sequence. Each MoS_2 layer constitutes a three-dimensional structure via strong S-Mo-S ionic/covalent bonds, and each adjacent layer of MoS_2 is separated by van der Waals forces, forming a spacing of 0.62 nm. This spacing is notably larger than the interlayer spacing of graphite (0.335 nm) [22,23]. The weaker van der Waals forces between the layers of MoS_2 facilitate the rapid diffusion of ions without significant changes in volume, making MoS_2 an ideal host material for electrochemical intercalation and deintercalation. Indeed, MoS_2 has been reported as an electrode material in LIBs, sodium-ion batteries (SIBs), zinc-ion batteries (ZIBs), and AIBs [24–29].

Li applied petal-shaped MoS_2 microspheres to AIBs for the first time. Electrochemical testing results revealed that the initial discharge specific capacity was 253.6 mAh·g^{-1} at a current density of 20 mA·g^{-1}, and the discharge specific capacity after 100 cycles was 66.7 mAh·g^{-1} at a current density of 40 mA·g^{-1}. The large ionic radius involved in the electrode reaction during charging and discharging resulted in a low discharge specific capacity and poor cycle stability [27]. Researchers have made structural designs and optimized the properties to improve their inherent defects and enhance their electrochemical performance [28,29]. Subsequently, Luo prepared MoS_2-RGO composite electrode materials using graphene as a carrier, and Lu designed MoS_2@CNFs composite electrodes using carbon nanotubes as a carrier [28,29]. The introduction of graphene and carbon nanotubes in the composite electrode effectively improved the electrical conductivity of the electrode material, increased the contact area between the electrode material and the electrolyte, and then effectively improved the electrochemical performance of the electrode. Therefore, in addition to designing the electrode structure reasonably, it is crucial to study the energy storage mechanism of MoS_2, as it has been demonstrated that MoS_2 is a suitable material for AIBs. Understanding the energy storage mechanism may help to successfully design stable electrode materials and optimize their performance.

Here, MoS_2/CNFs composites were synthesized using a simple electrospinning and subsequent hydrothermal method, with layered MoS_2 evenly distributed on the surface of the CNFs. Serving as a binder-free electrode for AIBs, the uniform nanosheet and layered structure of MoS_2/CNFs ensures complete contact with the electrolyte, improving its conductivity, thereby promoting the transport of Al^{3+} and the cycling performance. Additionally, the carbon substrate not only prevents the agglomeration of MoS_2, but also alleviates the volume expansion effects of the electrode during charging and discharging. As a result, the prepared AIBs exhibit good cycling stability, delivering a first discharge capacity of approximately 180 mAh g^{-1} at 100 mA g^{-1}, and maintaining their capacity at about 53 mAh g^{-1} after 260 cycles. Ex-situ XPS and XRD characterization were conducted to gain a deeper understanding of the energy storage mechanism of the MoS_2-based cathode. Consequently, the deintercalation energy of ions in different channels of MoS_2 was calculated, and the potential ion deintercalation sites and energy storage mechanism were inferred based on the magnitude of the deintercalation energy. Specifically, the layer distance of MoS_2 matches the size of Al^{3+}, and Al^{3+} is inserted into MoS_2 to generate AlxMoS$_2$.

2. Materials and Methods

2.1. *Synthesis of Binder-Free MoS_2/CNFs Cathode*

2.1.1. Electrospinning

All reagents were used without further purification. A weight of 3 g of PAN (PAN, average Mw = 150,000, ALDRICH, Shanghai, China) was dissolved in 27 mL of N,N-dimethyformamide (DMF, 99.9%, innochem) solution and vigorously stirred for 12 h until the solution became uniform and clear, and then it was poured into a needle tube. After that, the plastic needle tube was placed on the spinning machine injection pump for electrospinning. The parameters of the spinning process were: high-voltage power supply voltage of 15 kV, push rate of 0.5 mL h^{-1}, and needle diameter of Φ 18. After electrospinning, the fiber precursor was collected for thermal treatment. Under air atmosphere, the fibers were pre-treated at 280 °C with a temperature rise rate of 1 °C min^{-1}. The fiber obtained after pre-treatment was a brownish black. Finally, the pre-treated fibers were further carbonized, which needs to be carried out in Ar atmosphere at 800 °C with a temperature rise rate of 5 °C min^{-1}, as shown in Figure 1a.

2.1.2. Hydrothermal Method

Initially, 2 mmol $(NH_4)_6Mo_7O_{24}$ (99.95%, Sinopharm Chemical Reagent Co., Ltd., Shanghai, China) and 10 mmol CH_4N_2S (99.95%, Sinopharm Chemical Reagent Co., Ltd., Shanghai, China) were introduced into 80 mL of deionized water, followed by stirring at a speed of 600 r min^{-1} for a duration of 2 h until complete dissolution was achieved. The solution was then transferred to a 100 mL PTFE-lined autoclave. Subsequently, the CNFs prepared above were added to the solution. The PTFE-lined autoclave was reacted at 200 °C for 24 h with a heating rate of 2 °C min^{-1}. After completion of the reaction, the sample was thoroughly washed and dried overnight at 60 °C in a vacuum oven with a heating rate of 1 °C min^{-1}. Finally, the hybrid composite materials were calcined at 350 °C for 2 h with a heating rate of 2 °C min^{-1} under a nitrogen atmosphere to yield the desired MoS_2/CNFs composite (depicted in Figure 1b). The synthesis process for pure MoS_2 nanomaterials is comparable to the preparation method for MoS_2/CNFs, except that pure carbon fibers are not employed as the matrix material during the synthesis process.

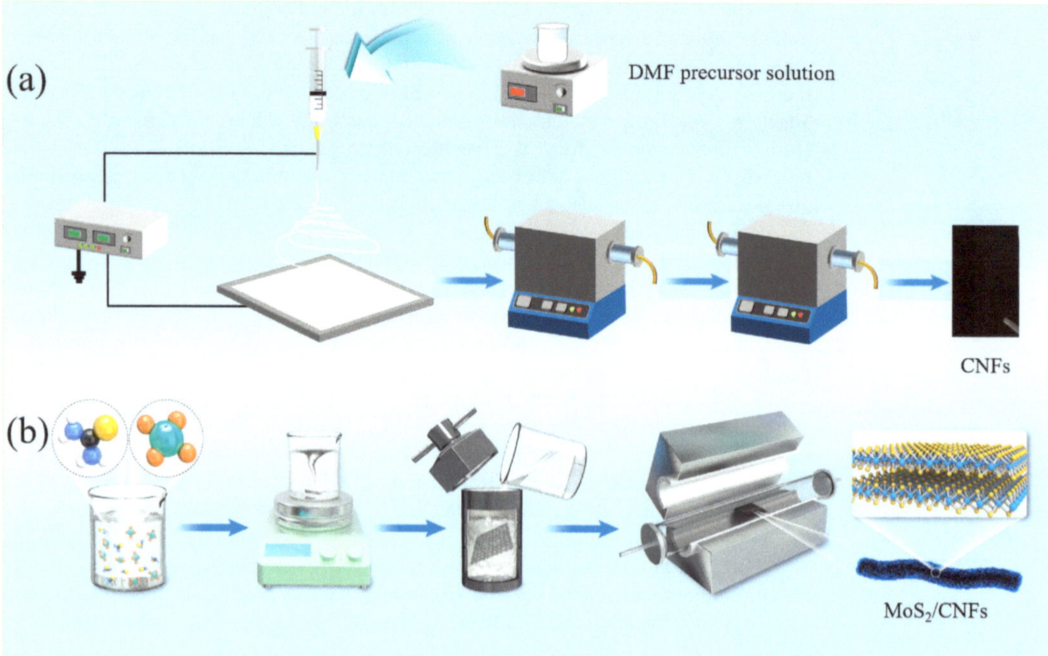

Figure 1. Schematic illustration of the preparation process: (**a**) CNFs; (**b**) MoS_2/CNFs.

2.2. Structural and Chemical Characterization

X-ray diffraction (XRD, SmartLab 9kw, Shanghai, China) analysis was employed for the phase characterization of the synthesized sample, operating within an angle range of 5–90° and a scan rate of 10° min^{-1}. The Thermo Fisher Nexsa (Waltham, MA, USA) instrument was utilized to analyze the surface element composition and valence states. The analysis chamber was operated at a vacuum level of approximately 5×10^{-9} mbar, utilizing a monochromatic Al-Kα X-ray source with an energy of 1486.6 eV and a voltage of 12 KV. The Thermo Fisher DXR instrument was employed to depict the molecular vibration and rotational energy level structure of the synthesized material, operating with a 523 nm laser and a power of 1 mW. To acquire a comprehensive understanding of the physical and chemical attributes, such as morphology, composition, and microstructure, the scanning electron microscopy (JSM-7800F, JEM-2100, JEOL, Tokyo, China) was researched.

2.3. Electrochemical Measurements

Considering the high corrosiveness of the ionic liquid electrolyte employed in the experiment towards stainless steel, soft-pack batteries were assembled for testing in this work. The batteries consist of binder-free electrodes made of MoS_2/CNFs, glass fiber separators, aluminum foils, and ionic liquid electrolytes. Specifically, the ionic liquid electrolyte was prepared by combining $AlCl_3$ (99.999%, Sigma Aldrich, St. Louis, MO, USA) and 1-ethyl-3-methylimidazolium chloride ([EMIm]Cl, 98%, Aladdin, Shanghai, China) in a molar ratio of 1:1.3. After assembly, the batteries were allowed to rest for 6 h to ensure the electrolyte fully contacted the electrode materials. All constant current charge–discharge tests were conducted on the Neware electrochemical testing system, with a test voltage range of 0.1–1.8 V (relative to Al/$AlCl_4^-$). In order to study the electrochemical activity and reaction process of the electrodes, CV tests were performed on the PARSTAT MCEIS instrument of the Princeton Electrochemical Workstation (AMETEK, Middleboro, MA, USA).

2.4. Computational Details

The simulations were conducted by the first-principles calculations based on density functional theory (DFT). For the computational simulation of MoS_2, the spatial symmetry group is P63/mcc (194). The computational model employed a 4 × 4 × 1 MoS_2 supercell structure, with a supercell size of 12.76 × 12.76 × 18.88 Å3. Considering the influence of van der Waals interactions, the DFT-D3 method was used for vdW correction during the calculation process. Prior to calculating the Al^{3+} insertion structure, the model was fully optimized for structural optimization.

3. Results

Figure 2a shows the XRD pattern of the MoS_2/CNFs composite material. The diffraction peaks of the synthesized sample match with the (002), (100), (102), (103), and (110) crystal planes of the standard card (JCPDS No. 37-1492) for 2H-MoS_2. From the figure, it can be seen that no impurity peaks of molybdenum or other impurities were observed, indicating that the synthesized product is pure-phase MoS_2. The peak at 2 θ = 25 ° can be attributed to amorphous carbon. Moreover, it can be concluded that the synthesized MoS_2 has a P63/mmc space group and belongs to the hexagonal crystal system. However, compared to the standard card, all diffraction peaks of MoS_2/CNFs are shifted to the left by a small angle, especially the diffraction peaks of the (002) crystal plane, which is most affected by CNFs. Raman spectroscopy was conducted to further confirm the consistency of the phase and chemical composition of the synthesized materials, and the results are shown in Figure 2b. The Raman spectrum of pure CNFs only has two distinct characteristic peaks, corresponding to the D and G peaks, respectively. However, the Raman spectrum of MoS_2/CNFs also has two additional characteristic peaks, located at 378 and 406 cm^{-1}, which correspond to the E_{2g} and A_{1g} of 2H-MoS_2 [30,31]. Therefore, through XRD and Raman characterization, MoS_2/CNFs composite materials were successfully synthesized.

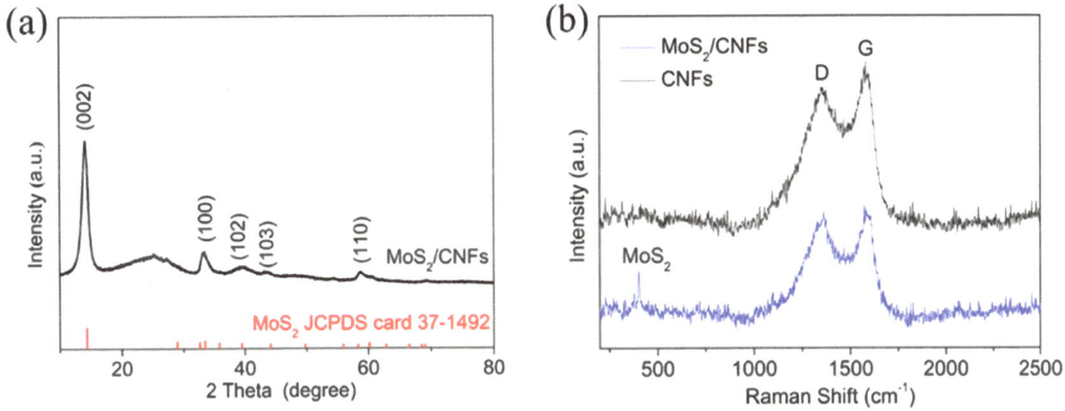

Figure 2. (**a**) Ex−situ XRD patterns and (**b**) Raman patterns of the MoS_2/CNFs.

To gain a better understanding of the morphology of the synthesized materials, SEM was performed. Figure 3a shows the corresponding SEM image of pure CNFs. It can be observed that the pure CNFs have a relatively smooth surface and uniform fiber diameter. Although the fibers have been calcined at high temperatures, they still maintain a long-range continuous morphology and form a three-dimensional (3D) network structure. After the hydrothermal synthesis process, the surface of the fiber becomes slightly rough, and layered MoS_2 nanosheets are uniformly grown on the fiber surface (Figure 3b,c). Due to the strong mechanical strength of CNFs, the 3D network structure of the composite material remains well after hydrothermal reaction. In order to further demonstrate the superiority of carbon nanofibers as a matrix material, SEM characterization was also performed on pure MoS_2. As shown in Figure 3d and Supplementary Figure S1, the MoS_2 nanomaterials prepared by one-step hydrothermal synthesis exhibit a clustered morphology formed by nanosheets. The main reason is that during the synthesis process, due to excessive reaction time, the layered structure gradually converges to form a spherical shape. However, the MoS_2 nanoballs are irregular and their size is large, exceeding 1 μm. Therefore, using carbon fibers as a substrate can effectively prevent the agglomeration of MoS_2 during formation. In addition, epitaxially grown MoS_2 nanosheets construct nanovoids on one-dimensional CNFs, which is beneficial for electrode/electrolyte interactions when used as battery electrodes.

Figure 3. (**a**) SEM image of pure CNFs; (**b**,**c**) SEM images of MoS_2/CNFs; (**d**) SEM image of pure MoS_2.

In addition, TEM characterization was studied to further analyze the morphology and microstructure of the prepared products. Figure 4a,b present TEM images of MoS_2/CNFs at varying magnifications. The observations in Figure 4a are comparable to those from the preceding SEM images, clearly showing the uniform growth of MoS_2 nanosheets on the fiber surface. From Figure 4b, the magnified TEM image reveals that the MoS_2 nanosheets on the fiber surface are relatively thin and have a clearly visible layered structure with a thickness of approximately 30 nm. Compared to MoS_2 nanoballs synthesized by a one-step hydrothermal method, small-sized MoS_2 nanosheets are more suitable as electrode materials for batteries, as smaller sizes of active materials can effectively improve electron transfer ability and shorten ion migration distance. Furthermore, the MoS_2 in the MoS_2/CNFs composite material is minimally aggregated and exposes a greater number of active sites. These factors indicate that the preparation process of growing active substances on the surface of CNFs using the hydrothermal method is superior. The polycrystalline nature of MoS_2/CNFs can be verified by a series of concentric rings in Figure 4c. The selected area electron diffraction pattern reveals that these rings correspond to the (103), (002), (110), and (101) planes of MoS_2. Figure 4d depicts the energy dispersive X-ray spectroscopy (EDS) of MoS_2/CNFs, revealing that the composite material comprises three elements of C, Mo, and S, and each element is evenly distributed, indicating that the

MoS_2/CNFs successfully synthesized by electrospinning combined with the hydrothermal method boast an optimal morphology, with MoS_2 predominantly distributed on the surface of the fibers.

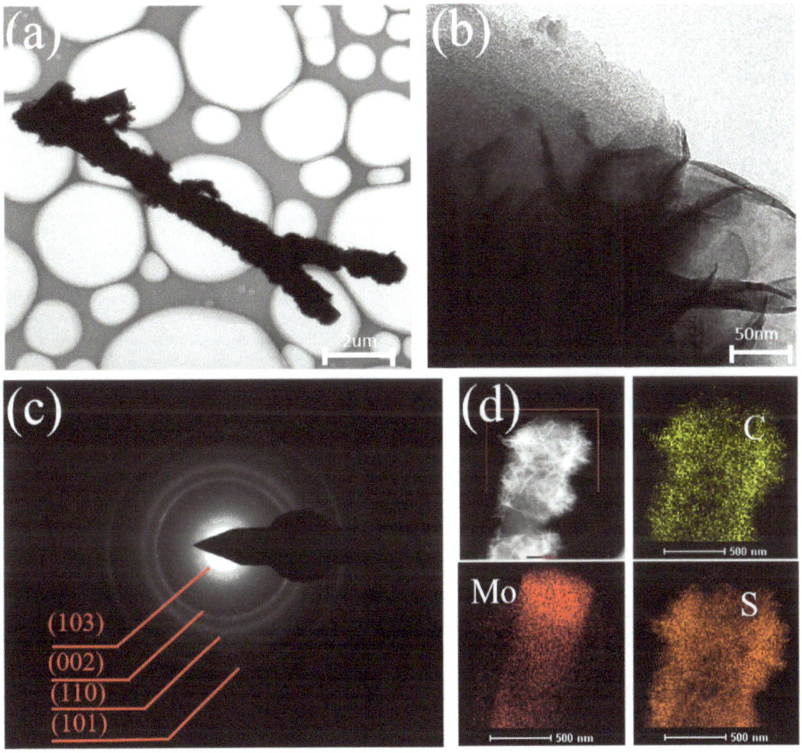

Figure 4. (**a**,**b**) TEM figures of MoS_2/CNFs; (**c**) SAED patterns of MoS_2/CNFs; (**d**) corresponding mapping patterns of MoS_2/CNFs.

An X-ray photoelectron spectroscopy (XPS) spectrum for MoS_2/CNFs composites was conducted to investigate the surface elemental composition and chemical state. Overall, the XPS full spectrum of MoS_2/CNFs composites showed the coexistence of Mo, C, and S elements, further supporting the composition of the composite consisting of these three elements (Figure 5a). Figure 5b shows a high-resolution C 1s spectrum consisting of three peaks, including the main C=C peak at 284.7 eV and two other peaks at 286.4 and 288.6 eV, corresponding to C-O and O-C=O, respectively. For the XPS high-resolution spectrum of Mo 3d (Figure 5c), two peaks located at 231.9 and 228.8 eV can be attributed to Mo $3d_{3/2}$ and Mo $3d_{5/2}$ of Mo^{4+} in MoS_2 [32,33]. In addition, the peak located at 235.8 eV corresponds to a satellite peak of Mo^{6+}, while another peak located at 226.2 eV is attributed to S 2s [34]. Figure 5d shows a high-resolution S 2p spectrum, where the peaks located at 163.1 and 161.9 eV belong to S $2p_{1/2}$ and S $2p_{3/2}$ [35,36]. Therefore, MoS_2 and CNFs are bound through C-O-Mo bonds, indicating a strong interaction between Mo species and functional groups present on CNFs, which originate from the peroxide and carbonization processes of PAN-[37].

Figure 6 shows the electrochemical performance of soft-pack batteries assembled with MoS_2/CNFs and CNFs as the electrodes. Since the MoS_2/CNFs electrodes tested in the experiment are composite materials, it is necessary to analyze the carbon substrate to understand its role in the composite electrode. Figure 6a depicts the CV curve of pure CNFs as the electrode. It is evident that the battery exhibits no electrochemical activity at this voltage window, indicating that the carbon material only plays a role as a self-supporting substrate for MoS_2. Furthermore, in comparison to its corresponding constant current charge–discharge curve, it can be clearly seen that the specific capacity released by the electrode at a current density of 100 mA g^{-1} is almost negligible (Figure 6b). Consequently, it can be inferred that using carbon nanofibers as the matrix to support the active substance has no impact on the electrochemical performance of the battery when operated under a voltage range of 0.1–1.8 V.

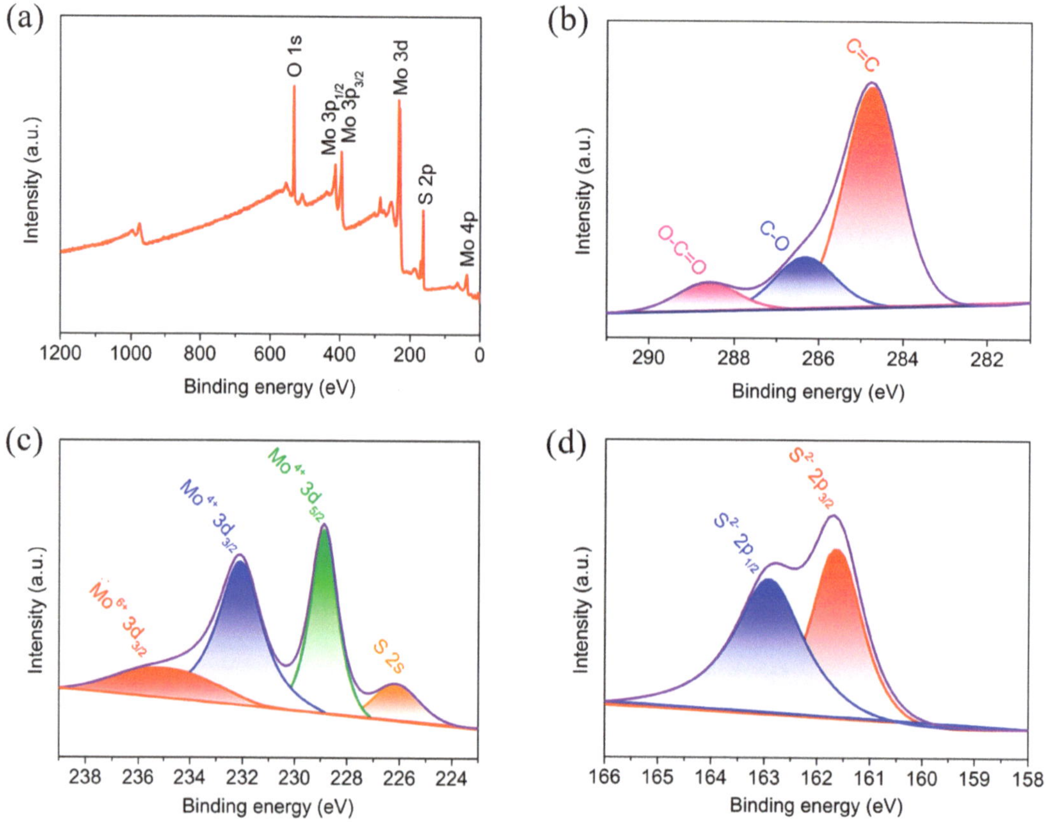

Figure 5. XPS spectra of (**a**) survey spectrum; (**b**) C 1s; (**c**) Mo 3d; (**d**) S 2p of MoS_2/CNFs.

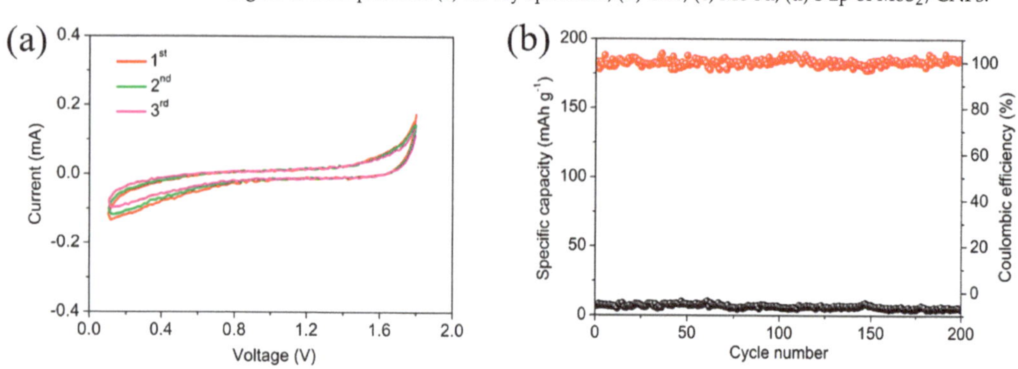

Figure 6. *Cont.*

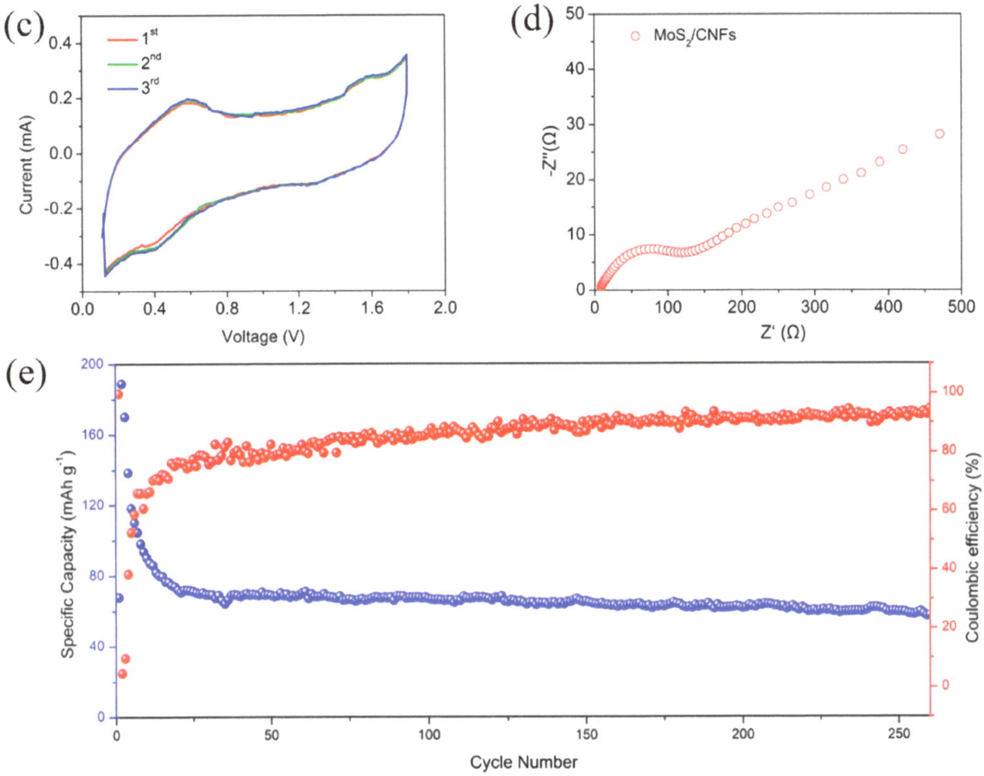

Figure 6. CV curves of: (**a**) CNFs, (**c**) MoS_2/CNFs; cycling performance at 100 mA g^{-1} of (**b**) CNFs, (**e**) MoS_2/CNFs; (**d**) Nyquist plots of MoS_2/CNFs.

Figure 6c displays the CV curve of the MoS_2/CNFs composite material serving as the cathode of AIBs. In contrast to the pure CNFs cathode, the battery demonstrates electrochemical activity within this voltage window, showcasing a subtle oxidation reduction peak. For the cathodic sweep peak, the oxidation peak is located at approximately 0.55 V and 1.58 V, while for the anodic sweep peak, the reduction peak positions of the curve are approximately 1.23 V and 0.42 V. Therefore, utilizing MoS_2 as the active substance in AIBs manifests battery characteristics, indicating the potential for cyclic performance. The Nyquist plot is shown in Figure 6d, revealing the kinetic characteristics of the MoS_2/CNFs electrode. It can be clearly observed that the impedance spectrum of the electrode consists of a semicircle and a slope, where the semicircle is related to the charge transfer dynamics between the electrode interface and the electrolyte, and the slope indicates the diffusion resistance of ions. The Nyquist curves are simulated by the modified Randles equivalent circuit (Figure S3, Supporting Information). From the fitted curve, it can be seen that the resistance value of the electrode is relatively small (128 Ω), indicating the high charge/ion conductivity of the electrode. After cycling, the hybrid electrode exhibited significantly enhanced charge transfer kinetics with a remarkably reduced charge transfer resistance of 66 Ω compared to the fresh cell (Figure S2, Supporting Information) [38]. The main reason is that the 3D network structure of the composite material is conducive to electrolyte penetration into the electrode, and the continuity of CNFs shortens the electron migration path. As illustrated in Figure 6e, since pure CNFs lack electrochemical activity within this voltage window, the curve reflects the capacity released by MoS_2. The initial capacity of the battery is approximately 180 mAh g^{-1}, but it experiences rapid capacity degradation in the initial cycles. After 30 cycles, the battery gradually tends to a stable state, reaching a capacity of about 60 mAh g^{-1}. In subsequent cycles, the battery displays good stability, and after 260 cycles of charging and discharging, the AIBs possess a capacity of approximately 53 mAh g^{-1}. To evaluate the rate performance of the MoS_2/CNFs electodes, charge/discharge cycles at different current densities from 100 to 500 mA g^{-1} were performed, as shown in Figure S4 (Supporting Information). Furthermore, we have compared the performance of MoS_2/CNFs for application in AIBs with several other transition metal dichalcogenides in Table S1 (Supporting Information).

MoS_2, serving as a host material, finds extensive applications in various metal secondary batteries (such as Li, Zn, Na, etc.). Its energy storage mechanism is mostly attributed to the deintercalation behavior of metal ions within MoS_2 crystals. However, AIBs have multiple ions during the charging and discharging processes, including Al^{3+}, $[Al_2Cl_7]^-$ and $[AlCl_4]^-$, among others. To further ascertain which ion undergoes the deintercalation behavior within MoS_2 crystals, in-depth studies were conducted on the energy storage mechanism of MoS_2/CNFs electrodes. Figure 7a,b present high-resolution TEM images of MoS_2/CNFs before and after cycling, respectively. For the pristine MoS_2/CNFs, the lattice spacing of MoS_2 is 0.62 nm, which aligns with its (002) crystal plane spacing (Figure 7a). However, as depicted in Figure 7b, the crystal plane spacing of the cycled electrode increased to 0.73 nm, indicating the presence of ion insertion in the MoS_2 crystal after cycling. The stability of the electrode was assessed by conducting SEM image analysis on the electrode post-cycling, and the obtained results are presented in Figure S5 (Supporting Information). To gain a more intuitive understanding of the composition and structure of the electrode material interface before and after cycling, EDS was also conducted. Upon EDS analysis of the pre- and post-cycling electrodes, it is evident that a distinct aluminum characteristic peak emerges in the EDS spectrum of the post-cycling electrode (Figure 7c,d). Therefore, it can be concluded that aluminum ions undergo a deintercalation process during the charge and discharge cycles, and this behavior can be summarized using the following equation:

$$\text{Cathode}: Al^{3+} + MoS_2 + 3xe^- \overset{\text{discharge/charge}}{\leftrightarrow} Al_xMoS_2$$

$$\text{Anode}: 7[Al\,Cl_4]^- + Al \overset{\text{discharge/charge}}{\leftrightarrow} 4[Al_2\,Cl_7]^- + 3e^-$$

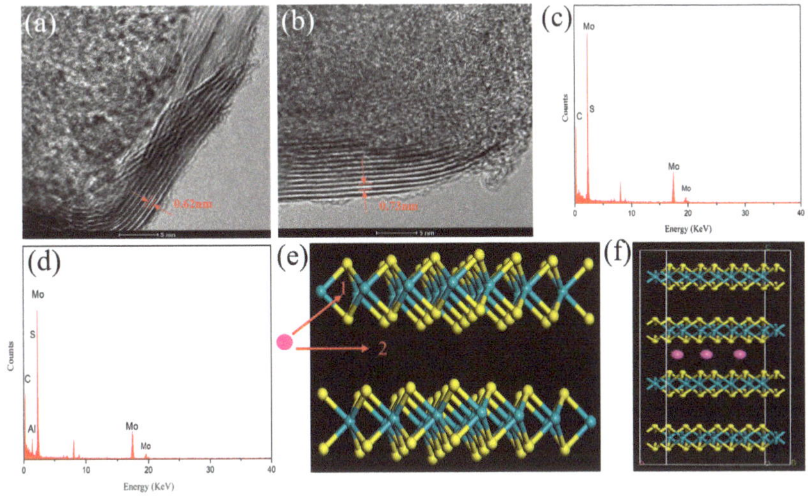

Figure 7. (**a**,**b**) HR-TEM images of the MoS_2/CNFs before and after cycled; (**c**,**d**) the EDS spectrum of the MoS_2/CNFs before and after cycled; (**e**) the crystal structure of MoS_2 and the possible entry methods of Al^{3+}; (**f**) schematic diagram of the structure of Al^{3+} in MoS_2 crystals. The red, green, and yellow circles represent Mo, Al, and S atoms.

However, due to the unique sandwich structure of MoS_2, molybdenum atoms and sulfur atoms form the spatial structure of S-Mo-S, and each adjacent MoS_2 layer is separated by van der Waals forces. To further investigate the deintercalation positions of aluminum ions, the experiment incorporated first-principles calculations to predict the potential entry pathways for aluminum ions. Figure 7e illustrates the crystal structure of MoS_2 and the possible entry pathways for aluminum ions. Initially, it was hypothesized that aluminum ions could be deintercalated in both channel 1 and channel 2 in MoS_2. However, the computational results revealed that the energy required for aluminum ions to be deintercalated in channel 1 is significantly high and nearly unattainable, while the energy required for deintercalation in channel 2 is relatively low, which can possibly attributed to the presence of minor van der Waals forces between layers. Based on this, it can be inferred that during the cycling process of the battery, aluminum ions predominantly undergo deintercalation in channel 2 of MoS_2. The corresponding computational results are shown in Table S2 (Supporting Information).

4. Conclusions

This study delves into the energy storage mechanism of AIBs, with a focus on the layered structure of MoS_2 as the host material. MoS_2/CNFs composite materials were synthesized through a combination of electrospinning and the hydrothermal method, serving as the self-supporting electrodes for AIBs. The electrode material boasts a 3D network structure, which not only enhances the conductivity of the electrode, but also effectively prevents the aggregation of MoS_2 during the synthesis process. As self-supporting electrodes, the AIBs exhibit a residual reversible capacity of 53 mAh g^{-1} at 100 mA g^{-1} after 260 cycles. The experimental and computational findings reveal the presence of Al^{3+} intercalation in MoS_2/CNFs electrodes during the charge and discharge processes. Based on first-principles calculations, it can be concluded that the energy required for Al^{3+} to be inserted in the interlayer gaps formed by S-Mo-S ion bonds is relatively high, indicating that this entry method is almost impossible. Meanwhile, the formation energy for deintercalation between each MoS_2 layer is relatively small, approximately 0.64 eV. Therefore, during the charging and discharging processes, Al^{3+} mainly undergoes deintercalation between each layer. The insights gained from this research are expected to guide the design and synthesis of next-generation high-capacity and durable AIBs electrodes, laying the foundation for a more sustainable energy future and facilitating the commercial application of AIBs.

Supplementary Materials: The following supporting information can be downloaded at: https://www.mdpi.com/article/10.3390/nano14050442/s1, Figure S1 Scanning electron microscopy (SEM) images of MoS_2. Figure S2 Nyquist plots of MoS_2/CNFs after cycling. Figure S3 Equivalent circuit model. Figure S4 Rate performance of MoS_2/CNFs electrode at different current density. Figure S5 Scanning electron microscopy (SEM) images of MoS_2/CNFs after cycling. Table S1 Performance comparison of MoS_2/CNFs with several other transition metal dichalcogenides for application in AIBs. Table S2 The formation energy of aluminum ions embedded in different channels of MoS_2. References [39–50] are cited in the Supplementary Materials.

Author Contributions: X.L. and J.L.: investigation, formal analysis. M.H. and P.S.: data curation. X.W.: methodology, writing—original draft preparation. R.Z.: conceptualization, validation, visualization, writing—reviewing and editing. J.W.: resources, software, writing—reviewing and editing. J.Y.: resources, supervision. Both authors contributed to the article and approved the submitted version. All authors have read and agreed to the published version of the manuscript.

Funding: This work was financially supported by the Science and Technology Bureau of Jiaxing (grant no. 2023AY11021), the Start-up Foundation of JiaXing NanHu University for Senior Talents (grant no. 70500000/047), and the JiaXing NanHu University Foundation (grant no. 62205ZL).

Data Availability Statement: Data are contained within the article.

Conflicts of Interest: The authors declare no conflicts of interest.

References

1. Shrestha, A.; Rajbhandari, Y.; Gonzalez-Longatt, F. Day-ahead energy-mix proportion for the secure operation of renewable energy-dominated power system. *Int. J. Electr. Power* **2024**, *155*, 109560. [CrossRef]
2. Wang, J.; Chen, L.; Tan, Z.; Du, E.; Liu, N.; Ma, J. Inherent spatiotemporal uncertainty of renewable power in China. *Nat. Commun.* **2023**, *14*, 5379. [CrossRef] [PubMed]
3. Zhang, J.W.; Zhang, Y.; Bao, S.; Huang, Y.Y.; Lu, J.L.; Yin, Y.S. Advances on layered transition-metal oxides for sodium-ion batteries: A mini review. *Front. Energy Res.* **2023**, *11*, 1246327.
4. Li, X.H.; Gan, Y.P.; Zhang, J.; Wu, Z.; Zhang, W.K.; Xia, Y. Interfaces in sulfide solid electrolyte-based all-solid-state lithium batteries: Characterization, mechanism, and strategy. *Electrochem. Energy Rev.* **2023**, *6*, 10.
5. Tu, J.G.; Song, W.L.; Lei, H.P.; Yu, Z.J.; Chen, L.L.; Wang, M.Y. Nonaqueous rechargeable aluminum batteries: Progresses, challenges, and perspectives. *Chem. Rev.* **2021**, *121*, 4903–4961. [CrossRef] [PubMed]
6. Ng, K.L.; Amrithraj, B.; Azimi, G. Nonaqueous rechargeable aluminum batteries. *Joule* **2022**, *6*, 134–170. [CrossRef]
7. Holleck, G.L.; Giner, J.J. The aluminum electrode in $AlCl_3$-Alkali-Halide melts. *Electrochem. Soc.* **1972**, *119*, 1161. [CrossRef]
8. Lin, M.C.; Gong, M.; Lu, B.G.; Wu, Y.P.; Wang, D.Y.; Guan, M.Y. An ultrafast rechargeable aluminum-ion battery. *Nature* **2015**, *520*, 324–328. [CrossRef]
9. Giraldo, D.A.; Almodóvar, P.; Álvarez-Serrano, I.; Chacón, J.; López, M.L. Electrochemical performance of tunnelled and layered MnO_2 electrodes in aluminum-ion batteries: A matter of dimensionality. *J. Electrochem. Soc.* **2022**, *169*, 100538. [CrossRef]
10. De, P.; Halder, J.; Priya, S.; Srivastava, A.K.; Chandra, A. Two-Dimensional V_2O_5 nanosheets as an advanced cathode material for realizing low-cost aqueous aluminum-ion batteries. *ACS Appl. Energy Mater.* **2023**, *6*, 753–762. [CrossRef]
11. Zhou, Q.; Zheng, Y.; Wang, D.; Lian, Y.; Ban, C.; Zhao, J.; Zhang, H. Cathode materials in non-aqueous aluminum-ion batteries: Progress and challenges. *Ceram. Int.* **2020**, *46*, 26454–26465. [CrossRef]
12. Liu, T.; Lv, G.; Liu, M.; Zhao, C.; Liao, L.; Liu, H. Cation-intercalation and conversion-type cathode materials for rechargeable aluminum batteries. *Mater. Chem. Front.* **2022**, *6*, 280–296. [CrossRef]
13. Nandi, S.; Pumera, M. Transition metal dichalcogenide-based materials for rechargeable aluminum-ion battery: A mini-review. *ChemSusChem* **2024**, e202301434. [CrossRef]
14. Huang, Z.; Du, X.; Ma, M.; Wang, S.; Xie, Y.; Meng, Y. Organic cathode materials for rechargeable aluminum-ion batteries. *ChemSusChem* **2023**, *16*, e202202358. [CrossRef] [PubMed]
15. Wei, T.T.; Peng, P.; Qi, S.Y.; Zhu, Y.R.; Yi, T.F. Advancement of technology towards high-performance non-aqueous aluminum-ion batteries. *J. Energy Chem.* **2021**, *57*, 169–188. [CrossRef]
16. Das, S.K. Graphene: A cathode material of choice for aluminum-ion batteries. *Angew. Chem. Int. Ed.* **2018**, *57*, 16606–16617. [CrossRef]
17. Ran, Q.; Zeng, S.P.; Zhu, M.H.; Wan, W.B.; Meng, H.; Shi, H. Uniformly MXene-grafted eutectic aluminum-cerium alloys as flexible and reversible anode materials for rechargeable aluminum-ion battery. *Adv. Funct. Mater.* **2023**, 33, 2211271. [CrossRef]

18. Yang, Z.; Huang, X.; Meng, P.; Jiang, M.; Wang, Y.; Yao, Z. Phenoxazine polymer-based p-type positive electrode for aluminum-ion batteries with ultra-long cycle life. *Angew. Chem. Int. Ed.* **2023**, *62*, e202216797. [CrossRef] [PubMed]
19. Hu, Z.; Zhang, H.; Wang, H.; Zhang, F.; Li, Q.; Li, H. Nonaqueous aluminum ion batteries: Recent progress and prospects. *ACS Mater. Lett.* **2020**, *2*, 887–904. [CrossRef]
20. Zhang, M.; Song, X.; Ou, X.; Tang, Y. Rechargeable batteries based on anion intercalation graphite cathodes. *Energy Storage Mater.* **2019**, *16*, 65–84. [CrossRef]
21. Lin, Z.; Mao, M.; Yang, C.; Tong, Y.; Li, Q.; Yue, J. Amorphous anion-rich titanium polysulfides for aluminum-ion batteries. *Sci. Adv.* **2021**, *7*, eabg6314. [CrossRef]
22. Vijayakumar, V.; Ghosh, M.; Asokan, K.; Sukumaran, S.B.; Kurungot, S.; Mindemark, J. 2D layered nanomaterials as fillers in polymer composite electrolytes for lithium batteries. *Adv. Funct. Mater.* **2023**, *13*, 2203326. [CrossRef]
23. Peng, K.; Fu, L.; Ouyang, J.; Yang, H. Emerging parallel dual 2D composites: Natural clay mineral hybridizing MoS_2 and interfacial structure. *Adv. Funct. Mater.* **2016**, *26*, 2666–2675. [CrossRef]
24. Wei, X.; Lin, C.C.; Wu, C.; Qaiser, N.; Cai, Y.; Lu, A.Y. Three-dimensional hierarchically porous MoS_2 foam as high-rate and stable lithium-ion battery anode. *Nat. Commun.* **2022**, *13*, 6006. [CrossRef] [PubMed]
25. Wen, X.; Feng, W.; Li, X.; Yang, J.; Du, R.; Wang, P. Diatomite-templated synthesis of single-atom cobalt-doped MoS_2/Carbon composites to boost sodium storage. *Adv. Mater.* **2023**, *35*, 2211690. [CrossRef]
26. Gao, S.; Chen, S.; Shi, F.; Jiang, W.; Chen, J. Synthesis of 1T-MoS_2 sheets with large space distance between layers for high-rate aqueous Zn-ion batteries cathode. *J. Power Sources* **2024**, *591*, 233866. [CrossRef]
27. Li, Z.; Niu, B.; Liu, J.; Li, J.; Kang, F. Rechargeable aluminum-ion battery based on MoS_2 microsphere cathode. *ACS Appl. Mater. Interfaces* **2018**, *10*, 9451–9459. [CrossRef] [PubMed]
28. Tan, B.; Han, S.; Luo, W.; Chao, Z.; Fan, J.; Wang, M. Synthesis of RGO-supported layered MoS_2 with enhanced electrochemical performance for aluminum ion batteries. *J. Alloys Compd.* **2020**, *841*, 155732. [CrossRef]
29. Yang, W.; Lu, H.; Cao, Y.; Xu, B.; Deng, Y.; Cai, W. Flexible free-standing MoS_2/carbon nanofibers composite cathode for rechargeable aluminum-ion batteries. *ACS Sustain. Chem. Eng.* **2019**, *7*, 4861–4867. [CrossRef]
30. Lukowski, M.A.; Daniel, A.S.; Meng, F.; Forticaux, A.; Li, L.; Jin, S. Enhanced hydrogen evolution catalysis from chemically exfoliated metallic MoS_2 nanosheets. *J. Am. Chem. Soc.* **2013**, *135*, 10274–10277. [CrossRef] [PubMed]
31. Zhao, C.; Yu, C.; Qiu, B.; Zhou, S.; Zhang, M.; Huang, H. Ultrahigh rate and long-life sodium-ion batteries enabled by engineered surface and near-surface reactions. *Adv. Mater.* **2018**, *30*, 1702486. [CrossRef] [PubMed]
32. Teng, Y.; Zhao, H.; Zhang, Z.; Li, Z.; Xia, Q.; Zhang, Y. MoS_2 nanosheets vertically grown on graphene sheets for lithium-ion battery anodes. *ACS Nano* **2016**, *10*, 8526–8535. [CrossRef]
33. Zhao, C.; Yu, C.; Zhang, M.; Huang, H.; Li, S.; Han, X. Ultrafine MoO_2-Carbon microstructures enable ultralong-life power-type sodium ion storage by enhanced pseudocapacitance. *Adv. Funct. Mater.* **2017**, *7*, 1602880. [CrossRef]
34. Shi, Y.; Wang, Y.; Wong, J.I.; Tan, A.Y.S.; Hsu, C.L.; Li, L.J. Self-assembly of hierarchical MoS_x/CNT nanocomposites ($2 < x < 3$): Towards high performance anode materials for lithium ion batteries. *Sci. Rep.* **2013**, *3*, 2169.
35. Tang, Y.J.; Wang, Y.; Wang, X.L.; Li, S.L.; Huang, W.; Dong, L.Z. Molybdenum disulfide/nitrogen-doped reduced graphene oxide nanocomposite with enlarged interlayer spacing for electrocatalytic hydrogen evolution. *Adv. Energy Mater.* **2016**, *6*, 1600116. [CrossRef]
36. Shi, Z.T.; Kang, W.; Xu, J.; Sun, L.L.; Wu, C.; Wang, L. In situ carbon-doped $Mo(Se_{0.85}S_{0.15})_2$ hierarchical nanotubes as stable Anodes for high-performance sodium-ion batteries. *Small* **2015**, *11*, 5667–5674. [CrossRef] [PubMed]
37. Li, W.; Bi, R.; Liu, G.; Tian, Y.; Zhang, L. 3D interconnected MoS_2 with enlarged interlayer spacing grown on carbon nanofibers as a flexible anode toward superior sodium-ion batteries. *ACS Appl. Mater. Interfaces* **2018**, *10*, 26982–26989. [CrossRef]
38. Baboukani, A.R.; Khakpour, I.; Adelowo, E.; Drozd, V.; Shang, W.; Wang, C.L. High-performance red phosphorus-sulfurized polyacrylonitrile composite by electrostatic spray deposition for lithium-ion batteries. *Electrochim. Acta* **2020**, *345*, 136227. [CrossRef]
39. Wu, L.; Sun, R.M.; Xiong, F.Y.; Pei, G.Y.; Han, K.; Peng, C.; Fan, Y.Q.; Yang, W.; An, Q.Y.; Mai, L.Q. A rechargeable aluminum-ion battery based on a VS_2 nanosheet cathode. *Phys. Chem. Chem. Phys.* **2018**, *20*, 22563–22568. [CrossRef] [PubMed]
40. Xing, L.L.; Owusu, K.A.; Liu, X.Y.; Meng, J.S.; Wang, K.; An, Q.Y.; Mai, L.Q. Insights into the storage mechanism of VS_4 nanowire clusters in aluminum-ion battery. *Nano Energy* **2021**, *79*, 105384. [CrossRef]
41. Wang, S.; Yu, Z.; Tu, J.; Wang, J.; Tian, D.; Liu, Y.; Jiao, S. A novel aluminum-ion battery: Al/$AlCl_3$-[EMIm]Cl/Ni_3S_2@Graphene. *Adv. Energy Mater.* **2016**, *6*, 1600137. [CrossRef]
42. Geng, L.; Lv, G.; Xing, X.; Guo, J. Reversible electrochemical intercalation of aluminum in Mo_6S_8. *Chem. Mater.* **2015**, *27*, 4926–4929.
43. Li, H.; Yang, H.; Sun, Z.; Shi, Y.; Cheng, H.-M.; Li, F. A highly reversible Co_3S_4 microsphere cathode material for aluminum-ion batteries. *Nano Energy* **2019**, *56*, 100–108. [CrossRef]
44. Liang, K.; Ju, L.; Koul, S.; Kushima, A.; Yang, Y. Self-supported Tin sulfide porous films for flexible aluminum-ion batteries. *Adv. Energy Mater.* **2019**, *9*, 1802543. [CrossRef]
45. Geng, L.; Scheifers, J.P.; Fu, C.; Zhang, J.; Fokwa, B.P.T.; Guo, J. Titanium sulfides as intercalation-type cathode materials for rechargeable aluminum batteries. *ACS Appl. Mater. Inter.* **2017**, *9*, 21251–21257. [CrossRef] [PubMed]
46. Zhang, X.; Wang, S.; Tu, J.; Zhang, G.; Li, S.; Tian, D.; Jiao, S. Flower-like vanadium suflide/reduced graphene oxide composite: An energy storage material for aluminum-ion batteries. *ChemSusChem* **2018**, *11*, 709–715. [CrossRef] [PubMed]

47. Yang, W.; Lu, H.; Cao, Y.; Jing, P. Single-/few-layered ultrasmall WS_2 nanoplates embedded in nitrogen-doped carbon nanofibers as a cathode for rechargeable aluminum batteries. *J. Power Sources* **2019**, *441*, 227173. [CrossRef]
48. Zhao, Z.C.; Hu, Z.Q.; Li, Q.; Li, H.S.; Zhang, X.; Zhuang, Y.D.; Wang, F.; Yu, G.H. Designing two-dimensional WS_2 layered cathode for high-performance aluminum-ion batteries: From micro-assemblies to insertion mechanism. *Nano Today* **2020**, *32*, 100870. [CrossRef]
49. Hu, Y.X.; Luo, B.; Ye, D.L.; Zhu, X.B.; Lyu, M.Q.; Wang, L.Z. An innovative freeze-dried reduced graphene oxide supported SnS_2 cathode active material for aluminum-ion batteries. *Adv. Mater.* **2017**, *29*, 1606132. [CrossRef]
50. Zhang, K.Q.; Lee, T.H.; Cha, J.H.; Jang, H.W.; Shokouhimehr, M.; Choi, J.-W. Properties of CoS_2/CNT as a cathode material of rechargeable aluminum-ion batteries. *Electron. Mater. Lett.* **2019**, *15*, 727–732. [CrossRef]

nanomaterials

Article

Mechanically Enhanced Detoxification of Chemical Warfare Agent Simulants by a Two-Dimensional Piezoresponsive Metal–Organic Framework

Yuyang Liu †, Shiyin Zhao †, Yujiao Li, Jian Huang, Xuheng Yang, Jianfang Wang * and Cheng-an Tao *

College of Science, National University of Defense Technology, Changsha 430083, China; liuyuyang0412@163.com (Y.L.); zhaoshiyin21@nudt.edu.cn (S.Z.); liyujiao15@nudt.edu.cn (Y.L.); huangjian2015@nudt.edu.cn (J.H.); ouyangxuheng@163.com (X.Y.)

* Correspondence: jianfangwang@nudt.edu.cn (J.W.); chengantao@nudt.edu.cn (C.-a.T.)

† These authors contributed equally to this work.

Abstract: Chemical warfare agents (CWAs) refer to toxic chemical substances used in warfare. Recently, CWAs have been a critical threat for public safety due to their high toxicity. Metal–organic frameworks have exhibited great potential in protecting against CWAs due to their high crystallinity, stable structure, large specific surface area, high porosity, and adjustable structure. However, the metal clusters of most reported MOFs might be highly consumed when applied in CWA hydrolysis. Herein, we fabricated a two-dimensional piezoresponsive UiO-66-F_4 and subjected it to CWA simulant dimethyl-4-nitrophenyl phosphate (DMNP) detoxification under sonic conditions. The results show that sonication can effectively enhance the removal performance under optimal conditions; the reaction rate constant *k* was upgraded 45% by sonication. Moreover, the first-principle calculation revealed that the band gap could be further widened with the application of mechanical stress, which was beneficial for the generation of $^1O_2$, thus further upgrading the detoxification performance toward DMNP. This work demonstrated that mechanical vibration could be introduced to CWA protection, but promising applications are rarely reported.

Keywords: piezoresponsive metal–organic frameworks; chemical warfare agent detoxification; mechanical energy harvesting

Citation: Liu, Y.; Zhao, S.; Li, Y.; Huang, J.; Yang, X.; Wang, J.; Tao, C.-a. Mechanically Enhanced Detoxification of Chemical Warfare Agent Simulants by a Two-Dimensional Piezoresponsive Metal–Organic Framework. *Nanomaterials* **2024**, *14*, 559. https://doi.org/10.3390/nano14070559

Academic Editor: Félix Zamora

Received: 6 February 2024
Revised: 8 March 2024
Accepted: 9 March 2024
Published: 22 March 2024

1. Introduction

Chemical warfare agents (CWAs) refer to various chemical substances with severe toxicity used for war purposes, capable of poisoning or killing enemy humans, animals, and plants on a large scale. In recent years, CWAs have been one of the crucial means for terrorists and extremist organizations to enforce terrorist activities and war activities due to their large killing ability, wide working range, strong concealment, and low cost [1–3]. Consequently, due to their terrible toxicity, the quick detoxification of CWAs has become one of the critical demands in preventing CWA attacks [4,5]. To deal with CWAs, conventional methods include adsorption and chemical reaction methods [6–8]. The adsorption method presents good universality; however, the removal performance largely relies on the adsorption capacity of the adsorbents (e.g., activated carbon). Moreover, after reaching saturation, there is a large degree of desorption, which is potentially dangerous, thus hindering its further application in protective equipment [9–11]. The chemical reaction method exhibits the great advantage of permanent detoxification. Recently, most of the commonly used disinfectants have included oxidizing chlorinated disinfectants, alkaline disinfectants, etc. [12–15]. However, various problems also exist such as low efficiency, strong corrosivity, storage difficulties over long periods of time, and serious environmental pollution. Moreover, the consumption of these disinfectants is always huge; it is difficult to carry them in military operations, and it is also not easy to integrate them into military personal protective equipment. Therefore, developing new detoxification materials is urgently needed to guard against potential threats from chemical warfare agents [16–18].

In recent years, researchers have made great efforts in developing new detoxification materials, especially materials with catalytic properties, such as biological enzymes, anionic polyoxometalates, nano metal oxides, amphiphilic polyoxometalates, etc. These materials show a certain degree of detoxification ability; however, a variety of demerits limit their further promotion and application. For instance, biological enzymes exhibit high cost; the pore structure of metal oxides is relatively inferior; simultaneously, the surface chemical properties of metal oxides are usually ultrastable and are hard to modify, thus leading to the active adsorption ability of metal oxides to chemical warfare agents being insufficient [17,19–21]. To achieve the ideal detoxification effect, the designed material needs the following key properties. Firstly, the material needs an outstanding pore structure to provide more active sites and larger space for further modification. Secondly, the chemical structure of the material is adjustable and easy to modify, so that the bandgap width, affinity, and application range can be controlled. Moreover, the material needs to have the characteristic of utilizing external driving energy to enhance its reusability and be able utilize external driving forces to achieve and accelerate the catalytic process during detoxification; then, the crystals can be well preserved even after many cycles of use.

Metal–organic frameworks (MOFs) are porous crystalline materials assembled by the coordination of metal clusters and organic ligands [22–25]. They have outstanding advantages of high crystallinity, stable structure, large specific surface area, high porosity, and adjustable structure, which give them broad application prospects in various fields, such as adsorption, drug delivery, sensing, and photocatalysis [26–29]. MOFs are also regarded as some of the ideal detoxification materials for CWAs and have had significant outcomes in the field of detoxification, greatly improving detoxification efficiency and largely shortening the half-life of the CWAs. However, the active sites are easy to inactivate in the detoxifying process, which severely hinders their further application. External driving energy is urgently required to be employed to transfer to chemical energy, which can detoxify CWAs. Most reported works developed photocatalytic MOFs to harvest solar energy, which were applied in the degradation of pollutants and CWA simulants, for instance, PCN-222, Cu-TCPP, NU-1000, etc. [30–33]. However, the working performance of photoresponsive MOFs is largely limited by the weather and daytime duration. Mechanical energy is one of the most common forms of energy in nature, and piezocatalysis has been emerging as a promising alternative technology for organic pollutant removal, as it can harvest various mechanical energies including vibrations, wind, and water waves from the surrounding environment [34–38]. However, the mechanically enhanced detoxification of CWAs by piezoresponsive MOFs has rarely been reported [39].

Herein, we fabricated a two-dimensional UiO-66-F_4 by the microwave-assisted hydrothermal method and subjected it to CWA simulant detoxification under mechanical vibration. The obtained UiO-66-F_4 presented ultrathin nanosheets with a thickness of ~5 nm and a large specific surface area of 331.827 m^2/g. It also showed outstanding piezoresponse capacity in an ultrasonic environment; the amplitude displacement of UiO-66-F_4 could reach ~50 mV under a 10 V DC bias field. Consequently, it could effectively harvest mechanical energy and transfer it to chemical energy, thus effectively detoxifying the CWAs. Under optimal conditions, the UiO-66-F_4 could detoxify almost all of the DMNP within 60 min, and the half-time of the DMNP was 10.58 min, which was much shorter than that under stirring conditions. The detoxification capacity was also excellent compared to other reported materials. Simultaneously, the reaction constant k was upgraded by 45% by the mechanical vibration. The first-principle calculation revealed the polarization behavior of the MOF crystals and the widening of the band gap under mechanical stress. This work demonstrated that CWAs could be mechanically detoxified by a piezoresponsive MOF, which provides great prospects for protection against CWAs of personnel in a toxic environment.

2. Materials and Methods

2.1. Chemicals

Tetrafluoroterephthalic acid (BDC-F_4) was bought from Yanshen Chemistry (Jilin, China). Zirconium chloride ($ZrCl_4$), methanol, absolute ethanol, N,Ndimethylformamide (DMF), and glacial acetic acid (HAc) were bought from Macklin (Shanghai, China). Thymolphthalein, sodium hydroxide (NaOH), zinc oxide (ZnO), and *N*-Ethylmorpholine were bought from Aladdin (Shanghai, China). Dimethyl-4-nitrophenyl phosphate (DMNP) was synthesized by our lab, and the chemical structure is displayed in Figure S1b.

2.2. Characterizations

The morphology of the UiO-66-F_4 was observed by field emission scanning electron microscopy (SEM, Zeiss Merlin, Oberkochen, Germany) with a working voltage of 10 kV and transmission electron microscopy (TEM, JEOL JEM F200, Tokyo Metropolitan, Japan) with a working voltage of 300 kV. The crystallinity was analyzed through a powder X-ray diffraction pattern which was recorded by a polycrystalline diffractometer (Rigaku MiniFlex 600, Tokyo, Japan) with Cu Kα radiation (λ = 0.154 nm). The Fourier transform infrared (FTIR) spectra were recorded by a PerkinElmer Spectrum-II spectrometer (Wellesley, MA, USA) with KBr pellets. The elements' valence status was investigated through X-ray photoelectronic spectroscopy (XPS) which was conducted by a Thermo ESCALAB 250XI (Waltham, MA, USA) fitted with Al Kα radiation, and all the binding energy was referenced as the C 1s peak at 284.8 eV. The surface area and porosity were analyzed by Brunauer–Emmett–Teller (BET) and Horvath–Kawazoe (HK) methods, respectively, using a Quantachrome Autosorb IQ (Boynton Beach, FL, USA) instrument. The thickness and piezoresponse capacity were measured through atom force microscopy (AFM) and piezoresponse force microscopy (PFM) by a Bruker-Icon (Billerica, MA, USA) with a working probe (MESP, coefficient of elasticity 2 N/m, resonance frequency 75 Hz) and polarization voltage of 10 V, respectively. The size distribution and zeta potential were collected by a Nanotrac Wave II (Microtrac MRB, Clearwater, FL, USA). The ultraviolet and visible spectrum (UV-Vis) was measured by a UV-5200 (Metash, Shanghai, China) with a working wavelength range of 200–800 nm.

2.3. Assembly of UiO-66-F_4

The UiO-66-F_4 was assembled by a previously described method [38]. Typically, 85.9 mg BDC-F_4 was added to 50 mL mixed solution (water/HAc = 30/20). After sonication for 5 min, 75 mg $ZrCl_4$ was added, followed by continuous sonication for 10 min. Right after, the mixed solution was placed in a microwave oven and heated at 100 °C for 4 h. The obtained precipitates were washed by DMF and ethanol, respectively, 3 times, followed by drying in a vacuum oven for 8 h at 80 °C. The resulting powder was referred to as UiO-66-F_4.

2.4. Removal Experiments

First 4.0 mL *N*-Ethylmorpholine and 13 mg as-prepared UiO-66-F_4 were added to a centrifugal tube (50 mL) and ultrasonicated for 5 min till the UiO-66-F_4 uniformly dispersed in the solution. Then, the tube was placed on a magnetic stirrer with a working speed of 600 rpm. Then, 16 µL of DMNP was injected in the mixture by a microinjector and the stirring speed was maintained. After certain time interval, the MOF particles were separated by a filter, then, 20 µL of the mixture was taken out and diluted to 10 mL to test the concentration of the DMNP. The time intervals were set as 0, 1, 3, 5, 10, 20, 30, 40, 60, and 100 min. When testing the piezoelectric degradation efficiency, the stirring was replaced by sonication with a frequency of 45 kHz. As the DMNP was decomposed into p-nitrophenol, the concentration of the DMNP was inferred by the concentration of the p-nitrophenol which can be detected by UV-Vis with characteristic wavelength of 407 nm.

2.5. Calculation Method

The calculations were performed using the first-principle calculation implementation of CASTEP (Materials Studio 2020) [40]. The generalized gradient approximation (GGA) with the Perdew–Burke–Ernzerhof (PBE) formula was employed for the exchange–correlation potential [41,42]. The Broyden–Fletcher–Goldfarb–Shanno (BFGS) method was used to search for the ground state of the supercells, and the convergence tolerance was set to energy change below 10^{-5} eV per atom, force less than 0.02 eV Å^{-1}, stress less than 0.05 GPa, and displacement change less than 0.001 Å. The cutoff energy of the atomic wave functions was set to 450 eV.

3. Results

3.1. Assembly and Characterizations

After 4 h of incubation through microwave-assisted heating, two-dimensional UiO-66-F_4 nanosheets were successfully obtained. Firstly, the morphology of the as-prepared sample was examined through the utilization of SEM, TEM, and AFM techniques. The SEM image demonstrates the sample's remarkable homogeneity (Figure 1a), while the enlarged SEM image reveals that nearly all UiO-66-F_4 clusters comprise interlaced nanosheets measuring approximately 200 nm in size (Figure 1b). The TEM image provides stronger evidence for this conclusion (Figure 1c). Furthermore, both the TEM image and the high-resolution TEM image reveal that the nanosheets appear transparent under fluorescent irradiation, indicating their ultrathin nature. However, upon observation using high-resolution TEM (Figure 1d), the crystal lattice was not clearly visible, indicating a relatively small crystal size. Consequently, this led to inferior crystallinity, which aligns with numerous reported UiO-66-type MOFs [43,44]. In order to further investigate the composition of the sample, the distribution of corresponding elements within a specific area was measured. The results exhibit a remarkable coincidence in the distributions of C, O, F, and Zr (Figure 1e,i), indicating a coordinated interaction between the Zr clusters and ligands. To accurately measure the thickness of the nanosheets, an atomic force microscope (AFM) was utilized. The results indicate that the sample is composed of multiple nanosheets, as depicted in Figure 1j. Subsequently, a specific region was selected for thickness measurement of the nanosheets, indicated by the white line in Figure 1j. The resulting height data reveal that the thickness of these nanosheets is approximately 5 nm, as shown in Figure 1k.

To further illustrate the compacted structure, a collection of size distributions was gathered. The results reveal that the peak on the curve begins at approximately 200 nm and attains its maximum at around 500 nm, indicating that the measured average size was approximately 500 nm (Figure 2a). This aggregation is primarily attributed to the remarkable tendency of the nanomaterial to aggregate, with multiple nanosheets coalescing into larger particles. When measuring, the equipment was capable of detecting only the particles and not single petals. Additionally, the zeta potential reveals that the weighted average potential is approximately 34.2 mV, indicating the nanosheet's remarkable hydrophilicity (Figure 2b). The porous structure was examined through the N_2 adsorption–desorption curve and the distribution of pore sizes, as depicted in Figure 2c. The results indicate that the specific surface area is as impressive as 331.827 m^2/g, aligning with our previously reported findings [38,44]. Furthermore, the weighted average pore diameter has been measured at 0.9 nm, as depicted in Figure 2d. To investigate the crystal structure of the sample, an XRD analysis was performed. The results demonstrate that the planes of the synthesized UiO-66-F_4 align well with the simulated data (Figure 2e). Notably, the plane of (002) merges with the plane of (111), resulting in a single crystal plane. This observation suggests that the crystals in UiO-66-F_4 exhibit a uniform crystal plane orientation. The XRD pattern further elucidates the disparities between UiO-66-F_4 synthesized in this study and those reported in numerous prior works [38]. To investigate the coordination mode of the carboxylic acid group, FTIR spectra were collected for the ligands of UiO-66-F_4, as carboxylate can exhibit distinct coordination modes that would result in variations in the corresponding spectrum. The broad peak from 2000–3200 cm^{-1} of the ligand is ascribed to the O−H and C=O stretching modes, and it disappears due to the full deprotonation and reduction of the ligand after the coordination process [45]. Moreover, the FTIR spectra of UiO-66-F_4 samples reveal two characteristic peaks located at around 1637 and 1407 cm^{-1}, which correspond to the C=O symmetric and asymmetric stretching vibration. The two peaks with $\Delta > 200$ cm^{-1} ($\nu_{as} - \nu_s = 230$ cm^{-1}, Figure 2f) indicate that the carboxylate ligands adopted a bridging bidentate coordination mode [46,47].

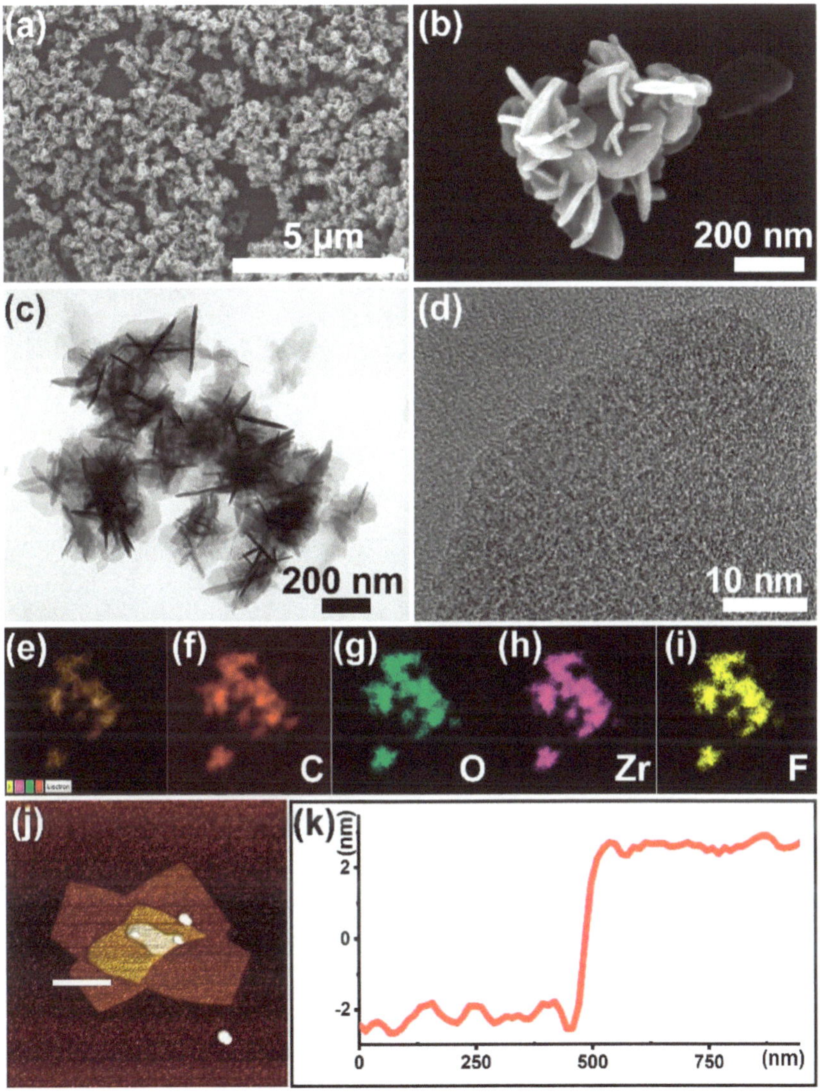

Figure 1. Characterizations of the UiO-66-F_4: (**a**,**b**) SEM and enlarged SEM images; (**c**,**d**) TEM and HR-TEM images; (**e**–**i**) corresponding element distribution images; (**j**) AFM image; and (**k**) corresponding height data.

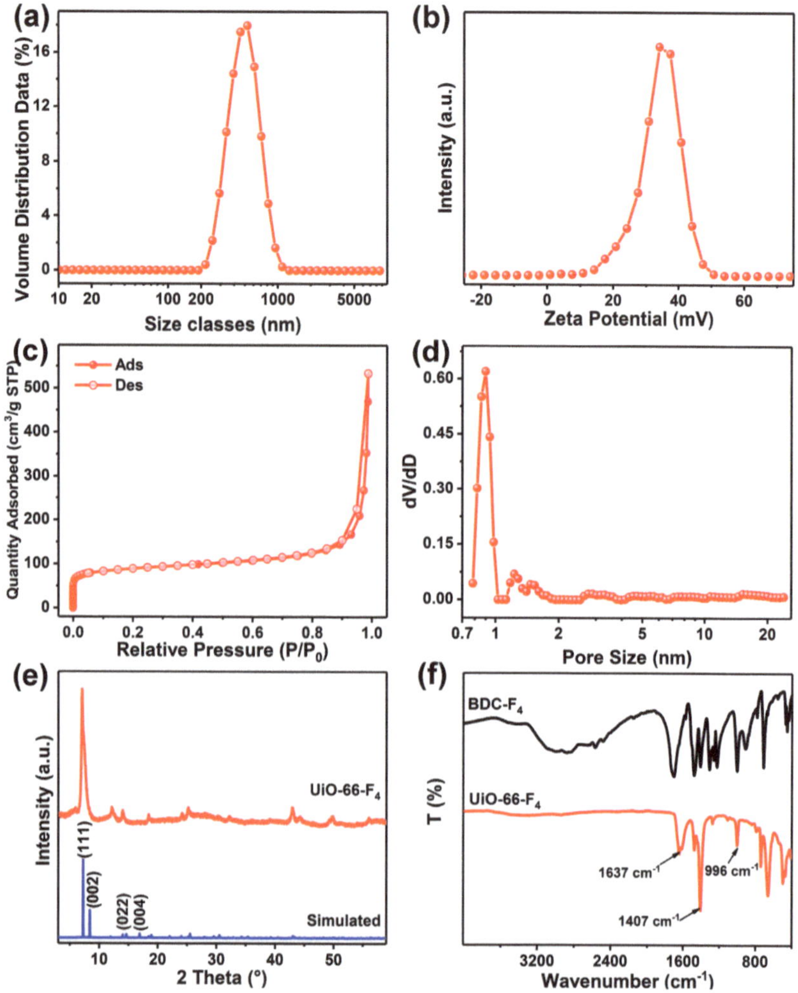

Figure 2. (**a**) The size distribution; (**b**) the zeta potential; (**c**) the N_2 adsorption and desorption curve; (**d**) the pore size distribution of the UiO-66-F_4; (**e**) the XRD pattern and the simulated data; (**f**) the FTIR spectra of BDC-F_4 and UiO-66-F_4.

To gain a deeper understanding of the coordination mode, XPS analysis was employed to investigate the valence states of the elements within the ligands and metal clusters, as depicted in Figure S2. The peak of the C species can be resolved into three distinct peaks, located at 284.7, 287.2, and 288.8 eV (Figure 3a), which are ascribed to C-C/C=C, C=O, and C-C=O, respectively [48,49]. The O atoms serve as a crucial constituent of clusters, with the deconvoluted peak of O 1s assigned to specific energy levels of 530.3, 531.6, and 533.2 eV, which are ascribed to C=O, O-metal (O-Zr), and O vacancy (Figure 3b), respectively [50,51]. And a distinct peak appears at 687.2 eV on the spectrum of F 1s (Figure 3c), which is attributed to the presence of C-F bonds, thus demonstrating the F did not attend the coordination. The Zr 3d spectra exhibits a spin–orbit doublet that splits into 182.5 and 184.5 eV, which is ascribed to $3d_{5/2}$ and $3d_{3/2}$ (Figure 3d) [43,52], respectively, strongly demonstrating the formation of $[Zr_6O_4(OH)_4(–COO)_{12}]$ in the frameworks [38,53–55].

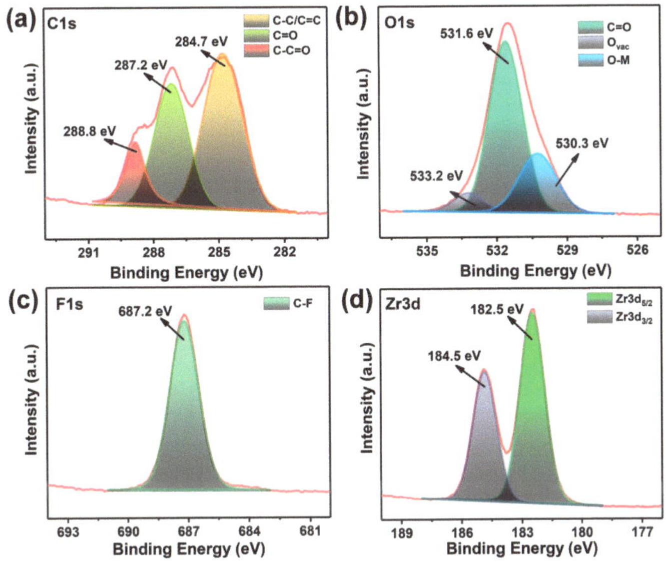

Figure 3. The high-resolution XPS spectra of UiO-66-F_4: (**a**) C 1s, (**b**) O 1s, (**c**) F 1s, and (**d**) Zr 3d.

Based on the aforementioned results, a putative coordination pathway can be inferred. BDC-F_4 is a is a strictly symmetric dicarboxylic acid, with two carboxyl groups situated on opposite sides of the benzene ring, thereby offering two coordination sites. The FTIR and XPS spectra obtained after the assembly process unambiguously reveal the presence of a bridging bidentate coordination mode between the ligand and the [$Zr_6O_4(OH)_4$(–COO)$_{12}$]. Utilizing these findings, we propose a mechanism for the formation of 2D UiO-66-F_4 nanosheets, which is schematically illustrated in Figure 4. Utilizing microwave heating at 100 °C, the carboxyl group of BDC-F_4 gradually interacted with Zr ions, as depicted in Figure 4a. This interaction led to the formation of [$Zr_6O_4(OH)_4$(–COO)$_{12}$] clusters, which are illustrated in Figure 4b. Subsequently, these clusters and ligands combined periodically to develop a network structure, shown in Figure 4c. Finally, through the addition of regulators and solvents to unsaturated coordination sites, the network was compacted and folded into 2D nanosheets, as seen in Figure 1b. Conventionally, the synthesis of UiO-66-type MOFs via hydrothermal methods entails prolonged reaction durations exceeding four days. However, our approach utilizes microwave irradiation to rapidly precipitate Zr ions with ligands, favoring the formation of nanosheets. This method offers a straightforward and remarkably swift pathway for the large-scale production of 2D UiO-66-F_4 MOFs.

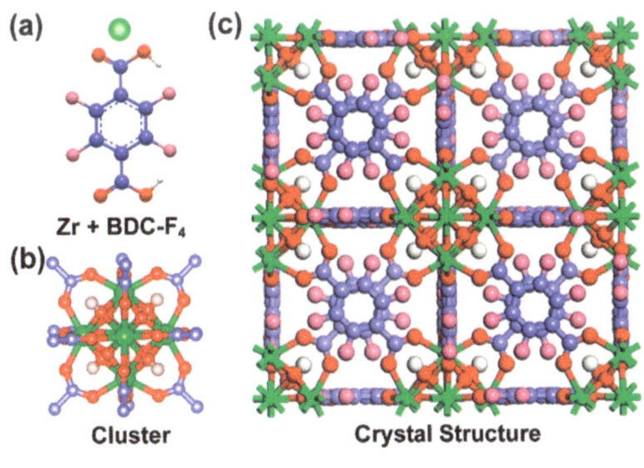

Figure 4. The structure of (**a**) Zr and BDC-F_4, (**b**) Zr-O cluster, and (**c**) crystal structure.

3.2. Mechanically Enhanced Detoxification Performance

The as-prepared UiO-66 nanosheets were introduced to detoxify the CWA simulant. Firstly, we investigated the mechanical property of the fabricated nanosheets by piezoresponse force microscopy (PFM). The phase hysteresis loop and amplitude loop were measured to evaluate the polarization degree of the MOF crystals. As reported, for the two hysteresis loops, the lower the overlap between the blue and red lines, the higher the degree of MOF crystal polarization [56,57]. As shown in Figure 5a, the sample exhibits well-defined 180° phase-reversal hysteresis, demonstrating a characteristic polarization switching behavior of the as-prepared MOF. Furthermore, a typical amplitude–voltage butterfly loop is obtained under a 10 V DC bias field (Figure 5b). The larger amplitude displacement of UiO-66-F_4 (~47 mV) indicates a stronger piezoelectric response than that of most reported UiO-66-type MOFs. Herein, the introduction of the F_4-BDC^{2-} linker that coordinated weakly to the Zr(IV) metal centers might have caused a strong polarity in the UiO-66-type MOFs. In addition, the strong hydrogen bonding between F and μ_3-OH can also lead to a large polarization of the nanosheets [37,38,58].

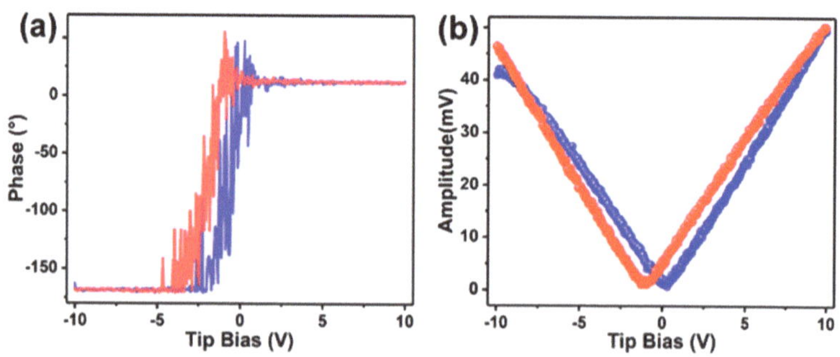

Figure 5. The piezoresponse capacity of the UiO-66-F_4: (**a**) phase hysteresis loop, (**b**) amplitude hysteresis loop. (Blue line: the change in phase or amplitude with the tip bias voltage gradually increasing; red line: the change in phase or amplitude with the tip bias voltage gradually reducing).

Subsequently, we subjected the fabricated MOF to CWA detoxification. Given the severe toxicity of sarin (as demonstrated in Figure S1a) and the necessity to conduct relevant experiments in high-standard laboratories, we chose DMNP as a representative of nerve agent simulants. As reported, DMNP can undergo hydrolysis to form p-Nitrophenol. Consequently, the detoxification performance of the prepared MOF can be evaluated by monitoring the concentration change in p-Nitrophenol [1]. The results indicate that, in a sonic environment, the conversion remains relatively unchanged after 100 min, indicating that the DMNP lacks self-degradation capabilities under stirring or ultrasonic conditions (Figure 6e). When UiO-66-F_4 is added to the catalytic system, the UV-Vis characteristic peak of DMNP gradually diminishes, while the peak of *p*-Nitrophenol concurrently increases (as shown in Figure 6a) under stirring conditions, which indicates that DMNP is efficiently hydrolyzed into *p*-Nitrophenol. In sonic conditions, UiO-66-F_4 demonstrates a superior hydrolysis efficiency compared to stirring conditions. Furthermore, the hydrolysis model aligns more closely with first-order reaction kinetics [59]. To more clearly investigate the hydrolysis performance under varying conditions, the introduction of the reaction rate constant, k, is employed [44]. The values are calculated as 0.040 and 0.058 min^{-1} for stirring and sonic conditions (Figure 6b,d), respectively, and the k is upgraded 45% by sonication with the addition of the UiO-66-F_4. Furthermore, the half-life time has been reduced from 13.61 min to 10.58 min, as demonstrated in Figure 6f. This significant reduction further underscores the ability of sonication to enhance the hydrolysis performance of UiO-66-F_4. ZnO is a prototypical inorganic material that exhibits remarkable piezocatalysis properties across diverse fields, making it a highly versatile and responsive material [60,61]. To further validate the distinct piezocatalysis characteristics of the UiO-66-F_4 material prepared,

ZnO was incorporated to assess the hydrolysis performance of alternative piezoresponse materials. However, the findings reveal that ZnO exhibits minimal catalytic activity for the hydrolysis of DMNP, even under sonication conditions (Figure 6e). Notably, the half-life of DMNP extends to 1487 min, indicating that conventional piezoelectric materials are ineffective at converting mechanical energy into chemical energy during the removal of DMNP. These results demonstrate that DMNP can be efficiently removed by piezoresponsive MOFs, aided by mechanical vibration. Additionally, the detoxification performance of UiO-66-F_4 was compared with various reported materials, revealing that the approach utilized in this study exhibits comparable detoxification efficiency (Table 1).

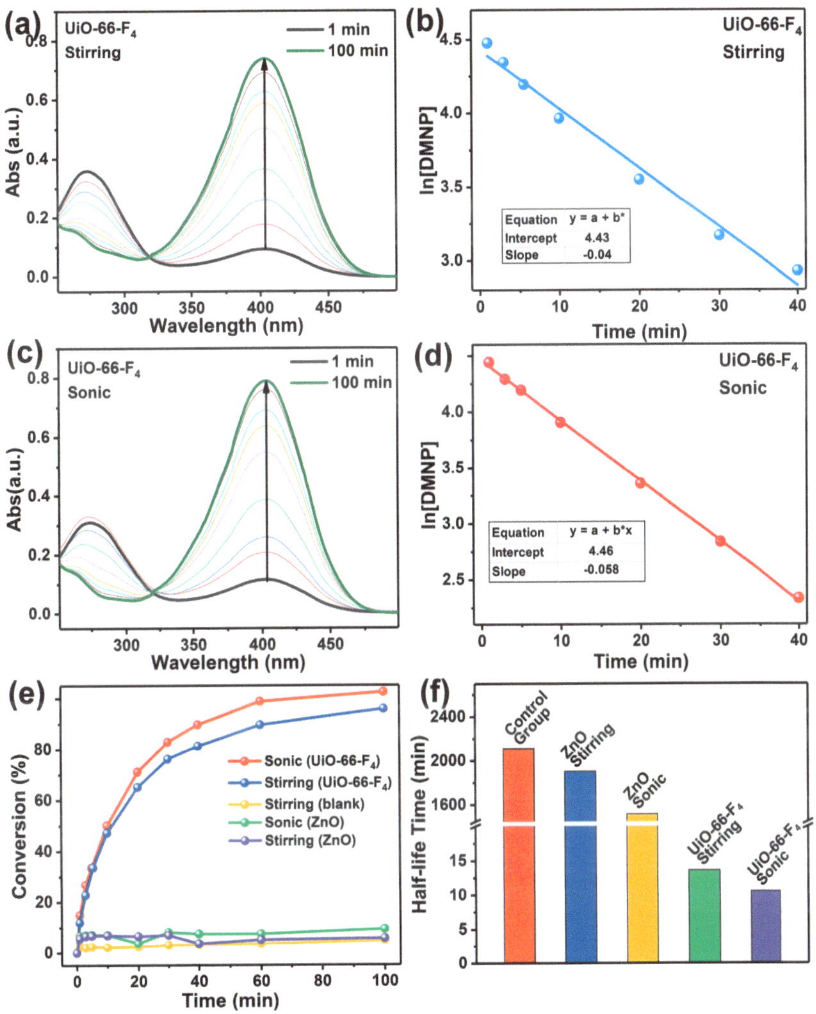

Figure 6. The detoxification experiments results: (**a**) degradation results and (**b**) first-pseudo-order kinetics under stirring conditions; (**c**) degradation results and (**d**) first-pseudo-order kinetics under sonication conditions; (**e**) comparison of degradation performance and (**f**) half-life time results of other control piezoresponse materials.

Table 1. The comparison with the reported materials.

Catalysts	$t_{1/2}$ (min)	k (min^{-1})	Ref.
PP/TiO_2/UiO-66-NH_2	15	--	[62]
$Zr(OH)_4$@PIM-1-Coat	12.6	0.055	[63]
UiO-66-NH_2/PAN	10	--	[64]
Ce-BDC	8.0	0.087	[65]
MOF@PDMAEA@LiCl@PNIPAM	>60	--	[66]
MIP/UiO-66-NH_2-0.5	9.4	--	[67]
UiO-66-F_4	10.58	0.058	This work

To gain a deeper comprehension of the alterations that occur under mechanical stress, a theoretical calculation was undertaken to emulate the responses exhibited by MOF crystals. The results initially reveal that the crystals maintain their integrity even when subjected to significant mechanical stress (Figure 7), indicating that MOF crystals possess remarkable stability, even under extreme conditions. Furthermore, the Z projection and side view of the MOFs' crystal structure reveal significant deformation. Specifically, the crystal has shrunk by approximately 5% in the direction of applied stress. Nevertheless, the metal cluster and the benzene ring of the ligand retain rigidity, resisting distortion. This deformation is primarily attributed to the rotation of carboxyl groups on the ligands' benzene rings, as evident from the side view of the crystal structure.

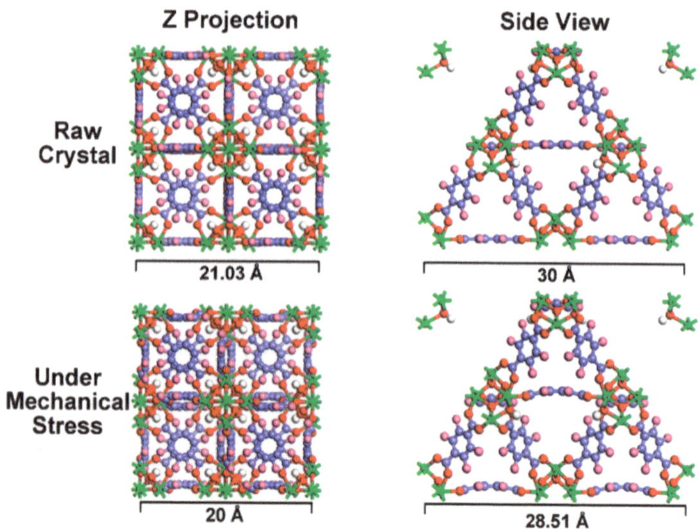

Figure 7. The first-principles calculation implementation of the crystal changes under stress.

Moreover, the band gap structure determined the generation of radical oxidation species (ROS) [68–70]. We also investigated the band gap changes in the MOFs under stress (Figure 8a,b). The results indicate that the band gap is 2.313 eV, which aligns closely with our prior research [38]. Upon the application of mechanical stress, the compressive strain notably altered the band gap structure, resulting in an increase to 2.405 eV. Simultaneously, the bands in the valence zone exhibit a significant increase under stress, as illustrated in Figure 8b. This observation underscores the fact that the band gap structure can be effectively modulated by mechanical stress [71,72]. The density of states (DOS) analysis offered deeper understanding of the distinct bands' origins, as illustrated in Figure 8c. Notably, the application of mechanical stress caused the peak value at the Fermi level to increase. Concurrently, the conduction band shifted slightly towards the high-energy level, leading to an expansion in the band gap. This observation aligns well with the calculated band gap results. The CB primarily aligns with the Zr 3d electrons or with orbitals that are delocalized throughout the entire linker (as shown in Figure S3), demonstrating a relatively low sensitivity to the functional group. Moreover, the application of mechanical stress causes the polarization to increase at an accelerated rate. Consequently, a novel VB' emerges, significantly broadening the band gap (Figure 9b).

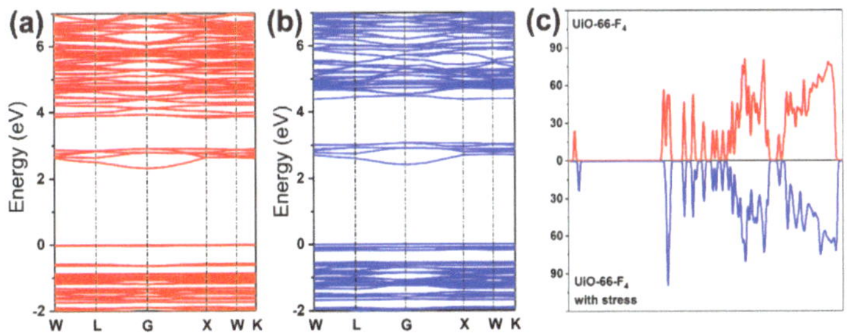

Figure 8. The band gap structure of (**a**) raw UiO-66-F_4 (2.313 eV) and (**b**) under stress (2.405 eV); (**c**) The DOS of UiO-66-F_4 and under stress.

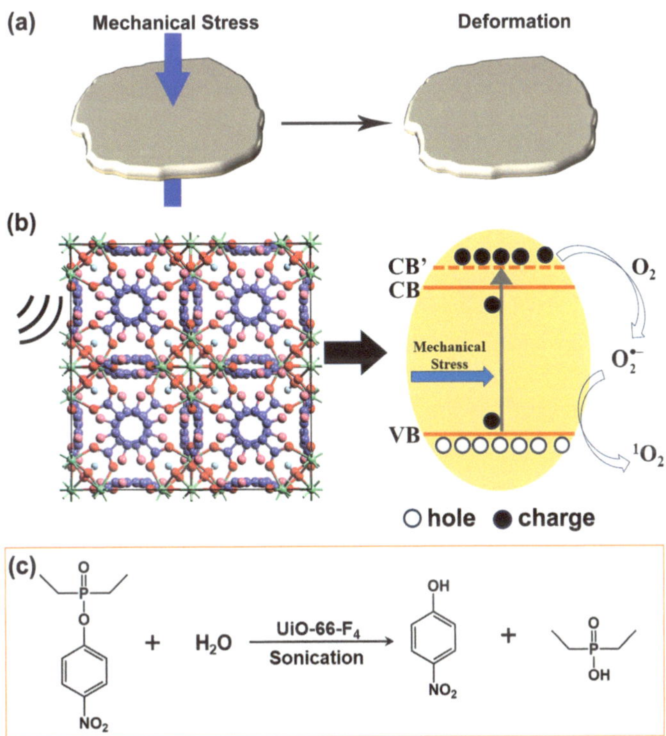

Figure 9. Schematic illustration of (**a**) mechanical deformation of UiO-66-F_4; (**b**) band diagram; (**c**) detoxification mechanism.

Drawing upon the aforementioned results and preceding research, a preliminary band diagram can be established (Figure 9). During the assembly process of UiO-66-F_4, the absence or unsaturated coordination of Zr-O clusters can lead to a significant number of metal node or ligand defects, ultimately endowing UiO-66-F_4 with remarkable catalytic capabilities, even without any external assistance. When the UiO-66-F_4 nanosheets are employed for the removal of CWA simulants, the unsaturated coordination sites can directly react with the ester group of DMNP, thereby achieving the aim of detoxification [1]. After the application of mechanical vibration, the band gap of the nanocrystals widens significantly, leading to the polarization effect. Consequently, a substantial number of free charges (e^-) are generated on the catalyst surface, efficiently reacting with O_2 to form $O_2^{\bullet-}$. On the one hand, partial $O_2^{\bullet-}$ can target the phosphoester bonds (O-P), thereby facilitating the hydrolysis of DMNP as depicted in Figure 9c. On the other hand, $O_2^{\bullet-}$ can expedite the generation of additional $^1O_2$ through interaction with residual holes (h^+), which can further oxidatively degrade the hydrolysate [73–77].

4. Conclusions

In this study, we successfully fabricated two-dimensional UiO-66-F_4 nanosheets using a microwave-assisted method. These nanosheets exhibit an ultrathin thickness of approximately 5 nm and possess a large specific surface area of 331.827 m^2/g. These nanosheets possess remarkable mechanical-energy-harvesting capabilities and have the potential to be utilized in the mechanically enhanced detoxification of CWA simulant DMNP. Under optimal conditions, the half-life of DMNP is 10.58 min, and mechanical vibration can significantly enhance its removal performance by approximately 45%. The first-principle simulation unveils the structural transformations within MOF crystals. The mechanical stress potentially influences the rotation angle of the carboxyl groups within the tetrafluoroterephthalic acid, potentially leading to slight deformations along the direction of pressure. Subsequently, this could broaden the band gap and reinstate the band gap structure of the crystals. All the stress-induced changes are advantageous for enhancing the detoxification capacity of UiO-66-F_4, thereby illustrating the efficacy of mechanical vibration in detoxifying CWAs. This study offers a novel perspective for defending against CWAs through innovative approaches.

Supplementary Materials: The following supporting information can be downloaded at: https://www.mdpi.com/article/10.3390/nano14070559/s1, Figure S1: The chemical structure of (a) sarin and (b) simulant DMNP; Figure S2: The XPS survey spectra of UiO-66-F_4; Figure S3: The valence XPS spectra of UiO-66-F_4.

Author Contributions: Conceptualization, Y.L. (Yuyang Liu) and S.Z.; methodology, Y.L. (Yuyang Liu) and S.Z.; software, Y.L. (Yuyang Liu) and S.Z.; validation, S.Z., J.H. and X.Y.; formal analysis, Y.L. (Yujiao Li); investigation, Y.L. (Yuyang Liu); resources, Y.L. (Yuyang Liu); data curation, Y.L. (Yuyang Liu); writing—original draft preparation, S.Z.; writing—review and editing, S.Z. and C.-a.T.; supervision, J.W. and C.-a.T.; project administration, J.W. and C.-a.T.; funding acquisition, S.Z. and C.-a.T. All authors have read and agreed to the published version of the manuscript.

Funding: This research was funded by the National Natural Science Foundation of China, grant No. 22075319, and Hunan Provincial Science and Technology Department, grant No. 2022JJ40545. Additional support was provided by the National University of Defense Technology (ZK-39).

Data Availability Statement: Dataset available on request from the authors.

Acknowledgments: We appreciate the *Shiyanjia Lab* (www.shiyanjia.com (accessed on 1 January 2018)) for their technical support on characterization and the first-principle calculation.

Conflicts of Interest: The authors declare no conflicts of interest.

References

1. Zhao, H.; Tao, C.A.; Zhao, S.; Zou, X.; Wang, F.; Wang, J. Porphyrin-Moiety-Functionalized Metal–Organic Layers Exhibiting Catalytic Capabilities for Detoxifying Nerve Agent and Blister Agent Simulants. *ACS Appl. Mater. Interfaces* **2023**, *15*, 3297–3306. [CrossRef] [PubMed]
2. Katz, M.J.; Mondloch, J.E.; Totten, R.K.; Park, J.K.; Nguyen, S.B.T.; Farha, O.K.; Hupp, J.T. Simple and Compelling Biomimetic Metal–Organic Framework Catalyst for the Degradation of Nerve Agent Simulants. *Angew. Chem. Int. Ed.* **2014**, *126*, 507–511. [CrossRef]
3. Liao, Y.; Sheridan, T.; Liu, J.; Farha, O.; Hupp, J. Product Inhibition and the Catalytic Destruction of a Nerve Agent Simulant by Zirconium-Based Metal–Organic Frameworks. *ACS Appl. Mater. Interfaces* **2021**, *13*, 30565–30575. [CrossRef] [PubMed]

4. Chen, R.; Tao, C.A.; Zhang, Z.; Chen, X.; Wang, J. Layer-by-Layer Fabrication of Core-Shell Fe_3O_4@UiO-66-NH_2 with High Catalytic Reactivity toward the Hydrolysis of Chemical Warfare Agent Simulants. *ACS Appl. Mater. Interfaces* **2019**, *11*, 43156–43165. [CrossRef] [PubMed]
5. Zhao, J.; Chen, R.; Huang, J.; Wang, F.; Wang, J. Facile Synthesis of Metal–Organic Layers with High Catalytic Performance toward Detoxification of a Chemical Warfare Agent Simulant. *ACS Appl. Mater. Interfaces* **2021**, *13*, 40863–40871. [CrossRef]
6. Nawala, J.; Jóźwik, P.; Popiel, S. Thermal and Catalytic Methods Used for Destruction of Chemical Warfare Agents. *Int. J. Environ. Sci. Technol.* **2019**, *16*, 3899–3912. [CrossRef]
7. Jouypazadeh, H.; Farrokhpour, H. DFT and TD-DFT Study of the Adsorption and Detection of Sulfur Mustard Chemical Warfare Agent by the C_{24}, $C_{12}Si_{12}$, $Al_{12}N_{12}$, $Al_{12}P_{12}$, $Be_{12}O_{12}$, $B_{12}N_{12}$ and $Mg_{12}O_{12}$ Nanocages. *J. Mol. Struct.* **2018**, *1164*, 227–238. [CrossRef]
8. Tomchenko, A.A.; Harmer, G.P.; Marquis, B.T. Detection of Chemical Warfare Agents Using Nanostructured Metal Oxide Sensors. *Sens. Actuators B Chem.* **2005**, *108*, 41–55. [CrossRef]
9. Liu, Y.; Howarth, A.J.; Vermeulen, N.A.; Moon, S.Y.; Hupp, J.T.; Farha, O.K. Catalytic Degradation of Chemical Warfare Agents and Their Simulants by Metal-organic Frameworks. *Coord. Chem. Rev.* **2016**, *346*, 101–111. [CrossRef]
10. Kiani, S.; Farooq, A.; Ahmad, M.; Irfan, N.; Nawaz, M.; Irshad, M. Impregnation on Activated Carbon for Removal of Chemical Warfare Agents (CWAs) and Radioactive Content. *Environ. Sci. Pollut. Res.* **2021**, *28*, 60477–60494. [CrossRef]
11. Matito-Martos, I.; Moghadam, P.Z.; Li, A.; Colombo, V.; Navarro, J.A.R.; Calero, S.; Fa, D. Discovery of an Optimal Porous Crystalline Material for the Capture of Chemical Warfare Agents. *Chem. Mater.* **2018**, *30*, 4571–4579. [CrossRef]
12. Li, J.; Singh, V.V.; Sattayasamitsathit, S.; Orozco, J.; Kaufmann, K.; Dong, R.; Gao, W.; Jurado-Sanchez, B.; Fedorak, Y.; Wang, J. Water-Driven Micromotors for Rapid Photocatalytic Degradation of Biological and Chemical Warfare Agents. *ACS Nano* **2014**, *8*, 11118–11125. [CrossRef]
13. Picard, B.; Chataigner, I.; Maddaluno, J.; Legros, J. Introduction to Chemical Warfare Agents, Relevant Simulants and Modern Neutralisation Methods. *Org. Biomol. Chem.* **2019**, *17*, 6528–6537. [CrossRef] [PubMed]
14. Xiao, L.; Bing, Y. Trace Detection of Organophosphorus Chemical Warfare Agents in Wastewater and Plants by Luminescent UIO-67(Hf) and Evaluating the Bioaccumulation of Organophosphorus Chemical Warfare Agents. *ACS Appl. Mater. Interfaces* **2018**, *10*, 14869–14876. [CrossRef]
15. Peterson, G.W.; Wagner, G.W. Detoxification of Chemical Warfare Agents by CuBTC. *J. Porous Mater.* **2014**, *21*, 121–126. [CrossRef]
16. Couzon, N.; Dhainaut, J.; Campagne, C.; Royer, S.; Loiseau, T.; Volkringer, C. Porous Textile Composites (PTCs) for the Removal and the Decomposition of Chemical Warfare Agents (CWAs)-A Review. *Coord. Chem. Rev.* **2022**, *467*, 214598. [CrossRef]
17. Snider, V.G.; Hill, C.L. Functionalized Reactive Polymers for the Removal of Chemical Warfare Agents: A Review. *J. Hazard. Mater.* **2023**, *442*, 130015. [CrossRef] [PubMed]
18. Chauhan, S.; Chauhan, S.; D'Cruz, R.; Faruqi, S.; Singh, K.K.; Varma, S.; Singh, M.; Karthik, V. Chemical Warfare Agents. *Environ. Toxicol. Pharmacol.* **2008**, *26*, 113–122. [CrossRef] [PubMed]
19. Black, R.M. History and Perspectives of Bioanalytical Methods for Chemical Warfare Agent Detection. *J. Chromatogr. B* **2010**, *878*, 1207–1215. [CrossRef]
20. Eubanks, L.M.; Dickerson, T.J.; Janda, K.D. Technological Advancements for the Detection of and Protection Against Biological and Chemical Warfare Agents. *Chem. Soc. Rev.* **2007**, *36*, 458–470. [CrossRef]
21. Balasubramanian, S.; Kulandaisamy, A.J.; Babu, K.J.; Das, A.; Balaguru Rayappan, J.B. Metal Organic Framework Functionalized Textiles as Protective Clothing for the Detection and Detoxification of Chemical Warfare Agents—A Review. *Ind. Eng. Chem. Res.* **2021**, *60*, 4218–4239. [CrossRef]
22. Hong-Cai, Z.; Long, J.R.; Omar, M.Y. Introduction to Metal–Organic Frameworks. *Chem. Rev.* **2012**, *112*, 673–674. [CrossRef]
23. Zhou, H.C.; Kitagawa, S. Metal–Organic Frameworks (MOFs). *Chem. Soc. Rev.* **2014**, *43*, 5415–5418. [CrossRef]
24. Furukawa, H.; Cordova, K.E.; O'Keeffe, M.; Yaghi, O.M. The Chemistry and Applications of Metal-organic Frameworks. *Science* **2013**, *341*, 974. [CrossRef] [PubMed]
25. Užarević, K.; Wang, T.C.; Moon, S.-Y.; Fidelli, A.M.; Hupp, J.T.; Farha, O.K.; Friščić, T. Mechanochemical and solvent-free assembly of zirconium-based metal–organic frameworks. *Chem. Commun.* **2016**, *52*, 2133–2136. [CrossRef]
26. Freund, R.; Zaremba, O.; Arnauts, G.; Ameloot, R.; Skorupskii, G.; Dincă, M.; Bavykina, A.; Gascon, J.; Ejsmont, A.; Goscianska, J.; et al. The Current Status of MOF and COF Applications. *Angew. Chem. Int. Ed.* **2021**, *60*, 23975–24001. [CrossRef]
27. Shekhah, O.; Liu, J.; Fischer, R.A.; Wöll, C. MOF thin films: Existing and future applications. *Chem. Soc. Rev.* **2011**, *40*, 1081–1106. [CrossRef]
28. Yu, J.; Mu, C.; Yan, B.; Qin, X.; Shen, C.; Xue, H.; Pang, H. Nanoparticle/MOF Composites: Preparations and Applications. *Mater. Horiz.* **2017**, *4*, 557–569. [CrossRef]
29. Ding, M.; Cai, X.; Jiang, H.-L. Improving MOF Stability: Approaches and Applications. *Chem. Sci.* **2019**, *10*, 10209–10230. [CrossRef]
30. Mondloch, J.E.; Katz, M.J.; Isley Iii, W.C.; Ghosh, P.; Liao, P.; Bury, W.; Wagner, G.W.; Hall, M.G.; DeCoste, J.B.; Peterson, G.W.; et al. Destruction of chemical warfare agents using metal–organic frameworks. *Nat. Mater.* **2015**, *14*, 512–516. [CrossRef]
31. De Koning, M.C.; Van Grol, M.; Breijaert, T. Degradation of Paraoxon and the Chemical Warfare Agents VX, Tabun, and Soman by the Metal–Organic Frameworks UiO-66-NH_2, MOF-808, NU-1000, and PCN-777. *Inorg. Chem.* **2017**, *56*, 11804–11809. [CrossRef]

32. Barton, H.F.; Jamir, J.D.; Davis, A.K.; Peterson, G.W.; Parsons, G.N. Doubly Protective MOF-Photo-Fabrics: Facile Template-Free Synthesis of PCN-222-Textiles Enables Rapid Hydrolysis, Photo-Hydrolysis and Selective Oxidation of Multiple Chemical Warfare Agents and Simulants. *Chem.—Eur. J.* **2021**, *27*, 1465–1472. [CrossRef]
33. Chen, H.; Snurr, R.Q. Insights into Catalytic Gas-Phase Hydrolysis of Organophosphate Chemical Warfare Agents by MOF-Supported Bimetallic Metal-Oxo Clusters. *ACS Appl. Mater. Interfaces* **2020**, *12*, 14631–14640. [CrossRef] [PubMed]
34. Cai, L.; Du, J.; Han, F.; Shi, T.; Zhang, H.; Lu, Y.; Long, S.; Sun, W.; Fan, J.; Peng, X. Piezoelectric Metal–Organic Frameworks Based Sonosensitizer for Enhanced Nanozyme Catalytic and Sonodynamic Therapies. *ACS Nano* **2023**, *17*, 7901–7910. [CrossRef] [PubMed]
35. Chen, Q.; Wu, H.; Hu, J.; Tian, Q.; He, D.; Lu, G.; Zhu, M. Fe-Metal Organic Framework Converts Mechanical Energy with Piezoelectric Polarization to Remove Carbamazepine in Water: Efficiency, Pathway and Mechanism. *Chem. Eng. J.* **2023**, *460*, 141839. [CrossRef]
36. He, J.; Yi, Z.; Chen, Q.; Li, Z.; Hu, J.; Zhu, M. Harvesting Mechanical Energy Induces Piezoelectric Polarization of MIL-100(Fe) for Cocatalyst-free Hydrogen Production. *Chem. Commun.* **2022**, *58*, 10723–10726. [CrossRef]
37. Zhang, C.; Lei, D.; Xie, C.; Hang, X.; He, C.; Jiang, H.-L. Piezo-Photocatalysis over Metal–Organic Frameworks: Promoting Photocatalytic Activity by Piezoelectric Effect. *Adv. Mater.* **2021**, *33*, 2106308. [CrossRef] [PubMed]
38. Zhao, S.; Liu, M.; Zhang, Y.; Zhao, Z.; Zhang, Q.; Mu, Z.; Long, Y.; Jiang, Y.; Liu, Y.; Zhang, J.; et al. Harvesting Mechanical Energy for Hydrogen Generation by Piezoelectric Metal–Organic Frameworks. *Mater. Horiz.* **2022**, *9*, 1978–1983. [CrossRef]
39. Kim, J.H.; Yun, H.; Kang, D.W.; Shin, J.; Kang, M.; Singh, N.; Jeong, J.-E.; Hong, C.S.; Kim, J.S. Isomeric sp2-C-conjugated porous organic polymer-mediated photo-and sono-catalytic detoxification of sulfur mustard simulant under ambient conditions. *Matter* **2021**, *4*, 3774–3785. [CrossRef]
40. Milman, V.; Winkler, B.; White, J.A.; Pickard, C.J.; Payne, M.C.; Akhmatskaya, E.V.; Nobes, R.H. Electronic Structure, Properties, and Phase Stability of Inorganic Crystals: A Pseudopotential Plane-wave Study. *Int. J. Quantum Chem.* **2000**, *77*, 895–910. [CrossRef]
41. Perdew, J.P.; Burke, K.; Ernzerhof, M. Generalized Gradient Approximation Made Simple. *Phys. Rev. Lett.* **1998**, *77*, 3865–3868. [CrossRef]
42. Perdew, J.P. Restoring the Density-Gradient Expansion for Exchange in a GGA for Solid and Surfaces. In Proceedings of the 2008 APS March Meeting, New Orleans, LA, USA, 10–14 March 2008.
43. Kandiah, M.; Nilsen, M.H.; Usseglio, S.; Jakobsen, S.; Olsbye, U.; Tilset, M.; Larabi, C.; Quadrelli, E.A.; Bonino, F.; Lillerud, K.P. Synthesis and stability of tagged UiO-66 Zr-MOFs. *Chem. Mater.* **2010**, *22*, 6632–6640. [CrossRef]
44. Zhao, S.; Long, Y.; Shen, X.; Wang, S.; Su, Y.; Zhang, X.; Zhang, Z. Regulation of Electronic Structures of MOF-derived Carbon via Ligand Adjustment for Enhanced Fenton-like Reactions. *Sci. Total Environ.* **2021**, *799*, 149497. [CrossRef]
45. Nyakuchena, J.; Ostresh, S.; Streater, D.; Pattengale, B.; Neu, J.; Fiankor, C.; Hu, W.; Kinigstein, E.D.; Zhang, J.; Zhang, X.; et al. Direct Evidence of Photoinduced Charge Transport Mechanism in 2D Conductive Metal Organic Frameworks. *J. Am. Chem. Soc.* **2020**, *142*, 21050–21058. [CrossRef]
46. Zhang, X.; Chen, Z.-K.; Loh, K.P. Coordination-Assisted Assembly of 1-D Nanostructured Light-Harvesting Antenna. *J. Am. Chem. Soc.* **2009**, *131*, 7210–7211. [CrossRef]
47. Deacon, G.B.; Phillips, R.J. Relationships Between the Carbon-Oxygen Stretching Frequencies of Carboxylato Complexes and the Type of Carboxylate Coordination. *Coord. Chem. Rev.* **1980**, *33*, 227–250. [CrossRef]
48. Chen, C.; Chen, D.; Xie, S.; Quan, H.; Luo, X.; Guo, L. Adsorption Behaviors of Organic Micropollutants on Zirconium Metal–Organic Framework UiO-66: Analysis of Surface Interactions. *ACS Appl. Mater. Interfaces* **2017**, *9*, 41043–41054. [CrossRef]
49. Min, X.; Wu, X.; Shao, P.; Ren, Z.; Ding, L.; Luo, X. Ultra-high capacity of lanthanum-doped UiO-66 for phosphate capture: Unusual doping of lanthanum by the reduction of coordination number. *Chem. Eng. J.* **2019**, *358*, 321–330. [CrossRef]
50. Qiu, H.; Ye, M.; Zeng, Q.; Li, W.; Fortner, J.; Liu, L.; Yang, L. Fabrication of Agricultural Waste Supported UiO-66 Nanoparticles With High Utilization in Phosphate Removal from Water. *Chem. Eng. J.* **2019**, *360*, 621–630. [CrossRef]
51. Peng, M.; You, D.; Shi, H.; Shao, P.; Ren, W.; Yang, L.; Sheng, X.; Shao, J.; Ding, X.; Ding, L.; et al. Disclosing the Role of Defective UiO-66 Over Sb(V) Removal: A Joint Experimental and Theoretical Study. *Chem. Eng. J.* **2022**, *448*, 137612. [CrossRef]
52. Banerjee, D.; Xu, W.; Nie, Z.; Johnson, L.E.; Coghlan, C.; Sushko, M.L.; Kim, D.; Schweiger, M.J.; Kruger, A.A.; Doonan, C.J. Zirconium-based metal–organic framework for removal of perrhenate from water. *Inorg. Chem.* **2016**, *55*, 8241–8243. [CrossRef]
53. Wang, J.; Ye, G.; Wang, H.; Chen, W.; Chu, H.; Wei, J.; Wang, D.; Li, Y. In Situ Implanting Single Tungsten Site Into Defective UiO-66(Zr) by Solvent-Free Route for Efficient Oxidative Desulfurization at Room Temperature. *Angew. Chem. Int. Ed.* **2021**, *133*, 20481–20487.
54. Geravand, E.; Farzaneh, F.; Ghiasi, M. Metalation and DFT Studies of Metal Organic Frameworks UiO-66(Zr) with Vanadium Chloride as Allyl Alcohol Epoxidation Catalyst. *J. Mol. Struct.* **2019**, *1198*, 126940. [CrossRef]
55. Driscoll, D.M.; Troya, D.; Usov, P.M.; Maynes, A.J.; Morris, A.J.; Morris, J.R. Characterization of Undercoordinated Zr Defect Sites in UiO-66 with Vibrational Spectroscopy of Adsorbed CO. *J. Phys. Chem. C* **2018**, *122*, 14582–14589. [CrossRef]
56. Sharma, P.; Xiang, F.-X.; Shao, D.-F.; Zhang, D.; Tsymbal, E.Y.; Hamilton, A.R.; Seidel, J. A room-temperature ferroelectric semimetal. *Sci. Adv.* **2019**, *5*, eaax5080. [CrossRef]
57. Li, L.; Wondraczek, L.; Peng, M.; Ma, Z.; Zou, B. Force-induced 1540 nm luminescence: Role of piezotronic effect in energy transfer process for mechanoluminescence. *Nano Energy* **2020**, *69*, 104413. [CrossRef]

58. Sun, Y.; Gao, J.; Cheng, Y.; Zhang, Y.-W.; Zeng, K. Design of the Hybrid Metal–Organic Frameworks as Potential Supramolecular Piezo-/Ferroelectrics. *J. Phys. Chem. C* **2019**, *123*, 3122–3129. [CrossRef]
59. Thomas, G.; Spitzer, D. Double-side microcantilevers as a key to understand the adsorption mechanisms and kinetics of chemical warfare agents on vertically-aligned TiO_2 nanotubes. *J. Hazard. Mater.* **2021**, *406*, 124672. [CrossRef] [PubMed]
60. Yuan, J.; Huang, X.; Zhang, L.; Gao, F.; Lei, R.; Jiang, C.; Feng, W.; Liu, P. Tuning piezoelectric field for optimizing the coupling effect of piezo-photocatalysis. *Appl. Catal. B Environ.* **2020**, *278*, 119291. [CrossRef]
61. Wen, Y.; Chen, J.; Gao, X.; Che, H.; Wang, P.; Liu, B.; Ao, Y. Piezo-enhanced photocatalytic performance of ZnO nanorod array for pollutants degradation in dynamic water: Insight into the effect of velocity and inner flow field. *Nano Energy* **2022**, *101*, 107614. [CrossRef]
62. Lee, D.T.; Zhao, J.; Oldham, C.J.; Peterson, G.W.; Parsons, G.N. UiO-66-NH_2 Metal–Organic Framework (MOF) Nucleation on TiO_2, ZnO, and Al_2O_3 Atomic Layer Deposition-Treated Polymer Fibers: Role of Metal Oxide on MOF Growth and Catalytic Hydrolysis of Chemical Warfare Agent Simulants. *ACS Appl. Mater. Interfaces* **2017**, *9*, 44847–44855. [CrossRef]
63. Aina, P.O.; Mondal, S.K.; Rownaghi, A.A.; Rezaei, F. Assessing Detoxification Performance of Metal Hydroxide@Polymer Composites Prepared via Matrix-Incorporation and Spray-Coating Techniques. *ACS Appl. Eng. Mater.* **2024**, *2*, 454–466. [CrossRef]
64. Yao, A.; Jiao, X.; Chen, D.; Li, C. Photothermally Enhanced Detoxification of Chemical Warfare Agent Simulants Using Bioinspired Core–Shell Dopamine–Melanin@Metal–Organic Frameworks and Their Fabrics. *ACS Appl. Mater. Interfaces* **2019**, *11*, 7927–7935. [CrossRef]
65. Islamoglu, T.; Atilgan, A.; Moon, S.-Y.; Peterson, G.W.; DeCoste, J.B.; Hall, M.; Hupp, J.T.; Farha, O.K. Cerium(IV) vs Zirconium(IV) Based Metal–Organic Frameworks for Detoxification of a Nerve Agent. *Chem. Mater.* **2017**, *29*, 2672–2675. [CrossRef]
66. Wang, X.; Yang, J.; Zhang, M.; Hu, Q.; Li, B.-X.; Qu, J.; Yu, Z.-Z.; Yang, D. Spontaneously super-hygroscopic MOF-gel microreactors for efficient detoxification of nerve agent simulant in atmospheric environments. *Appl. Catal. B Environ.* **2023**, *328*, 122516. [CrossRef]
67. Jiang, P.; Niu, Y.; Cao, J.; Xie, D.; Li, J.; Guo, T. A MOF-doped molecularly imprinted polymer/MOF hybrid gel incorporating with pH-buffering sodium acrylate for practical detoxification of organophosphorus nerve agents. *Chem. Eng. J.* **2024**, *481*, 148377. [CrossRef]
68. George, S.; Pokhrel, S.; Ji, Z.; Henderson, B.L.; Xia, T.; Li, L.; Zink, J.I.; Nel, A.E.; Mädler, L. Role of Fe Doping in Tuning the Band Gap of TiO_2 for the Photo-Oxidation-Induced Cytotoxicity Paradigm. *J. Am. Chem. Soc.* **2011**, *133*, 11270–11278. [CrossRef] [PubMed]
69. Wang, C.; Luo, Y.; Liu, X.; Cui, Z.; Zheng, Y.; Liang, Y.; Li, Z.; Zhu, S.; Lei, J.; Feng, X.; et al. The Enhanced Photocatalytic Sterilization of MOF-Based Nanohybrid for Rapid and Portable Therapy of Bacteria-Infected Open Wounds. *Bioact. Mater.* **2022**, *13*, 200–211. [CrossRef] [PubMed]
70. Li, X.; Wang, J.; Xue, F.; Wu, Y.; Xu, H.; Yi, T.; Li, Q. An Imine-Linked Metal–Organic Framework as a Reactive Oxygen Species Generator. *Angew. Chem. Int. Ed.* **2021**, *60*, 2534–2540. [CrossRef] [PubMed]
71. Zhang, M.; Zhao, S.; Zhao, Z.; Li, S.; Wang, F. Piezocatalytic Effect Induced Hydrogen Production from Water over Non-noble Metal Ni Deposited Ultralong GaN Nanowires. *ACS Appl. Mater. Interfaces* **2021**, *13*, 10916–10924. [CrossRef] [PubMed]
72. Zhao, G.; Sun, Y.; Zhou, W.; Wang, X.; Chang, K.; Liu, G.; Liu, H.; Kako, T.; Ye, J. Superior Photocatalytic H_2 Production with Cocatalytic Co/Ni Species Anchored on Sulfide Semiconductor. *Adv. Mater.* **2017**, *29*, 1703258. [CrossRef]
73. Tan, G.; Wang, S.; Zhu, Y.; Lei, Z.; Peng, Y.; Wang, X.; He, T.; Chen, J.; Mao, C.; Ning, C. Surface-Selective Preferential Production of Reactive Oxygen Species on Piezoelectric Ceramics for Bacterial Killing. *ACS Appl. Mater. Interfaces* **2016**, *8*, 24306–24309. [CrossRef]
74. Zhang, M.; Guo, W.; Chen, Y.; He, D.; Isaev, A.B.; Zhu, M. Dissolved Oxygen in Aeration-Driven Piezo-catalytic for Antibiotics Pollutants Removal in Water. *Chin. Chem. Lett.* **2023**, *34*, 108229. [CrossRef]
75. Nie, G.; Xiao, L.; Bi, J.; Wang, S.; Duan, X. New Insight to Piezocatalytic Peroxymonosulfate Activation: The Critical Role of Dissolved Oxygen in Mediating Radical and Nonradical Pathways. *Appl. Catal. B Environ.* **2022**, *315*, 121584. [CrossRef]
76. Dai, J.; Shao, N.; Zhang, S.; Zhao, Z.; Long, Y.; Zhao, S.; Li, S.; Zhao, C.; Zhang, Z.; Liu, W. Enhanced Piezocatalytic Activity of $Sr_{0.5}Ba_{0.5}Nb_2O_6$ Nanostructures by Engineering Surface Oxygen Vacancies and Self-Generated Heterojunctions. *ACS Appl. Mater. Interfaces* **2021**, *13*, 7259–7267. [CrossRef] [PubMed]
77. Wang, Y.; Xu, Y.; Dong, S.; Wang, P.; Gao, G. Ultrasonic Activation of Inert Poly(tetrafluoroethylene) Enables Piezocatalytic Generation of Reactive Oxygen Species. *Nat. Commun.* **2021**, *12*, 3508. [CrossRef] [PubMed]

materials

MDPI

Article

Antifouling and Antioxidant Properties of PVDF Membrane Modified with Polyethylene Glycol Methacrylate and Propyl Gallate

Ting Wang [1,2], Jun Hu [1], Zhengchi Hou [3,*] and Haijun Yang [3]

1 Shanghai Institute of Applied Physics, Chinese Academy of Sciences, Shanghai 201800, China; wangting@sinap.ac.cn (T.W.); hujun@sinap.ac.cn (J.H.)
2 University of Chinese Academy of Sciences, Beijing 100049, China
3 Shanghai Synchrotron Radiation Facility (SSRF), Shanghai Advanced Research Institute, Chinese Academy of Sciences, Shanghai 201204, China; yanghj@sari.ac.cn
* Correspondence: houzc@sari.ac.cn; Tel.: +86-13818279326

Abstract: In this study, molecules of propyl gallate (PG) and polyethylene glycol methacrylate (PEGMA) were covalently bonded via a transesterification reaction and subsequently grafted onto polyvinylidene fluoride substrates using a homogeneous radiation grafting technique. The enhancement of the membranes' hydrophilicity with the increment of the grafting rate was corroborated by scanning electron microscopy imaging and measurements of the water contact angle. At a grafting degree of 10.1% and after a duration of 4 min, the water contact angle could decrease to as low as 40.1°. Cyclic flux testing demonstrated that the membranes modified in this manner consistently achieved a flux recovery rate exceeding 90% across varying degrees of grafting, indicating robust anti-fouling capabilities. Furthermore, these modified membranes exhibited significant antioxidant ability while maintaining antifouling performance over 30 days. The ability of the modified membranes to scavenge 2,2-diphenyl-1-picrylhydrazyl (DPPH) and 2,2′-Azino-bis(3-ethylbenzothiazoline-6-sulfonic acid) diammonium salt ($ABTS^+$) free radicals remained nearly unchanged after being stored in pure water for 30 days, and the flux recovery rate remained above 95% after immersion in sodium hypochlorite solution for 30 days. Among the tested membranes, the PVDF-g-PEGMAG modified membrane with a grafting degree of 7.2% showed the best antioxidant effect.

Keywords: polyvinylidene fluoride; antifouling; antioxidation; homogeneous radiation grafting; ultrafiltration membrane

Citation: Wang, T.; Hu, J.; Hou, Z.; Yang, H. Antifouling and Antioxidant Properties of PVDF Membrane Modified with Polyethylene Glycol Methacrylate and Propyl Gallate. *Materials* **2024**, *17*, 1867. https://doi.org/10.3390/ma17081867

Academic Editors: Guangxu Lan and Yi-Nan Wu

Received: 30 March 2024
Revised: 13 April 2024
Accepted: 17 April 2024
Published: 18 April 2024

1. Introduction

As economies grow and societies progress, the role of high-performance polymer filtration membranes gains prominence across sectors like water treatment and biomedicine [1–8]. Among these membranes, ultrafiltration membranes are indispensable components that face limitations due to fouling [9–12], which impairs performance and reduces lifespan due to oxidation during the cleaning phases [12–14]. Moreover, the demand for these membranes to possess antioxidant properties is critical, especially in the separation of bioactive substances to curb or eliminate oxidative stress triggered by reactive oxygen species in procedures like hemodialysis [15–18]. Therefore, the development of ultrafiltration membranes with enhanced antifouling and antioxidant features is of paramount importance.

In recent years, numerous studies have proposed various modification methods to alleviate membrane fouling issues, such as modification with inorganic ion exchangers, oxidation of graphene, metal–organic frameworks (MOFs), and others [19–24]. Polyvinylidene fluoride (PVDF) is a commonly used material for ultrafiltration membranes. There are many research works that highlight the significant improvement in fouling resistance when PVDF undergoes PEGylation, enhancing the performance of membranes fabricated from this material [25–28]. Wang et al. [25] employed the oxidative strength of ozone to activate PVDF, creating peroxides. This step was crucial for initiating the free radical graft polymerization process, which facilitated the grafting of polyethylene glycol methacrylate (PEGMA) onto the PVDF structure. Such a modification markedly improved the membrane's capacity to repel protein adsorption. Through the utilization of Atom Transfer Radical Polymerization (ATRP), Hester [26] and Liu et al. [27] developed PVDF-g-PEGMA. Additionally, Liu et al. [27] successfully constructed a PVDF/PVDF-g-PEGMA filtration membrane with exemplary antifouling characteristics and devoid of any defects by incorporating the grafted polymer as a functional additive. Previous explorations by our research group [28] have demonstrated the efficacious grafting of methoxy polyethylene glycol acrylate (mPEGA) onto PVDF, employing it as a PEG monomer via a homogeneous radiation grafting technique. This grafted composite was directly utilized in membrane formation, showcasing a water flux recovery rate that consistently remained above 90% across three fouling cycles. Furthermore, there are studies that have also delved into improving the hydrophilicity and antifouling capabilities of PVDF membranes through the strategic incorporation of PEGylation during the surface grafting [29–32] and the membrane fabrication process [33], thus underscoring the importance of PEGylation in enhancing membrane functionality.

Enhancing a membrane's hydrophilicity significantly improves its resistance to fouling [9,34]. However, maintaining these membranes through periodic cleaning is equally critical for extending their operational longevity and reducing expenses associated with their use [35–40]. In the realm of maintenance, chemical cleaning stands alongside physical methods as a vital approach, employing potent agents like hydrochloric acid (HCl), sodium hydroxide (NaOH), and sodium hypochlorite (NaClO), notable for their robust acidic, alkaline, and oxidizing effects, respectively [13,38,41–50]. NaClO is one of the most commonly used chemical agents in membrane cleaning processes [51]. The durability and functional lifespan of these membranes are heavily influenced by their capacity to withstand oxidation, making oxidation resistance a pivotal characteristic [52]. Investigative efforts by Gao et al. [43] have shed light on how different concentrations of sodium hypochlorite impact PVDF membranes, establishing guidelines for optimal cleaning practices to ensure filtration efficiency. Furthermore, Zhang et al. [34] have provided insights through comparative studies on the susceptibility of PVDF and polysulfone ultrafiltration membranes to damage and performance degradation when exposed to sodium hypochlorite solutions.

Derived from tannin hydrolysis, gallic acid (GA) is a polyphenolic compound with a spectrum of biological activities, including anti-inflammatory, antioxidant, and antibacterial properties [53–59]. In the realm of membrane technology, GA has been the focus of various studies. Employing a one-step dip-coating technique, Yang et al. [60] fabricated a hydrophilic PVDF membrane with GA and gamma-aminopropyltriethoxysilane, which exhibited remarkable hydrophilicity and antifouling characteristics. Through biomimetic coating, Cheng et al. [61] demonstrated the creation of a porous nanofiltration (NF) membrane via the Schiff base reaction, linking GA's phenolic group with the amino group of polyethyleneimine. Additionally, Zhao et al. [62] synthesized a GA-modified polyethyleneimine layer on a hydrolyzed polyacrylonitrile base using a co-deposition method.

To enhance the antioxidative capabilities of filter membranes, numerous studies have integrated a variety of antioxidants directly into the membrane material [63]. Addressing the challenges associated with the migration of antioxidant molecules and their extraction by solvents during application, several researchers have explored chemically bonding antioxidants to the membrane surface [64,65]. However, reports on antioxidative modifications applied to PEGylated filter membranes remain absent from the literature.

In this study, a dual-functional PVDF membrane endowed with both antifouling and antioxidative properties was successfully developed. The initial phase involved a transesterification process between propyl gallate (PG) and PEGMA. Following this reaction, the incorporation of PVDF into the mixture led to the formation of PVDF-g-PEGMAG through a homogeneous radiation grafting method. The fabrication of ultrafiltration membranes was achieved using the technique of non-solvent-induced phase separation (NIPS). The degree of grafting (DG) was determined using the gravimetric method. Furthermore, the grafted product underwent molecular structural analysis through FTIR spectroscopy. Assessments of the membrane's surface morphology, hydrophilicity, and filtration capacity were conducted utilizing SEM, water contact angle measurements, and water flux tests, respectively. The antifouling properties of the membrane were rigorously evaluated by filtering a solution of BSA and subsequently measuring the recovery rate of water flux following a cleansing process. The membrane's antioxidant efficacy was determined through its ability to scavenge DPPH and $APTS^+$ free radicals, with additional tests conducted to ascertain the longevity of these antioxidative characteristics by submerging the membrane in distilled water for an extended duration. Lastly, the resilience of the modified PVDF membrane against oxidative challenges was rigorously assessed by exposing it to a sodium hypochlorite solution over a specified timeframe, thereby evaluating its long-term oxidation resistance. This comprehensive study illustrates the potential of PVDF-g-PEGMAG membranes in applications requiring robust antifouling and antioxidative functionalities.

2. Materials and Methods

2.1. Materials

The PVDF powder (Solef 6020, Mn = 670,000) was purchased from Solvay Co., Ltd. (Qingdao, China). Polyethylene glycol methyl acrylate (PEGMA), 2,2-diphenyl-1-picrylhydrazyl (DPPH), and 2,2′-Azino-bis(3-ethylbenzothiazoline-6-sulfonic acid) diammonium salt ($ABTS^+$) were obtained from Sigma-Aldrich Trading Co., Ltd. (Shanghai, China). Propyl gallate (PG), p-toluenesulfonic acid (TsOH), N,N-dimethylformamide (DMF), polyvinylpyrrolidone (PVP-K30), sodium chloride (NaCl), disodium hydrogen phosphate (Na_2HPO_4), potassium chloride (KCl), potassium dihydrogen phosphate (KH_2PO_4), hydrochloric acid (HCl), sodium hypochlorite solution (NaClO), bovine serum albumin (BSA), anhydrous ethanol, and potassium persulfate ($K_2S_2O_8$) were purchased from Sinopharm Chemical Reagent Co., Ltd. (Shanghai, China). The PVDF powder was washed three times with ultrapure water before use and dried at 80 °C for 24 h in a vacuum oven. All chemical reagents were used without further purification. Additionally, water purified through a Milli-Q system from Millipore was employed.

2.2. Synthesis of Graft Copolymer

A mixture of PEGMA and TsOH was placed into a reaction flask, subsequently heated to 120 °C in an oil bath with ongoing stirring. Following this, PG was incrementally introduced to the mixture while stirring persisted. This reaction was conducted over a span of four hours, with vacuum application for one minute at each hour to eliminate volatiles. Once the reaction phase concluded and the mixture had cooled, PVDF and DMF were added, followed by stirring at 70 °C for a full day to achieve a homogenous solution. After cooling to ambient temperature, nitrogen gas was purged into the flask for thirty minutes before sealing. The samples were then subjected to γ-radiation from a ^{60}Co source, with a dose of 20 kGy for 18 h at room temperature. After irradiation, the solution was slowly transferred into an ethanol solution for precipitation. Through repeated rinsing and washing in ethanol, unreacted substances and homopolymers were removed. The final step involved drying the precipitate in a vacuum oven at 80 °C until reaching a constant weight to yield PVDF-g-PEGMAG. The reagent quantities used are specified in Table 1. The schematic diagram of the study is shown in Figure 1.

Table 1. The formula for synthesizing PVDF-g-PEGMAG.

No.	PEGMA/g	TsOH/g	PG/g	PVDF/g	DMF/g	Total/g
1	3	0.08	4.24	5	37.68	50
2	5	0.08	7.07	5	32.85	50
3	9	0.08	12.72	5	24	50

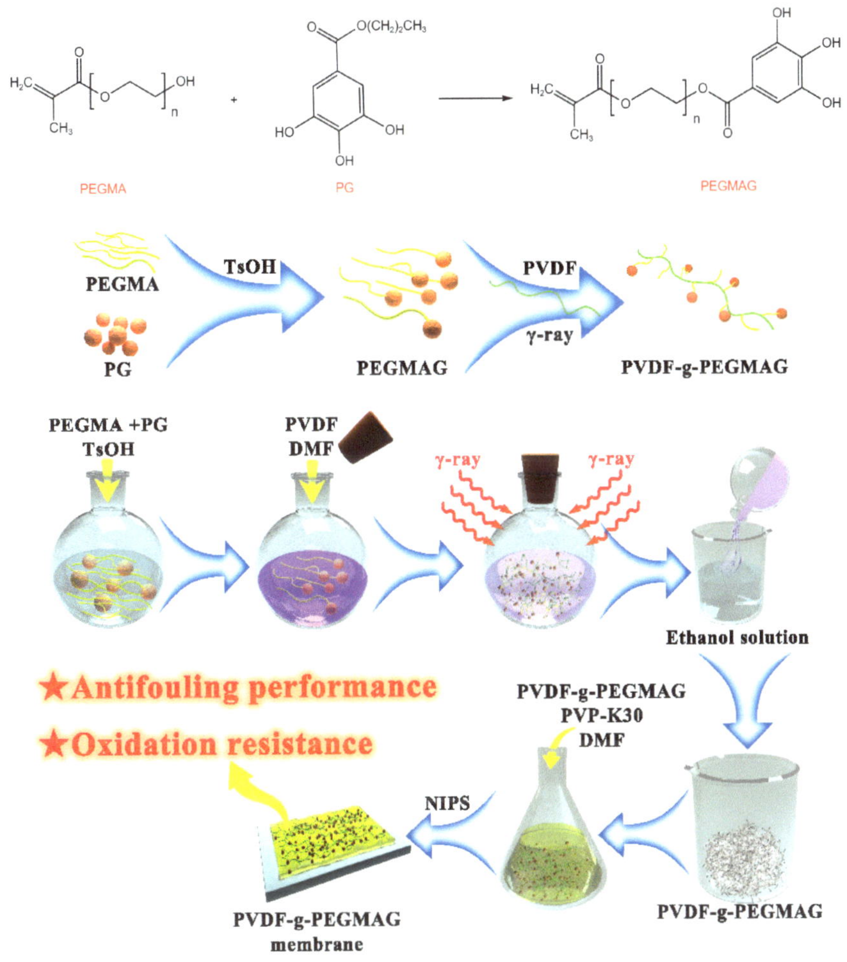

Figure 1. Schematic diagram of the synthesis/fabrication process.

The degree of grafting (DG) of PVDF-g-PEGMAG is defined in Equation (1):

$$DG = \frac{W_1 - W_0}{W_0} \times 100\% \quad (1)$$

where W_1 denotes the mass of PVDF-g-PEGMAG and W_0 is the mass of pristine PVDF.

2.3. Fourier Transform Infrared Spectroscopy Characterization

ATR-IR spectroscopy analyses were conducted for pristine polyvinylidene fluoride (PVDF), PVDF-g-PEGMA, and PVDF-g-PEGMAG samples utilizing a Nicolet Avatar 370 infrared spectrometer (Thermo Nicolet Instrument Corporation, Madison, WI, USA). The spectral data collection covered a wavelength range from 400 cm^{-1} to 4000 cm^{-1} with a resolution set at 4 cm^{-1}. Each spectrum acquisition involved 32 repetitive scans to ensure accuracy. Given the challenges associated with pulverizing the grafted polymer, a membrane approach was adopted for preparing the samples, consistent with methodologies described in preceding research [28].

2.4. Preparation of Membrane

The membranes were fabricated using the non-solvent-induced phase separation (NIPS) method. The composition of the casting solution encompassed a polymer concentration of 15 wt.%, a porogen PVP-K30 at 4 wt.%, and the solvent DMF constituting 81 wt.%. Utilization of a casting knife facilitated the even distribution of the casting solution across a glass plate, targeting a uniform film thickness of 200 μm. The solution underwent brief evaporation in the ambient atmosphere for 15 s, followed by immersion in deionized water. To assure the thorough elimination of any residual solvents, the membrane was meticulously preserved in water for a duration of two days, during which the water was refreshed multiple times before any testing commenced.

2.5. Dynamic Water Contact Angle Measurement

Contact angle measurements of the membrane were carried out using the Attension Theta system (KSV Instruments Ltd., Helsinki, Finland). For this process, droplets, each measuring 5 μL, were carefully dispensed onto the membrane's surface from a needle's tip. Over a 4 min duration, a digital camera captured the interaction between the droplet and the membrane surface in video format. Subsequently, specialized software in the system analyzed these video recordings to determine the contact angle values.

2.6. Analysis of Scanning Electron Microscope

Morphological analysis of the membrane was conducted using a LEO1530vp SEM (Zeiss, Jena, Germany). To prepare the membrane for examination, it was first subjected to rapid freezing in liquid nitrogen, followed by a brittle fracture to reveal the cross-section. The images obtained were then analyzed with ImageJ software (National Institutes of Health, Bethesda, MD, USA, v1.8.0), allowing for the quantification of pores, as well as the measurement of pore size and distribution across a given unit area.

2.7. Filtering Experiment

The membrane's performance was assessed using the identical cross-flow apparatus utilized in prior experiments, as depicted in Figure 2 [28]. Each membrane was fashioned into a circular specimen boasting a 15.9 cm^2 surface area and underwent a pre-compression process at 0.1 MPa for a duration of 20 min. For each type of membrane, a set of three independent trials was conducted. The flux measurement involved the quantification of the solution volume passing through the membrane over a specified period, subsequently normalized to the membrane's surface area.

The water flux (F_0) was calculated according to Equation (2):

$$F_0 = \frac{Q}{A\Delta T} \tag{2}$$

where Q refers to the volume of the filtered water; A, the area of the membrane; and ΔT, the time of the filtration.

In this study, a BSA buffer solution at a concentration of 1 g/L was selected as the model for pollutant testing, employing the same buffer preparation protocol established in preceding experiments. Samples were collected both before and after the filtration process to facilitate subsequent analysis. The quantification of BSA concentration within the solution was accurately performed using a UV-1100 visible ultraviolet spectrophotometer (Beijing Ruili Instrument Co., Ltd., Beijing, China), in accordance with methodologies cited in the reference literature [66].

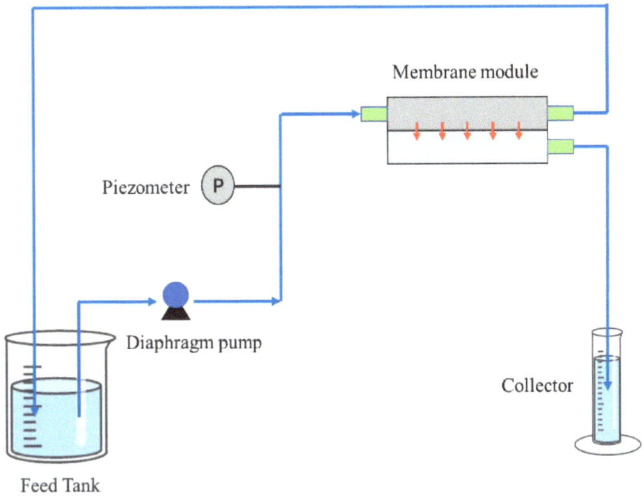

Figure 2. The schematic diagram of the cross-flow device used in the filtration experiment [28].

2.8. Evaluation of Antifouling Performance of Membrane

The evaluation of the membrane's antifouling capability was conducted by measuring the pure water flux recovery rate following the filtration of a BSA solution. In the cross-flow filtration setup, the feed solution was alternated between pure water and the BSA buffer solution every 60 min. After completing three cycles of this process, the recovery rate of water flux through the membrane was determined.

The flux recovery ratio (*FRR*) is calculated according to Equation (3):

$$FRR = \frac{F_1}{F_0} \times 100\% \quad (3)$$

where F_0 and F_1 denote the water flux prior to and following fouling, respectively.

2.9. Evaluation of the Oxidation Resistance of the Membrane

The antioxidative properties of the membrane were evaluated from two perspectives. Firstly, the membrane's ability to scavenge DPPH and $ABTS^+$ radicals was assessed. Secondly, the membrane's resistance to oxidation in NaClO solution was evaluated.

2.9.1. Free Radical Scavenging Effect

A solution of DPPH (0.1 mM) was formulated using anhydrous ethanol. Subsequently, membrane samples, each measuring 1×1 cm^2, were submerged in 3 mL of this DPPH solution and left to incubate for 1 h in a dark environment at room temperature. As a control, a DPPH solution without any membrane sample underwent identical treatment. The absorbance of these solutions was then recorded at a wavelength of 517 nm.

To prepare the $ABTS^+$ radical cation solution (0.1 mM), a mixture of $ABTS^+$ solution (2 mL, 7 mM) and potassium persulfate ($K_2S_2O_8$) solution (2 mL, 2.45 mM), diluted with pure water, was allowed to stand for 12 h in the dark at 4 °C. This solution was subsequently adjusted with anhydrous ethanol to achieve an absorbance of 0.700 ± 0.025. Each membrane sample, again measuring 1×1 cm^2, was immersed in 4 mL of the $ABTS^+$ solution and left for 20 min in a dark environment at room temperature. The absorbance of this solution was measured at 734 nm.

Equation (4) for calculating the scavenging rate of the membranes against these two free radicals is presented as follows:

$$Scavenging = \frac{A_0 - A_1}{A_0} \times 100\% \quad (4)$$

where A_0 represents the absorbance of the control sample devoid of the membrane, while A_1 signifies the absorbance recorded for the solution containing various membrane samples.

The enduring stability of antioxidant capabilities holds considerable relevance for the practical application of antioxidant membranes. To assess this, membrane samples measuring 1×1 cm^2 were submerged in distilled water for a duration of 30 days. The efficiency of these membranes in scavenging free radicals was determined at intervals—on the 1st, 7th, 14th, and 30th days, to gauge the long-term retention of their antioxidant attributes.

2.9.2. Oxidation Resistance of Membrane to NaClO Solution

Initially, a variety of membranes were submerged in a NaClO solution with a concentration of 0.2 wt% for a duration of 30 days, with the NaClO solution being refreshed daily to maintain consistent concentration levels. The performance of each membrane, in terms of water flux and BSA rejection rates, was documented on the 1st, 7th, 14th, and 30th days. Furthermore, a modified membrane, possessing a DG of 8.6%, was also immersed in the NaClO solution for 30 days. Following this period, it was thoroughly rinsed with pure water, and a circulation flux test was conducted. The resistance of the membrane to oxidation by NaClO solution was assessed based on these two investigative approaches.

3. Results and Discussion

3.1. FTIR Spectra of Graft Copolymer

Figure 3 illustrates the infrared spectra of the pristine PVDF alongside various grafted polymers, following vector normalization. Among these, PVDF-g-PEGMA, with a DG of 8.9%, was synthesized by directly grafting PEGMA onto PVDF using the previously described homogeneous radiation grafting technique [28]. Notably, both grafted polymers displayed a new characteristic absorption peak at 1712 cm^{-1}, indicative of the carbonyl (C=O) group present in PEGMA [67]. When comparing PVDF-g-PEGMA to PVDF-g-PEGMAG, the latter exhibited subtle signals corresponding to phenolic hydroxyl groups and benzene rings, characteristic of PG molecules, in the range of 3300–3600 cm^{-1} and at around 1610 cm^{-1}, respectively [68]. These spectral features confirm the successful grafting of PEGMA and PG onto the PVDF chain, resulting in the synthesis of PVDF-g-PEGMAG.

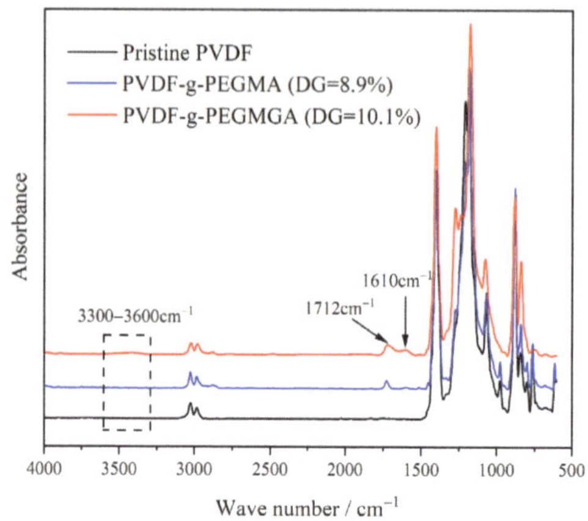

Figure 3. FTIR spectra of pristine PVDF and different grafted polymers.

3.2. Dynamic Water Contact Angle Test

Figure 4 displays the dynamic water contact angles of both the pristine PVDF and the PVDF-g-PEGMAG membranes with varying degrees of grafting. The water contact angle for the unmodified PVDF remained consistently above 95°, indicating pronounced hydrophobicity. For the PVDF-g-PEGMAG membranes, an increase in the degree of grafting correlated with an increase in the water contact angle, signifying enhanced hydrophilicity and, consequently, improved surface wetting. Notably, at a grafting degree of 10.1% and after a duration of 4 min, the water contact angle could decrease to as low as 40.1°. This demonstrates a substantial enhancement in hydrophilicity for the grafted polymers in comparison to the pristine PVDF. It was observed that the addition of PG segments to the PEGylated PVDF materials did not alter their hydrophilic properties.

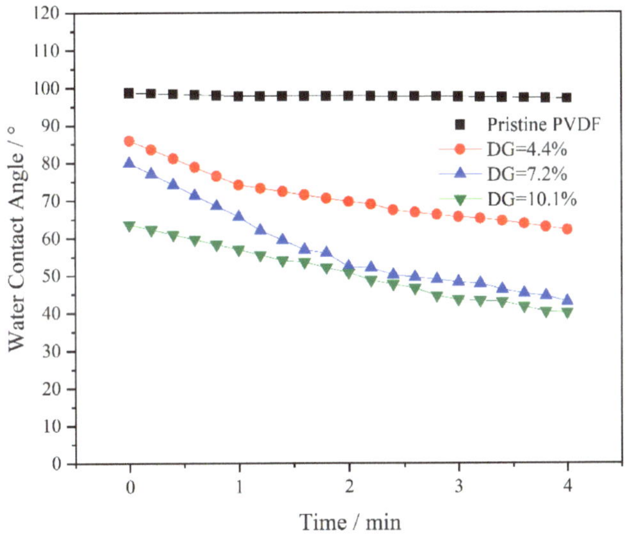

Figure 4. Dynamic water contact angles of pristine PVDF and PVDF-g-PEGMAG membranes with different DG.

3.3. Morphology Characterization of Membrane

Figure 5 presents the surface and cross-sectional images of membranes with varying grafting ratios, alongside a pore size distribution histogram derived from the surface images using ImageJ software (v1.8.0). The images reveal that each membrane exhibits an asymmetric structure characterized by finger-like pores. The membrane surface pore sizes are 27.650 nm, 28.432 nm, 30.923 nm, and 32.072 nm in Figure 5a–d, respectively. With an increase in grafting ratio, there is a noticeable enlargement in pore size on the membrane surface, and the cross-sectional views begin to show larger pores more prominently [69,70]. In general, an NIPS process generates a dense skin and porous sub-layer. The dense surface is formed due to the solidification of the polymer top layer induced by a fast solvent outflow, and the porous sub-layer is induced from the liquid–liquid demixing where the solution phase separates into a polymer-lean phase and a polymer-rich phase [71]. The pore formation in the NIPS process may be attributed to the rapid diffusion of solvent into the non-solvent (water), while the non-solvent does not permeate as swiftly through the interface between the casting solution and water, generating significant tension within the final membrane matrix, thus forming the distinctive finger-like pores. As the grafting ratio rises, the diffusion rate of the non-solvent (water) accelerates, making the macroscopic pore structure in the membrane matrix more pronounced in the cross-sectional area [72]. Moreover, at the same polymer weight percentage, a higher grafting ratio implies a greater content of grafting segments and, consequently, a lower concentration of pure PVDF in the polymer. This results in a quicker solvent diffusion rate and, subsequently, larger pores in the modified membrane [73].

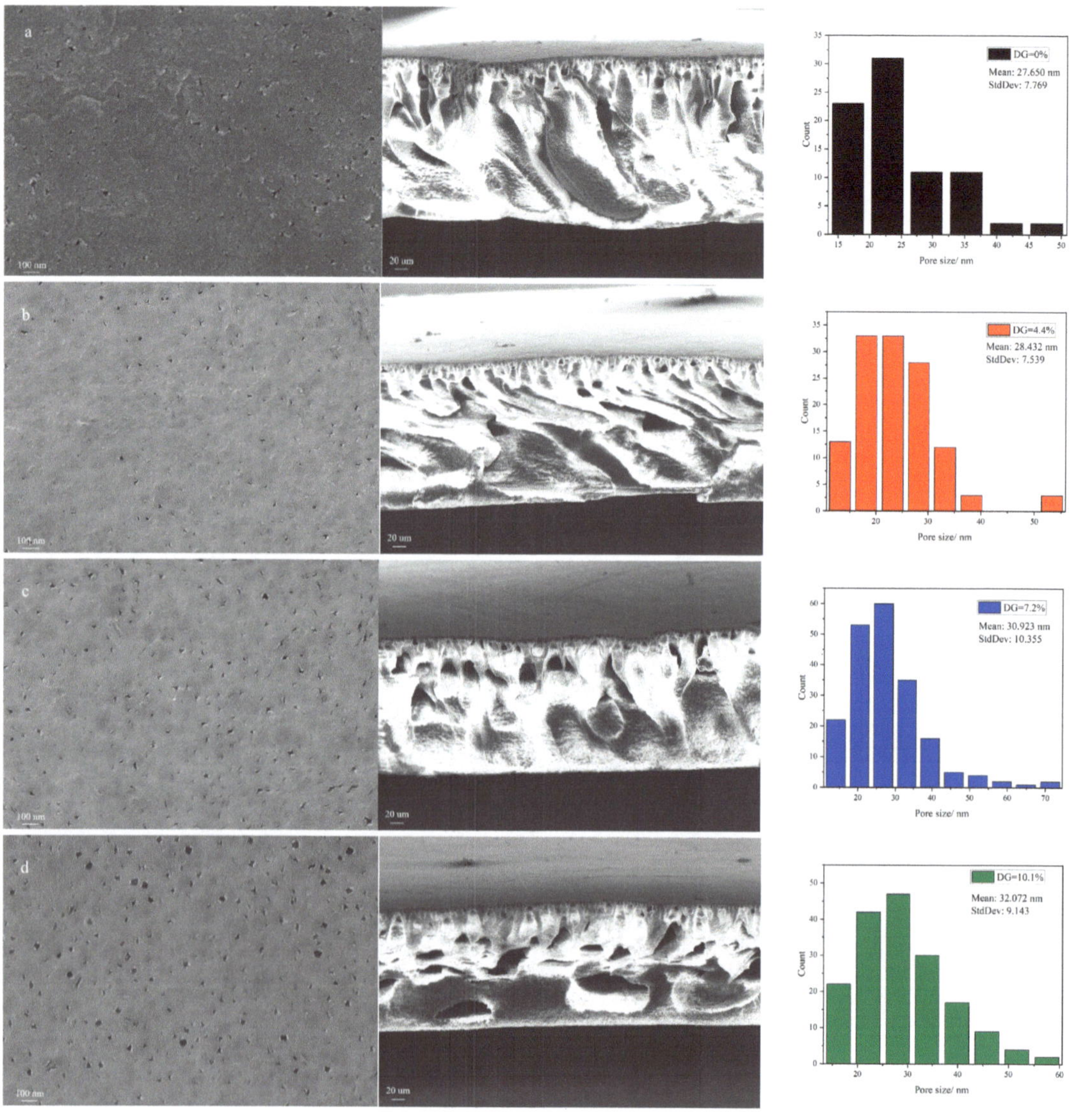

Figure 5. SEM images of membrane surface and cross-section with different grafting ratios: (**a**) DG = 0, (**b**) DG = 4.4%, (**c**) DG = 7.2%, (**d**) DG = 10.1%. (The rightmost is the square distribution of the membrane pore size calculated by ImageJ based on the SEM membrane surface image).

3.4. Evaluation of Filtration Performance and Antifouling Performance of Membrane

Figure 6 illustrates the comparison of water flux and BSA rejection between the pristine PVDF membranes and those modified with PVDF-g-PEGMAG at various DG. The graph demonstrates that, as the DG increases, so does the water flux through the membrane. This observation aligns with findings from previous contact angle tests, which indicated an enhancement in membrane hydrophilicity concurrent with rising DG levels. Enhanced hydrophilicity leads to diminished filtration resistance, thereby facilitating increased flux. Additionally, this elevation in water flux corresponds with the morphological attributes identified in the membrane, such as the expansion in pore size on the membrane surface and the emergence of larger pores within the membrane cross-section. The BSA rejection rates for all the membranes hovered around 90%, indicating the effectiveness of the modified membranes, crafted using grafted polymers, in maintaining robust filtration performance.

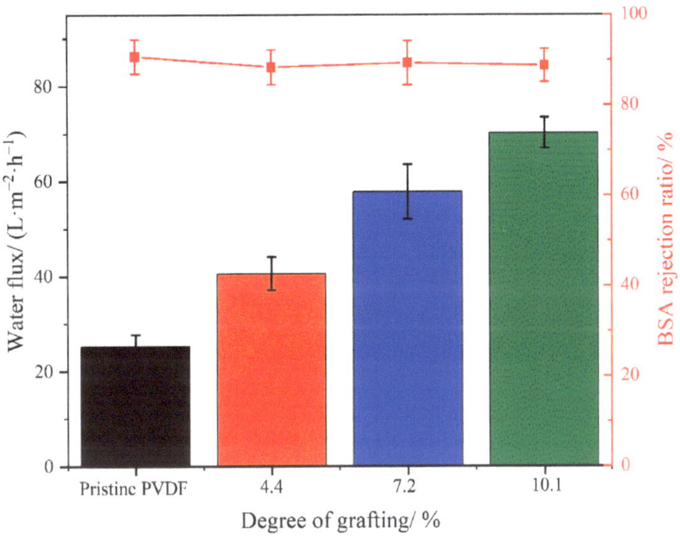

Figure 6. Water flux and BSA rejection of polymer membranes with different DG.

In this study, an in situ measurement technique was employed, integrating online cleaning of the membrane via a cross-flow filtration method. This process entailed alternating the fluid composition, specifically switching between water and a BSA solution. Figure 7 illustrates the results from antifouling cycle tests on membranes with varying DG. Specifically, Figure 7a depicts the fouling and cleaning flux of the membrane across multiple cycles, whereas Figure 7b presents the FRR derived from the data in Figure 7a. During the cleaning phase, consistent trends in flux variation were observed among different modified membranes, with a notable increase in flux upon switching the feed to pure water. After three cycles (420 min) of testing, the flux of pure water post-cleaning remained relatively unchanged, with only a minor reduction observed during the initial cycle. Conversely, the pristine PVDF membrane exhibited a significant flux decline upon the initial switch to pure water. Furthermore, the FRR values calculated from the flux data indicated a significant decrease in the FRR of the pristine PVDF membrane during the first cycle, with a continued decline over the subsequent two cycles. In contrast, the FRRs for the modified membranes consistently exceeded 90% and remained stable through later cycles, suggesting that higher FRR values correlate with enhanced membrane stability and superior antifouling performance. The antifouling cycle tests comprehensively demonstrate the superior antifouling capabilities of the modified membranes.

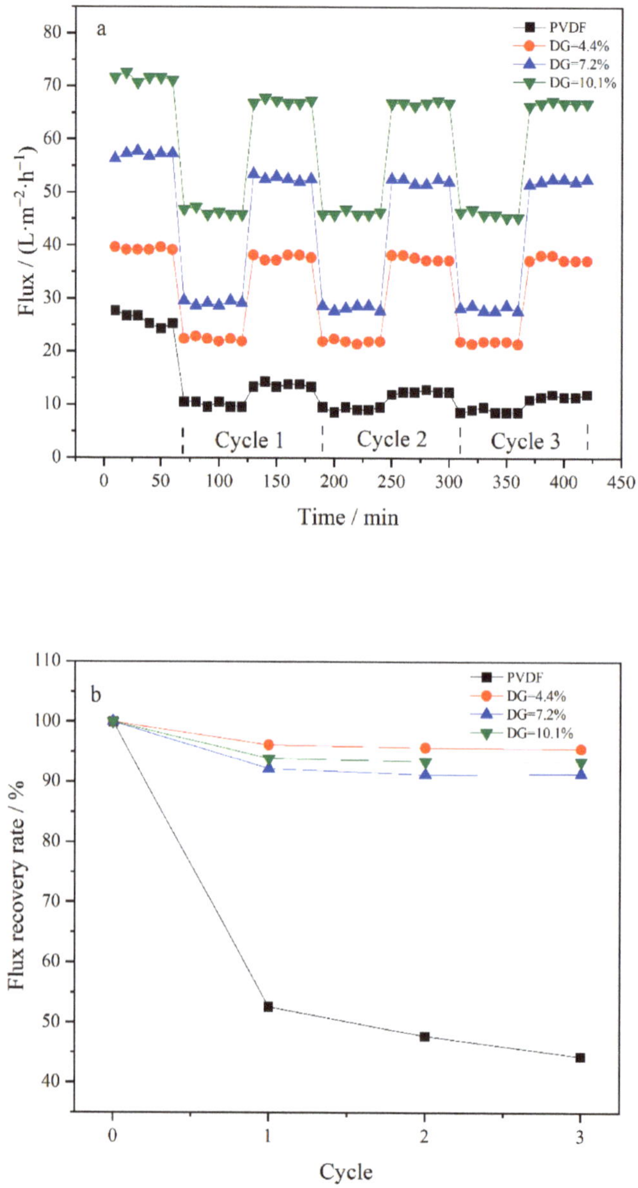

Figure 7. Antifouling cycle test of membrane: (**a**) water and BSA flux of membrane with different DG under multiple cycles, (**b**) FRR of membrane with different DG under multiple cycles.

3.5. Oxidation Resistance of the Membrane

3.5.1. Free Radical Scavenging Effect of Membrane

Figure 8 demonstrates the efficiency of different types of membranes in scavenging DPPH and $ABTS^+$ free radicals. The membranes modified with PG segments, PVDF-g-PEGMAG, showcased a significantly higher free radical scavenging capacity compared to both the pristine PVDF and the PVDF-g-PEGMA. This superior scavenging ability highlights the impactful role of the PG segments in enhancing the membrane's antioxidative properties. Furthermore, variations in the grafting rate appeared to have a minimal impact on the membranes' ability to neutralize free radicals. This observation could be attributed to the fact that the grafting rate mentioned in this study reflects the combined grafting of both PG and PEGMA segments onto the membrane. During the synthesis of the grafted polymer with a DG of 10.1%, a larger quantity of PEGMA segments was introduced onto the PVDF chain compared to PEGMAG, due to the higher proportion of PEGMA, thereby leading to a slight decrease in the membrane's free radical scavenging rate compared to the membrane with DG = 7.2%.

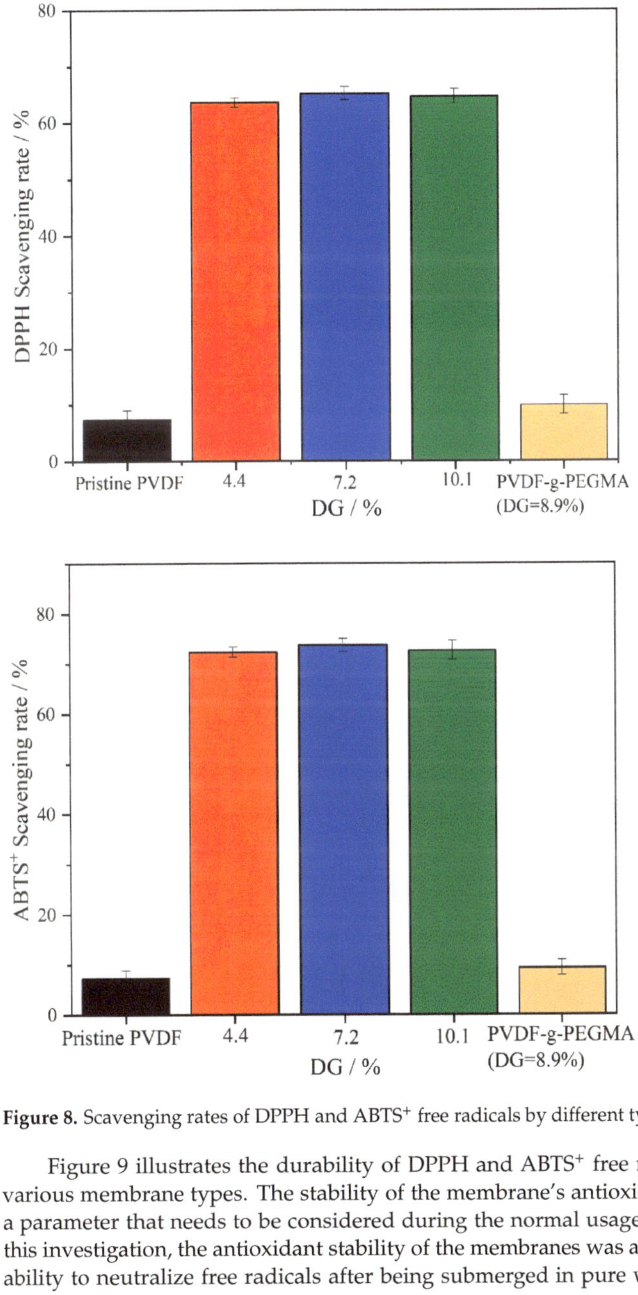

Figure 8. Scavenging rates of DPPH and $ABTS^+$ free radicals by different types of membranes.

Figure 9 illustrates the durability of DPPH and $ABTS^+$ free radical scavenging by various membrane types. The stability of the membrane's antioxidative performance is a parameter that needs to be considered during the normal usage of the membrane. In this investigation, the antioxidant stability of the membranes was assessed based on their ability to neutralize free radicals after being submerged in pure water for an extended duration. The data indicate that the scavenging efficiency for both types of free radicals diminished only marginally after a 30-day immersion period in pure water. This experiment underscores that the modified membranes possess not just a measure of resistance to oxidation, but also maintain considerable stability in their antioxidative properties over time.

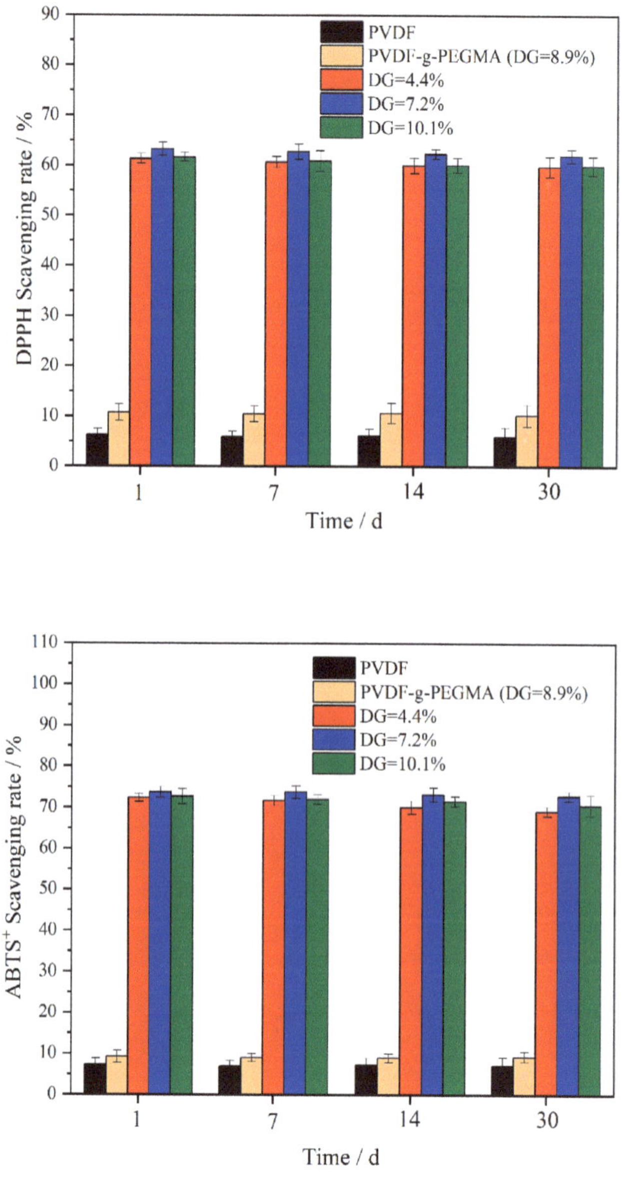

Figure 9. Scavenging stability of DPPH and $ABTS^+$ free radicals by different types of membranes. (Membrane samples were submerged in distilled water for a duration of 30 days).

3.5.2. Oxidation Resistance of Membrane to NaClO Solution

To investigate the influence of an oxidizing agent on membranes post-grafting of PG segments, membranes, including PVDF-g-PEGMA (with a DG of 8.9%) and PVDF-g-

PEGMAG with various DG, were immersed in a sodium hypochlorite (NaClO) solution (2 wt.%) for a set duration. The impact on membrane filtration performance was assessed through variations in water flux and bovine serum albumin (BSA) rejection rate over time. Previous reports on the effect of hypochlorite treatment on UF membranes indicated a flux increase in NaClO-treated membranes [12,14]. Specifically, the PVDF-g-PEGMAG membrane with a DG of 8.6% was analyzed to compare its circulation flux before and after a 30-day immersion in NaClO solution. The antifouling efficacy of the membrane was appraised by comparing the FRR before and after the 30-day exposure to oxidation. Figure 10 reveals that, after 30 days of exposure to the oxidizing environment, the water flux through the PVDF-g-PEGMA membrane increased, while its BSA rejection rate dropped below 80%, indicating a significant detriment to its filtration performance due to oxidation. Conversely, the water flux and BSA rejection rates for PVDF-g-PEGMAG membranes with various DG remained largely stable after the 30-day oxidation period. This suggests that the PG segment grafting effectively mitigated the oxidation impact on PEGylated PVDF membranes. Figure 11 displays the antifouling cycle test results for the PVDF-g-PEGMAG membrane with a DG of 8.6%, both before and after undergoing oxidation treatment for 30 days. It is observed that the flux of the membrane post-oxidation closely mirrors the FRR prior to oxidation, maintaining stability across three cycles. This outcome indicates that the antifouling properties of the modified membrane were well-preserved even after 30 days of exposure to oxidizing conditions.

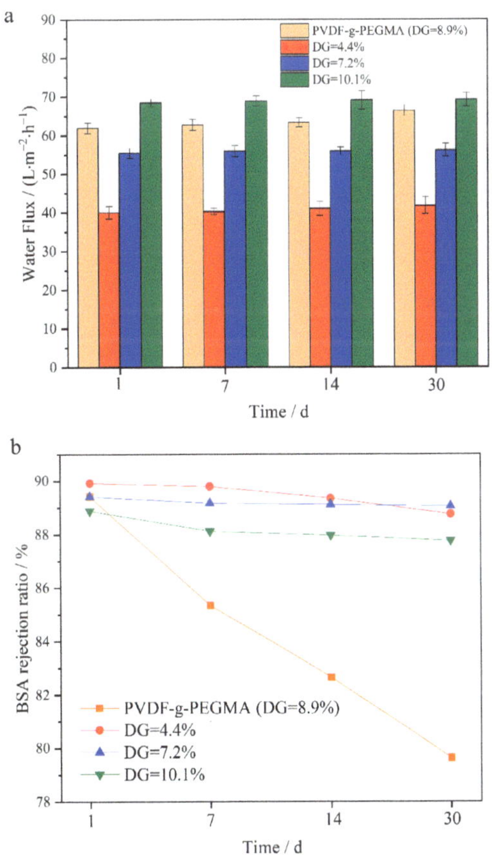

Figure 10. Water flux (**a**) and BSA rejection (**b**) of different types of membranes immersed in NaClO (2 wt.%) solution for a period of time.

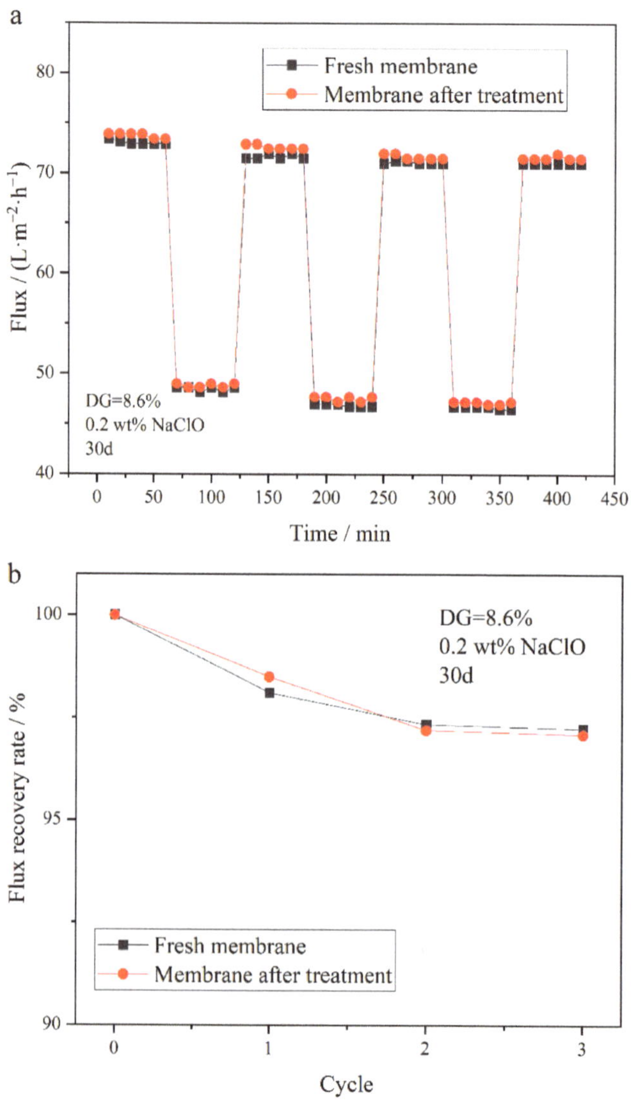

Figure 11. Antifouling cycle test of PVDF-g-PEGMAG (DG = 8.6%) membrane before and after immersion in NaClO (0.2 wt.%) solution for 30 days: (**a**) water and BSA flux under multiple cycles before and after oxidation, (**b**) FRR under multiple cycles before and after oxidation.

4. Conclusions

In this study, the PG segment was initially bonded to the PEGMA molecule through a transesterification process. Subsequently, this complex was grafted onto PVDF using a homogeneous radiation grafting technique to create PVDF-g-PEGMAG, which incorporates both antioxidant (PG segment) and antifouling (PEGMA segment) functionalities.

The membrane of the grafted polymer was fabricated by employing the NIPS method. Analysis of the membrane's surface and cross-section via SEM alongside water contact angle measurements revealed that the hydrophilicity of the membrane enhanced as the DG increased. The pore size of the membrane increased with the increase in grafting rate. Specifically, for the modified membrane with a grafting rate of 10.1%, the pore size expands from 27.650 nm in the original PVDF membrane to 32.072 nm. Additionally, after a duration of 4 min, the water contact angle of the membrane could decrease to as low as 40.1°. Cyclic flux testing of the membrane indicated that introducing the PG segment to the modified membrane did not detract from the excellent antifouling capabilities inherent to PEGylated PVDF membranes. In tests of circulation flux across membranes with varying DG, the FRR consistently exceeded 90%. Moreover, the antioxidant performance tests demonstrated that the modified membrane retained its capacity to scavenge DPPH and $ABTS^+$ free radicals, with its efficacy largely unaffected by prolonged immersion in water. After soaking in pure water for 30 days, the modified membrane exhibited only a slight decrease in scavenging rates for both DPPH and $ABTS^+$ free radicals, maintaining them at above 60% and 70%, respectively. Additionally, the PG grafting significantly bolstered the membrane's resistance to oxidation, enabling it to maintain antifouling performance; even after a 30-day immersion in a NaClO solution, the flux recovery rate remained above 95%, thus illustrating the membrane's durability against oxidation during chemical cleaning processes.

This study successfully developed a membrane that exhibits both antifouling and antioxidative properties, highlighting its long-term effectiveness and stability. Although further research is needed to assess the membrane's tolerance to other chemical cleaning methods, such as acid–base treatments, to differentiate the proportion of functional components in the grafted product, and to elucidate the mechanism of antioxidation. However, the simplicity and potential scalability of the membrane fabrication process make it a viable option from an industrial perspective. The membrane exhibits promising potential for applications in water treatment and the membrane-based separation of bioactive substances.

Author Contributions: Conceptualization, T.W. and Z.H.; methodology, Z.H.; validation, T.W. and Z.H.; data curation, T.W., Z.H., and J.H.; writing—original draft preparation, T.W.; writing—review and editing, Z.H., H.Y. and J.H.; project administration, Z.H.; funding acquisition, Z.H., H.Y. and J.H. All authors have read and agreed to the published version of the manuscript.

Funding: This work was financially supported by the National Natural Foundation of China (NSFC, No. 11375252 and No. U1632135) and the National Key Research and Development Program (2022YFA1603600).

Institutional Review Board Statement: Not applicable.

Informed Consent Statement: Not applicable.

Data Availability Statement: Data are contained within the article.

Conflicts of Interest: The authors declare no conflicts of interest.

References

1. Li, Q.; Bi, Q.; Lin, H.-H.; Bian, L.-X.; Wang, X.-L. A Novel Ultrafiltration (UF) Membrane with Controllable Selectivity for Protein Separation. *J. Membr. Sci.* **2013**, *427*, 155–167. [CrossRef]
2. Bai, T.; Zhao, K.; Lu, Z.; Liu, X.; Lin, Z.; Cheng, M.; Li, Z.; Zhu, D.; Zhang, L. Simple Fabrication of Cu^{2+} Doped Calcium Alginate Hydrogel Filtration Membrane with Excellent Anti-Fouling and Antibacterial Properties. *Chin. Chem. Lett.* **2021**, *32*, 1051–1054. [CrossRef]
3. Imahase, H.; Sakamoto, Y.; Kusunose, M.; Koami, H.; Nishimura, Y.; Goto, A.; Yamashita, T.; Nakashima, A.; Iwamura, T.; Kutsukata, N. Examination of Blood Filtration Membrane Removal Ability of HMGB1. *Crit. Care* **2012**, *16*, P10. [CrossRef]
4. Abdelhamid, H.N.; Mathew, A.P. Cellulose-Based Nanomaterials Advance Biomedicine: A Review. *Int. J. Mol. Sci.* **2022**, *23*, 5405. [CrossRef] [PubMed]
5. Venugopal, J.; Ramakrishna, S. Applications of Polymer Nanofibers in Biomedicine and Biotechnology. *ABAB* **2005**, *125*, 147–158. [CrossRef] [PubMed]
6. Liu, Y.; Liu, H.; Shen, Z. Nanocellulose Based Filtration Membrane in Industrial Waste Water Treatment: A Review. *Materials* **2021**, *14*, 5398. [CrossRef] [PubMed]
7. Pronk, W.; Ding, A.; Morgenroth, E.; Derlon, N.; Desmond, P.; Burkhardt, M.; Wu, B.; Fane, A.G. Gravity-Driven Membrane Filtration for Water and Wastewater Treatment: A Review. *Water Res.* **2019**, *149*, 553–565. [CrossRef] [PubMed]

8. Madsen, H.T. Membrane Filtration in Water Treatment—Removal of Micropollutants. In *Chemistry of Advanced Environmental Purification Processes of Water*; Elsevier: Amsterdam, The Netherlands, 2014; pp. 199–248. ISBN 978-0-444-53178-0.
9. Subasi, Y.; Cicek, B. Recent Advances in Hydrophilic Modification of PVDF Ultrafiltration Membranes—A Review: Part I. *Membr. Technol.* **2017**, *2017*, 7–12. [CrossRef]
10. Al Aani, S.; Mustafa, T.N.; Hilal, N. Ultrafiltration Membranes for Wastewater and Water Process Engineering: A Comprehensive Statistical Review over the Past Decade. *J. Water Process Eng.* **2020**, *35*, 101241. [CrossRef]
11. Ahmad, T.; Guria, C.; Mandal, A. A Review of Oily Wastewater Treatment Using Ultrafiltration Membrane: A Parametric Study to Enhance the Membrane Performance. *J. Water Process Eng.* **2020**, *36*, 101289. [CrossRef]
12. Shi, X.; Tal, G.; Hankins, N.P.; Gitis, V. Fouling and Cleaning of Ultrafiltration Membranes: A Review. *J. Water Process Eng.* **2014**, *1*, 121–138. [CrossRef]
13. Arkhangelsky, E.; Kuzmenko, D.; Gitis, V. Impact of Chemical Cleaning on Properties and Functioning of Polyethersulfone Membranes. *J. Membr. Sci.* **2007**, *305*, 176–184. [CrossRef]
14. Regula, C.; Carretier, E.; Wyart, Y.; Gésan-Guiziou, G.; Vincent, A.; Boudot, D.; Moulin, P. Chemical Cleaning/Disinfection and Ageing of Organic UF Membranes: A Review. *Water Res.* **2014**, *56*, 325–365. [CrossRef] [PubMed]
15. Mahlicli, F.Y.; Altinkaya, S.A. Immobilization of Alpha Lipoic Acid onto Polysulfone Membranes to Suppress Hemodialysis Induced Oxidative Stress. *J. Membr. Sci.* **2014**, *449*, 27–37. [CrossRef]
16. Chen, Q.; He, Y.; Zhao, Y.; Chen, L. Intervening Oxidative Stress Integrated with an Excellent Biocompatibility of Hemodialysis Membrane Fabricated by Nucleobase-Recognized Co-Immobilization Strategy of Tannic Acid, Looped PEtOx Brush and Heparin. *J. Membr. Sci.* **2021**, *625*, 119174. [CrossRef]
17. Hasegawa, T.; Iwasaki, Y.; Ishihara, K. Preparation and Performance of Protein-Adsorption-Resistant Asymmetric Porous Membrane Composed of Polysulfone/Phospholipid Polymer Blend. *Biomaterials* **2001**, *22*, 243–251. [CrossRef]
18. Qi, X.; Yang, N.; Luo, Y.; Jia, X.; Zhao, J.; Feng, X.; Chen, L.; Zhao, Y. Resveratrol as a Plant Type Antioxidant Modifier for Polysulfone Membranes to Improve Hemodialysis-Induced Oxidative Stress. *Mater. Sci. Eng. C* **2021**, *123*, 111953. [CrossRef]
19. Myronchuk, V.; Dzyazko, Y.; Zmievskii, Y.; Ukrainets, A.; Bildukevich, A.; Kornienko, L.; Rozhdestvenskaya, L.; Palchik, A. Organic-Inorganic Membranes for Filtration of Corn Distillery. *Acta Period. Technol.* **2016**, *47*, 153–165. [CrossRef]
20. Marjani, A.; Nakhjiri, A.T.; Adimi, M.; Jirandehi, H.F.; Shirazian, S. Effect of Graphene Oxide on Modifying Polyethersulfone Membrane Performance and Its Application in Wastewater Treatment. *Sci. Rep.* **2020**, *10*, 2049. [CrossRef]
21. Jaber, L.; Almanassra, I.W.; Backer, S.N.; Kochkodan, V.; Shanableh, A.; Atieh, M.A. A Comparative Analysis of the Effect of Carbonaceous Nanoparticles on the Physicochemical Properties of Hybrid Polyethersulfone Ultrafiltration Membranes. *Membranes* **2022**, *12*, 1143. [CrossRef]
22. Almanassra, I.W.; Jaber, L.; Backer, S.N.; Chatla, A.; Kochkodan, V.; Al-Ansari, T.; Shanableh, A.; Atieh, M.A. Oxidized Carbide-Derived Carbon as a Novel Filler for Improved Antifouling Characteristics and Permeate Flux of Hybrid Polyethersulfone Ultrafiltration Membranes. *Chemosphere* **2023**, *313*, 137425. [CrossRef] [PubMed]
23. Almanassra, I.W.; Jaber, L.; Chatla, A.; Abushawish, A.; Shanableh, A.; Ali Atieh, M. Unveiling the Relationship between MOF Porosity, Particle Size, and Polyethersulfone Membranes Properties for Efficient Decontamination of Dye and Organic Matter. *Chem. Eng. J.* **2023**, *471*, 144616. [CrossRef]
24. Jaber, L.; Almanassra, I.W.; AbuShawish, A.; Chatla, A.; Ihsanullah, I.; Ali, M.M.; Manawi, Y.; Shanableh, A.; Atieh, M.A. Pioneering Biofouling Resistant PES UF Membrane with MnFe2O4/g-C3N4 Nanocomposite: Insight into Mechanisms and Fouling Dynamics. *J. Membr. Sci.* **2024**, *691*, 122259. [CrossRef]
25. Wang, P.; Tan, K.L.; Kang, E.T.; Neoh, K.G. Synthesis, Characterization and Anti-Fouling Properties of Poly(Ethylene Glycol) Grafted Poly(Vinylidene Fluoride) Copolymer Membranes. *J. Mater. Chem.* **2001**, *11*, 783–789. [CrossRef]
26. Hester, J.F.; Banerjee, P.; Won, Y.-Y.; Akthakul, A.; Acar, M.H.; Mayes, A.M. ATRP of Amphiphilic Graft Copolymers Based on PVDF and Their Use as Membrane Additives. *Macromolecules* **2002**, *35*, 7652–7661. [CrossRef]
27. Liu, B.; Chen, C.; Li, T.; Crittenden, J.; Chen, Y. High Performance Ultrafiltration Membrane Composed of PVDF Blended with Its Derivative Copolymer PVDF-g-PEGMA. *J. Membr. Sci.* **2013**, *445*, 66–75. [CrossRef]
28. Wang, T.; Hou, Z.; Yang, H.; Hu, J. A PEGylated PVDF Antifouling Membrane Prepared by Grafting of Methoxypolyethylene Glycol Acrylate in Gama-Irradiated Homogeneous Solution. *Materials* **2024**, *17*, 873. [CrossRef] [PubMed]
29. Chang, Y.; Shih, Y.-J.; Ruaan, R.-C.; Higuchi, A.; Chen, W.-Y.; Lai, J.-Y. Preparation of Poly(Vinylidene Fluoride) Microfiltration Membrane with Uniform Surface-Copolymerized Poly(Ethylene Glycol) Methacrylate and Improvement of Blood Compatibility. *J. Membr. Sci.* **2008**, *309*, 165–174. [CrossRef]
30. Chang, Y.; Ko, C.-Y.; Shih, Y.-J.; Quémener, D.; Deratani, A.; Wei, T.-C.; Wang, D.-M.; Lai, J.-Y. Surface Grafting Control of PEGylated Poly(Vinylidene Fluoride) Antifouling Membrane via Surface-Initiated Radical Graft Copolymerization. *J. Membr. Sci.* **2009**, *345*, 160–169. [CrossRef]
31. Wang, P.; Tan, K.L.; Kang, E.T.; Neoh, K.G. Plasma-Induced Immobilization of Poly(Ethylene Glycol) onto Poly(Vinylidene Fluoride) Microporous Membrane. *J. Membr. Sci.* **2002**, *195*, 103–114. [CrossRef]
32. Liu, F.; Du, C.-H.; Zhu, B.-K.; Xu, Y.-Y. Surface Immobilization of Polymer Brushes onto Porous Poly(Vinylidene Fluoride) Membrane by Electron Beam to Improve the Hydrophilicity and Fouling Resistance. *Polymer* **2007**, *48*, 2910–2918. [CrossRef]
33. Guo, H.; Li, C.; Zhu, L.; Wang, Y.; Zeng, Z.; Wang, L.; Xue, Q. Preparation of PEGylated Poly(Vinylidene Fluoride) Porous Membranes with Improved Antifouling Property. *Surf. Topogr. Metrol. Prop.* **2019**, *7*, 014002. [CrossRef]

34. Zhang, M.; Nguyen, Q.T.; Ping, Z. Hydrophilic Modification of Poly (Vinylidene Fluoride) Microporous Membrane. *J. Membr. Sci.* **2009**, *327*, 78–86. [CrossRef]
35. Gao, F.; Wang, J.; Zhang, H.; Hang, M.A.; Cui, Z.; Yang, G. Interaction Energy and Competitive Adsorption Evaluation of Different NOM Fractions on Aged Membrane Surfaces. *J. Membr. Sci.* **2017**, *542*, 195–207. [CrossRef]
36. Li, S.; Chen, H.; Zhao, X.; Lucia, L.A.; Liang, C.; Liu, Y. Impact Factors for Flux Decline in Ultrafiltration of Lignocellulosic Hydrolysis Liquor. *Sep. Purif. Technol.* **2020**, *240*, 116597. [CrossRef]
37. Robinson, S.; Abdullah, S.Z.; Bérubé, P.; Le-Clech, P. Ageing of Membranes for Water Treatment: Linking Changes to Performance. *J. Membr. Sci.* **2016**, *503*, 177–187. [CrossRef]
38. Rabuni, M.F.; Nik Sulaiman, N.M.; Aroua, M.K.; Yern Chee, C.; Awanis Hashim, N. Impact of in Situ Physical and Chemical Cleaning on PVDF Membrane Properties and Performances. *Chem. Eng. Sci.* **2015**, *122*, 426–435. [CrossRef]
39. Kumar, A.P.; Depan, D.; Singh Tomer, N.; Singh, R.P. Nanoscale Particles for Polymer Degradation and Stabilization—Trends and Future Perspectives. *Prog. Polym. Sci.* **2009**, *34*, 479–515. [CrossRef]
40. Liu, W.-W.; Chai, S.-P.; Mohamed, A.R.; Hashim, U. Synthesis and Characterization of Graphene and Carbon Nanotubes: A Review on the Past and Recent Developments. *J. Ind. Eng. Chem.* **2014**, *20*, 1171–1185. [CrossRef]
41. Yu, T.; Meng, L.; Zhao, Q.-B.; Shi, Y.; Hu, H.-Y.; Lu, Y. Effects of Chemical Cleaning on RO Membrane Inorganic, Organic and Microbial Foulant Removal in a Full-Scale Plant for Municipal Wastewater Reclamation. *Water Res.* **2017**, *113*, 1–10. [CrossRef]
42. Hajibabania, S.; Antony, A.; Leslie, G.; Le-Clech, P. Relative Impact of Fouling and Cleaning on PVDF Membrane Hydraulic Performances. *Sep. Purif. Technol.* **2012**, *90*, 204–212. [CrossRef]
43. Gao, F.; Wang, J.; Zhang, H.; Zhang, Y.; Hang, M.A. Effects of Sodium Hypochlorite on Structural/Surface Characteristics, Filtration Performance and Fouling Behaviors of PVDF Membranes. *J. Membr. Sci.* **2016**, *519*, 22–31. [CrossRef]
44. Abdullah, S.Z.; Bérubé, P.R. Assessing the Effects of Sodium Hypochlorite Exposure on the Characteristics of PVDF Based Membranes. *Water Res.* **2013**, *47*, 5392–5399. [CrossRef] [PubMed]
45. Awanis Hashim, N.; Liu, Y.; Li, K. Stability of PVDF Hollow Fibre Membranes in Sodium Hydroxide Aqueous Solution. *Chem. Eng. Sci.* **2011**, *66*, 1565–1575. [CrossRef]
46. Rouaix, S.; Causserand, C.; Aimar, P. Experimental Study of the Effects of Hypochlorite on Polysulfone Membrane Properties. *J. Membr. Sci.* **2006**, *277*, 137–147. [CrossRef]
47. Causserand, C.; Rouaix, S.; Lafaille, J.-P.; Aimar, P. Ageing of Polysulfone Membranes in Contact with Bleach Solution: Role of Radical Oxidation and of Some Dissolved Metal Ions. *Chem. Eng. Process. Process Intensif.* **2008**, *47*, 48–56. [CrossRef]
48. Kwon, Y.; Leckie, J. Hypochlorite Degradation of Crosslinked Polyamide membranesII. Changes in Hydrogen Bonding Behavior and Performance. *J. Membr. Sci.* **2006**, *282*, 456–464. [CrossRef]
49. Gaudichetmaurin, E.; Thominette, F. Ageing of Polysulfone Ultrafiltration Membranes in Contact with Bleach Solutions. *J. Membr. Sci.* **2006**, *282*, 198–204. [CrossRef]
50. Wolff, S.H.; Zydney, A.L. Effect of Bleach on the Transport Characteristics of Polysulfone Hemodialyzers. *J. Membr. Sci.* **2004**, *243*, 389–399. [CrossRef]
51. Mohammadi, T.; Madaeni, S.S.; Moghadam, M.K. Investigation of Membrane Fouling. *Desalination* **2003**, *153*, 155–160. [CrossRef]
52. Trägårdh, G. Membrane Cleaning. *Desalination* **1989**, *71*, 325–335. [CrossRef]
53. Cota, D.; Patil, D. Antibacterial Potential of Ellagic Acid and Gallic Acid against IBD Bacterial Isolates and Cytotoxicity against Colorectal Cancer. *Nat. Prod. Res.* **2023**, *37*, 1998–2002. [CrossRef] [PubMed]
54. Xu, Y.; Tang, G.; Zhang, C.; Wang, N.; Feng, Y. Gallic Acid and Diabetes Mellitus: Its Association with Oxidative Stress. *Molecules* **2021**, *26*, 7115. [CrossRef] [PubMed]
55. Bai, J.; Zhang, Y.; Tang, C.; Hou, Y.; Ai, X.; Chen, X.; Zhang, Y.; Wang, X.; Meng, X. Gallic Acid: Pharmacological Activities and Molecular Mechanisms Involved in Inflammation-Related Diseases. *Biomed. Pharmacother.* **2021**, *133*, 110985. [CrossRef] [PubMed]
56. Yilmaz, Y.; Toledo, R.T. Major Flavonoids in Grape Seeds and Skins: Antioxidant Capacity of Catechin, Epicatechin, and Gallic Acid. *J. Agric. Food Chem.* **2004**, *52*, 255–260. [CrossRef]
57. Kumaran, A.; Joel Karunakaran, R. Antioxidant and Free Radical Scavenging Activity of an Aqueous Extract of Coleus Aromaticus. *Food Chem.* **2006**, *97*, 109–114. [CrossRef]
58. Ferrari, C.K.B.; Torres, E.A.F.S. Biochemical Pharmacology of Functional Foods and Prevention of Chronic Diseases of Aging. *Biomed. Pharmacother.* **2003**, *57*, 251–260. [CrossRef] [PubMed]
59. Locatelli, C.; Rosso, R.; Santos-Silva, M.C.; De Souza, C.A.; Licínio, M.A.; Leal, P.; Bazzo, M.L.; Yunes, R.A.; Creczynski-Pasa, T.B. Ester Derivatives of Gallic Acid with Potential Toxicity toward L1210 Leukemia Cells. *Bioorg. Med. Chem.* **2008**, *16*, 3791–3799. [CrossRef] [PubMed]
60. Yang, X.; Sun, H.; Pal, A.; Bai, Y.; Shao, L. Biomimetic Silicification on Membrane Surface for Highly Efficient Treatments of Both Oil-in-Water Emulsion and Protein Wastewater. *ACS Appl. Mater. Interfaces* **2018**, *10*, 29982–29991. [CrossRef]
61. Cheng, X.Q.; Wang, Z.X.; Zhang, Y.; Zhang, Y.; Ma, J.; Shao, L. Bio-Inspired Loose Nanofiltration Membranes with Optimized Separation Performance for Antibiotics Removals. *J. Membr. Sci.* **2018**, *554*, 385–394. [CrossRef]
62. Zhao, S.; Wang, Z. A Loose Nano-Filtration Membrane Prepared by Coating HPAN UF Membrane with Modified PEI for Dye Reuse and Desalination. *J. Membr. Sci.* **2017**, *524*, 214–224. [CrossRef]

63. Rajesh, S.; Murthy, Z.V.P. Synthesis, Characterization and Application of Antioxidants Nanoparticles Incorporated Polymeric Membranes. *Sep. Sci. Technol.* **2019**, *54*, 247–257. [CrossRef]
64. Teotia, R.; Verma, S.K.; Kalita, D.; Singh, A.K.; Dahe, G.; Bellare, J. Porosity and Compatibility of Novel Polysulfone-/Vitamin E-TPGS-Grafted Composite Membrane. *J. Mater. Sci.* **2017**, *52*, 12513–12523. [CrossRef]
65. Buchmüller, Y.; Wokaun, A.; Gubler, L. Polymer-Bound Antioxidants in Grafted Membranes for Fuel Cells. *J. Mater. Chem. A* **2014**, *2*, 5870–5882. [CrossRef]
66. Noble, J.E.; Bailey, M.J.A. Chapter 8 Quantitation of Protein. In *Methods in Enzymology*; Elsevier: Amsterdam, The Netherlands, 2009; Volume 463, pp. 73–95. ISBN 978-0-12-374536-1.
67. Uğuzdoğan, E.; Denkbaş, E.B.; Öztürk, E.; Tuncel, S.A.; Kabasakal, O.S. Preparation and Characterization of Polyethyleneglycolmethacrylate (PEGMA)-Co-Vinylimidazole (VI) Microspheres to Use in Heavy Metal Removal. *J. Hazard. Mater.* **2009**, *162*, 1073–1080. [CrossRef] [PubMed]
68. Gálico, D.A.; Nova, C.V.; Guerra, R.B.; Bannach, G. Thermal and Spectroscopic Studies of the Antioxidant Food Additive Propyl Gallate. *Food Chem.* **2015**, *182*, 89–94. [CrossRef] [PubMed]
69. Smolders, C.A.; Reuvers, A.J.; Boom, R.M.; Wienk, I.M. Microstructures in Phase-Inversion Membranes. Part 1. Formation of Macrovoids. *J. Membr. Sci.* **1992**, *73*, 259–275. [CrossRef]
70. Lee, J.; Park, B.; Kim, J.; Park, S.B. Effect of PVP, Lithium Chloride, and Glycerol Additives on PVDF Dual-Layer Hollow Fiber Membranes Fabricated Using Simultaneous Spinning of TIPS and NIPS. *Macromol. Res.* **2015**, *23*, 291–299. [CrossRef]
71. Jung, J.T.; Kim, J.F.; Wang, H.H.; Di Nicolo, E.; Drioli, E.; Lee, Y.M. Understanding the Non-Solvent Induced Phase Separation (NIPS) Effect during the Fabrication of Microporous PVDF Membranes via Thermally Induced Phase Separation (TIPS). *J. Membr. Sci.* **2016**, *514*, 250–263. [CrossRef]
72. Yu, L.; Yang, F.; Xiang, M. Phase Separation in a PSf/DMF/Water System: A Proposed Mechanism for Macrovoid Formation. *RSC Adv.* **2014**, *4*, 42391–42402. [CrossRef]
73. Ying, L.; Kang, E.T.; Neoh, K.G. Covalent Immobilization of Glucose Oxidase on Microporous Membranes Prepared from Poly(Vinylidene Fluoride) with Grafted Poly(Acrylic Acid) Side Chains. *J. Membr. Sci.* **2002**, *208*, 361–374. [CrossRef]

materials

MDPI

Article

Preparation of Fe-HMOR with a Preferential Iron Location in the 12-MR Channels for Dimethyl Ether Carbonylation

Wenrong Liu, Yaquan Wang *, Lingzhen Bu, Kailiang Chu, Yitong Huang, Niandong Guo, Liping Qu, Juncai Sang, Xuemei Su, Xian Zhang and Yaoning Li

Key Laboratory for Green Chemical Technology of Ministry of Education, School of Chemical Engineering and Technology, Tianjin University, Tianjin 300072, China; liuwr19@tju.edu.cn (W.L.)
* Correspondence: yqwang@tju.edu.cn

Abstract: As the Brønsted acid sites in the 8-membered ring (8-MR) of mordenite (MOR) are reported to be the active center for dimethyl ether (DME) carbonylation reaction, it is of great importance to selectively increase the Brønsted acid amount in the 8-MR. Herein, a series of Fe-HMOR was prepared through one-pot hydrothermal synthesis by adding the EDTA–Fe complex into the gel. By combining XRD, FTIR, UV–Vis, Raman and XPS, it was found that the Fe atoms selectively substituted for the Al atoms in the 12-MR channels because of the large size of the EDTA–Fe complex. The NH_3-TPD and Py-IR results showed that with the increase in Fe addition from Fe/Si = 0 to 0.02, the Brønsted acid sites derived from Si-OH-Al in the 8-MR first increased and then decreased, with the maximum at Fe/Si = 0.01. The Fe-modified MOR with Fe/Si = 0.01 showed the highest activity in DME carbonylation, which was three times that of HMOR. The TG/DTG results indicated that the carbon deposition and heavy coke formation in the spent Fe-HMOR catalysts were inhibited due to Fe addition. This work provides a practical way to design a catalyst with enhanced catalytic performance.

Keywords: mordenite; Fe modification; acid distribution; DME carbonylation

Citation: Liu, W.; Wang, Y.; Bu, L.; Chu, K.; Huang, Y.; Guo, N.; Qu, L.; Sang, J.; Su, X.; Zhang, X.; et al. Preparation of Fe-HMOR with a Preferential Iron Location in the 12-MR Channels for Dimethyl Ether Carbonylation. *Materials* **2024**, *17*, 2417. https://doi.org/10.3390/ma17102417

Academic Editors: Yi-Nan Wu and Guangxu Lan

Received: 27 March 2024
Revised: 9 May 2024
Accepted: 10 May 2024
Published: 17 May 2024

1. Introduction

Ethanol can be used as fuel or fuel additive and has broad application prospects in both the pharmaceutical industry and chemical industry [1,2]. In recent years, ethanol synthesis from syngas ($CO + H_2$) has attracted great interest. Among the different technologies, the route involving dimethyl ether (DME) carbonylation to methyl acetate (MA) followed by hydrogenation is considered the most promising. Syngas can be easily obtained via coal gasification and then be converted to DME. DME can react through a carbonylation reaction to form MA, and finally, MA can be hydrogenated to form ethanol (Equations (1) and (2)) [3]. The key step in syngas-to-ethanol conversion is DME carbonylation catalyzed by solid acids. Among the candidates, H-form mordenite (MOR) was found to possess high MA selectivity and good DME conversion in a carbonylation reaction [4]. The specific catalytic performance lies in the unique pore structure of MOR.

$$CH_3OCH_3 + CO \rightarrow CH_3COOCH_3 \quad (1)$$

$$CH_3COOCH_3 + 2H_2 \rightarrow CH_3OH + C_2H_5OH \quad (2)$$

A typical MOR structure seen along the c-axis is shown in Scheme 1. MOR exhibits 12-membered ring (MR) channels (0.65 × 0.70 nm) along the c-axis, which are interconnected with 8-MR channels (0.34 × 0.48 nm) along the b-axis [5]. The latter are usually referred to as side pockets. There are also 8-MR channels (0.26 × 0.57 nm) parallel to the 12-MR channels, which are too small for most molecules. In the DME carbonylation reaction, DME molecules first adsorb on the Brønsted acid sites (Z-OH) to form methoxy groups

($Z\text{-}OCH_3$) and release methanol. Methanol can also adsorb on Z-OH to form $Z\text{-}OCH_3$. Then, CO molecules insert into methoxy groups to form acetyl species ($Z\text{-}OCOCH_3$) [6]. Finally, the acetyl species react with DME, leading to the reformation of methoxy groups and the formation of MA [7].

$$CH_3OCH_3 + Z\text{-}OH \rightarrow Z\text{-}OCH_3 + CH_3OH \quad (3)$$

$$CH_3OH + Z\text{-}OH \rightarrow Z\text{-}OCH_3 + H_2O \quad (4)$$

$$CO + Z\text{-}OCH_3 \rightarrow Z\text{-}OCOCH_3 \quad (5)$$

$$CH_3OCH_3 + Z\text{-}OCOCH_3 \rightarrow Z\text{-}OCH_3 + CH_3COOCH_3 \quad (6)$$

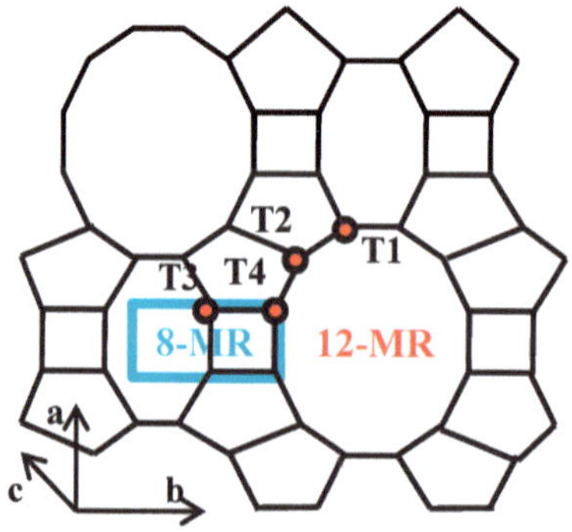

Scheme 1. A typical MOR structure seen along the c-axis.

The MA formation rate was found to increase linearly with CO pressure, but independently of DME pressure, indicating that CO insertion was the rate-determining step [3]. The formation of methoxy groups and CO insertion (Equations (3)–(5)) were found to be more energetically feasible in the 8-MR [8]. Rasmussen et al. claimed that the reaction intermediate of DME carbonylation was ketene, which was formed by CO reacting with methoxy groups in the 8-MR (Equations (7) and (8)) [9]. In the 8-MR, ketene underwent rapid protonation to form acylium ion, followed by MA formation [10]. However, in the 12-MR, the dimerization of ketene led to the formation of carbon deposit [11]. Therefore, the Brønsted acid sites in the 8-MR side pockets were the active centers of DME carbonylation to MA, while those in the 12-MR were favorable for side reactions [12,13]. To improve the catalytic properties of MOR catalysts, it is critical to adjust the acid distribution of Brønsted acid sites, especially to selectively increase the Brønsted acid amount in the 8-MR.

$$CO + Z\text{-}OCH_3 \rightarrow Z\text{-}O^{-}\text{-}CH_3CO^{+} \quad (7)$$

$$Z\text{-}O^{-}\text{-}CH_3CO^{+} \rightarrow Z\text{-}O\text{-}COCH_2 \quad (8)$$

Many efforts have been made to tune the acid distribution of MOR. Post-treatment, such as acid/alkaline leaching, and steam treatment are extensively used to remove or introduce framework Al, but they will lead to a decrease in crystallinity and the blockage of pore channels [14–16]. Trivalent and tetravalent metal ions can substitute for framework atoms during one-pot synthesis, which can adjust the distribution of Brønsted acid sites through the incorporation of heteroatom. Ce-incorporated MOR was reported to exhibit higher activity thanks to increased Brønsted acid sites in the 8-MR [17]. Zhou [18] reported that the DME carbonylation performance of Fe-incorporated MOR was improved because the acid amount in the 12-MR was reduced. However, the location of Fe ions and the detailed influence of Fe species on acid distribution had not been studied. Zhang [19] synthesized template-free Fe-MOR with oxalic acid complexed iron. The Fe atoms were mainly located at the 8-MR; thus, the Brønsted acid sites in the 8-MR decreased, which led to better stability. However, the activity was decreased compared to HMOR because the acid amount and strength in the 8-MR of the reported Fe-MOR catalysts were not improved.

In this work, Fe atoms were selectively incorporated into the 12-MR, which led to an increase in the Brønsted acid sites derived from Si-OH-Al in the 8-MR. A series of Fe-modified mordenite zeolites were synthesized through the one-pot hydrothermal synthesis method by adding ferric complex into the gel. The complexing agent was ethylenediaminetetraacetic acid (EDTA). The effects of Fe addition on the structure and acid properties of the MOR catalysts were investigated in detail. The location of the framework Fe atoms was found to be mainly in the 12-MR channels, which enhanced the amount and strength of the Brønsted acid sites in the 8-MR. The catalytic performances for the DME carbonylation reaction were studied and correlated with the properties of the zeolites.

2. Materials and Methods

2.1. Synthesis of Fe-Modified Mordenite

The details of the chemicals are provided in the Supplementary Materials. The synthesis procedure of Fe-modified mordenite is shown in Scheme 2. In a typical synthesis procedure, sodium hydroxide and sodium meta-aluminate were dissolved in deionized water under magnetic stirring. When a clear solution was formed, tetraethylammonium hydroxide was added to the flask. Then, colloidal silica was added dropwise under vigorous stirring, followed by 2 h of aging. $Fe(NO_3)_3{\cdot}9H_2O$ and ethylenediaminetetraacetic acid (EDTA) were dissolved in NaOH aqueous solution. The clear yellow solution formed was added to the gel and stirred for another 2 h. Then, the mixture was transferred to a Teflon-lined stainless-steel autoclave and crystallized at 170 °C for 120 h. The molar composition of the mixture was $0.23Na_2O$: $1SiO_2$: $0.033Al_2O_3$: 0.10TEAOH: $40H_2O$: xFe: xEDTA, (x = 0.005, 0.01, 0.015, 0.02). After crystallization, the solid product was separated by filtration and washed with deionized water until the pH of the supernatant liquid was less than 8. The resulting solid was dried at 120 °C overnight and calcined at 500 °C for 6 h. The obtained Na-form mordenite was then ion-exchanged three times with 1 mol/L NH_4NO_3 aqueous solution at 80 °C for 2 h. After ion exchange, the product was dried and calcined. The prepared Fe-HMOR zeolites are referred to as M-Fe-x.

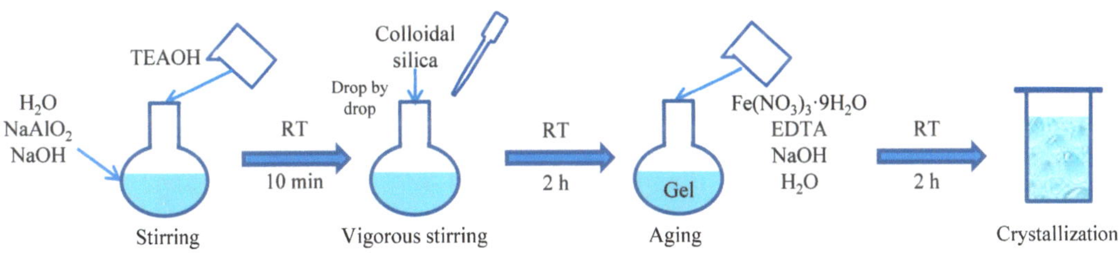

Scheme 2. Synthesis procedure for Fe-modified mordenite.

An HMOR zeolite without Fe modification was prepared. The synthesis process for this sample was the same as that for M-Fe-x, but no $Fe(NO_3)_3{\cdot}9H_2O$ or EDTA were added. The final composition of the gel was 0.23 Na_2O:1 SiO_2:0.033 Al_2O_3:0.10 TEAOH:40 H_2O. The ion-exchange and calcination procedures were also the same as described above. This sample is referred to as HM.

For comparison, a Fe-containing sample was prepared without the complexing agent. The synthesis procedure was similar to M-Fe-x, but ferric nitrate aqueous solution was added to the gel instead of ferric complex. The final gel contained 0.23 Na_2O:1 SiO_2:0.033

Al_2O_3:0.10 TEAOH:40 H_2O:0.01 $Fe(NO_3)_3$. The ion-exchange and calcination procedures were the same as described above. The obtained sample is referred to as 0.01Fe/M.

2.2. Characterization

Powder X-ray diffraction (XRD) patterns were collected using a Bruker D8 ADVANCE X-ray diffractometer (Bruker AXS GMBH, Karlsruhe, Germany) with Cu Kα radiation (λ = 0.15418 nm) operating at 40 kV and 40 mA. The scanning speed was 8 °/min in the range of 5–50° with a step size of 0.01°. The relative crystallinity was calculated using the sum of the heights of the five characteristic peaks (2θ = 9.9°, 19.8°, 22.5°, 25.9°, 26.5°) relative to that of the reference sample.

Elemental analysis of the samples was performed on Bruker S4 Pioneer X-ray fluorescence (XRF).

The solid state nuclear magnetic resonance (NMR) experiments were performed on a JEOL JNM ECZ600R spectrometer (JEOL, Tokyo, Japan). Chemical shifts were referenced versus excess aluminum nitrate aqueous solution.

Field-emission scanning electron microscopy (SEM) images were recorded using a Hitachi S-4800 instrument (Hitachi High-Technologies Corporation, Ibaraki, Japan) operated at 3 kV. The samples were coated with Pt.

The textural properties of the samples were measured in nitrogen adsorption–desorption experiments on a Micromeritics TriStar 3000 automated physisorption instrument (Norcross, GA, USA) at liquid nitrogen temperature (77 K). Before the nitrogen adsorption–desorption experiments, the samples were heated to 350 °C for 5 h under vacuum to remove adsorbed species. The specific surface area (S_{BET}) was calculated based on the Brunauer–Emmett–Teller equation. The external surface area (S_{ext}) and micropore volume (V_{mic}) were determined using the t-plot method. The total pore volume (V_{total}) was determined from the quantity adsorbed at relative pressure of $P/P_0 = 0.97$.

FTIR spectra were recorded on a Nicolet iS 50 spectrometer (Thermo Fisher Scientific Inc., Waltham, MA, USA). For pyridine-adsorbed FTIR (Py-IR) measurements, 18 mg of H-form mordenite samples was pressed into self-supported disk wafers, and background spectra were recorded before pyridine adsorption. The wafers were saturated with pyridine at 300 °C under vacuum for 4 h, followed by evacuation to remove physically adsorbed pyridine. The Py-IR spectra were taken at 4 cm^{-1} resolution from 2000 to 1000 cm^{-1} when the self-supported wafers were cooled down to room temperature. For NH_3-adsorbed FTIR (NH_3-IR), 18 mg of H-form mordenite samples was pressed into self-supported disk wafers, and background spectra were recorded before NH_3 adsorption. The samples were subjected to pretreatment at 400 °C for 30 min and cooled down to 100 °C to adsorb NH_3 for 10 min. Afterward, the samples were flushed with flowing He for 2 h to remove physically adsorbed NH_3. Then, NH_3-IR spectra were taken at 4 cm^{-1} resolution from 2000 to 1000 cm^{-1}.

UV–Vis spectra were collected using a PE Lambda 750 spectrophotometer (Perkin Elmer, Shanghai, China) in the range of 190–600 nm with $BaSO_4$ as reference.

Raman spectra were taken on a Horiba Confocal Raman Microscopy (Kyoto, Japan) using a 532 nm laser.

X-ray photoelectron spectroscopy (XPS) was performed using a Thermo Fisher Scientific ESCALAB 250Xi instrument (Thermo Fisher Scientific, MA, USA). The X-ray source was Al Kα. The binding energy was calibrated relative to C 1 s at 284.8 eV.

NH_3-TPD experiments were performed on a TP-5076 chemical adsorption instrument (Xianquan Industrial and Trading Co., Ltd., Tianjin, China). In a typical test, a 100 mg H-MOR sample was loaded in a quartz tube and purged with flowing He at 400 °C for 1 h. After cooling down to 100 °C, NH_3 was introduced for 10 min, followed by purging with He for 2 h to remove physically adsorbed NH_3. The TPD profiles were recorded from 100 to 700 °C at a heating rate of 10 °C/min.

TG analysis was conducted on a Shimadzu TGA-50 instrument (Shimadzu, Kyoto, Japan). The spent catalysts after 24 h of reaction were used to analyze the carbon deposits. The samples were combusted from room temperature to 750 °C at a heating rate of 10 °C/min in oxygen.

2.3. Catalyst Tests

DME carbonylation reactions were performed with a tubular fixed-bed reactor under 2.0 MPa. An amount of 0.50 g of the catalyst (40–60 mesh) was loaded in the central part of the reactor. The catalyst was heated to 500 °C for 2 h under a nitrogen atmosphere (60 mL/min) to remove the moisture. After the temperature dropped down to 200 °C, the gas feed was then changed to the reactant mixture (3%DME/30%CO/67%Ar) with a flow rate of 30 mL/min. The outlet gas was analyzed in an Agilent 4890 gas chromatograph (Santa Clara, CA, USA) equipped with a flame ionization detector. DME conversion and product selectivity were calculated as reported [20].

3. Results and Discussion

3.1. Fe-Modified Mordenite Zeolites

The XRD patterns of all samples are shown in Figure 1a. The typical diffraction peaks of MOR zeolite can be observed, indicating that well-crystallized mordenite zeolites are successfully synthesized either with or without Fe modification. There are no characteristic peaks of iron oxide species in the XRD patterns, probably due to the good dispersion of the iron species. All the peaks for Fe-modified MOR zeolites shift to lower angles compared with HM. The peaks positioned at around 22.5° of the Fe-HMOR samples (Figure 1b)—one of the most prominent peaks for the MOR structure—also shift to lower angles, which is ascribed to the increases in d values and unit cell parameters [21]. This is evidence that Fe^{3+} has been incorporated into the mordenite framework because Fe^{3+} has a larger ionic radius than Al^{3+} [18]. To confirm the accuracy of the XRD measurements, the average crystal sizes of these samples are calculated based on the Scherrer equation and compared with the average particle sizes obtained from SEM images (Figure S2). As shown in Table S1, the crystal sizes from the Scherrer equation are similar to those obtained from SEM images. The relative crystallinity is calculated according to the sum of the heights of the five characteristic peaks at 2θ = 9.9°, 19.8°, 22.5°, 25.9° and 26.5° relative to that of M-Fe-0. As shown in Table 1, an increase in relative crystallinity is observed with the increasing Fe/Si ratio, suggesting that the substitution of Al by Fe does not affect the zeolite framework.

The iron contents, Si/Al and Si/(Al+Fe) ratios of the products are determined by XRF and listed in Table 1. As the Fe/Si ratio in the raw materials increases from 0.005 to 0.02, the iron content in the Fe-HMOR product increases from 1.03 wt% to 3.55 wt%. The iron content in 0.01Fe/M is 2.38 wt%, which is slightly higher than that in M-Fe-0.01 (2.06 wt%). The Si/Al ratios of M-Fe-x increase from Fe/Si = 0 to 0.02, but the Si/(Al+Fe) ratios of these samples are similar. This indicates that the substitution of Fe for Al increases with the Fe addition increasing. The Si/Al ratio and Si/(Al+Fe) ratio of 0.01Fe/M are slightly lower than those of M-Fe-0.01. ^{27}Al NMR spectra were employed to determine the fraction of different Al species and are shown in Figure 2. The peaks at 58 ppm and 0 ppm correspond to the framework Al (Al_F) and extraframework Al (Al_{EF}), respectively [22]. The Al_F and Al_{EF} fractions are also listed in Table 1. HM exhibits a high Al_{EF} fraction of 21.4%. When Fe/Si = 0.005 and 0.01, the Al_{EF} fractions of the Fe-HMOR samples decrease slightly. When the Fe/Si ratio increases to 0.015 and 0.02, the Al_{EF} fractions of the samples apparently decrease. The lower fraction of Al_{EF} might result from the competitive incorporation of Fe and Al. With the increase in Fe/Si, the large size of the complexed iron (III) inhibits the access of excessive Al species to the aluminosilicate network. Additionally, Al is preferred to be incorporated rather than Fe [21]; thus, most of the Al species have been successfully incorporated into the zeolite framework, which results in a decrease in Al_{EF}. The Al_{EF} fraction of 0.01Fe/M is also lower than that of M-Fe-0.

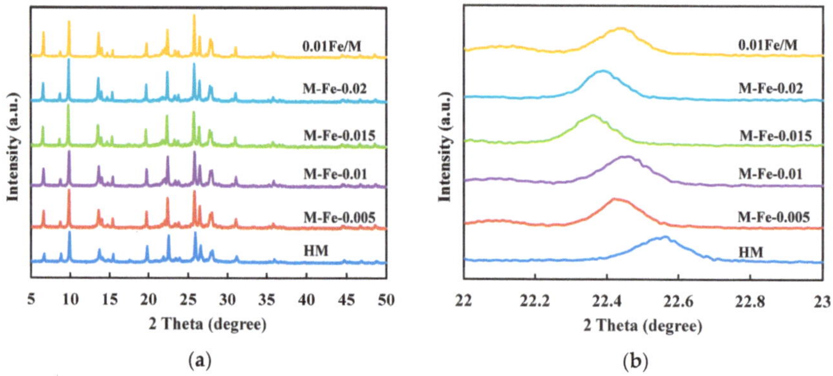

Figure 1. (**a**) XRD patterns of the samples; (**b**) XRD patterns in the 2θ region of 22–23°.

Table 1. Relative crystallinity and chemical compositions of HM, M-Fe-x and 0.01Fe/M.

Sample	Relative Crystallinity (%)	Iron Content [a] (wt%)	Si/Al [a]	Si/(Al+Fe) [a]	^{27}Al NMR (%) [b]	
					Al_F	Al_{EF}
HM	100	-	8.6	-	78.6	21.4
M-Fe-0.005	129	1.03	10.1	8.9	80.9	19.1
M-Fe-0.01	125	2.06	11.3	8.8	80.5	19.5
M-Fe-0.015	131	2.69	11.8	8.6	89.3	10.7
M-Fe-0.02	134	3.55	12.4	8.1	87.1	12.9
0.01Fe/M	142	2.38	10.3	7.9	81.8	18.2

[a] Determined by XRF. [b] The fractions of framework Al (Al_F) and extraframework Al (Al_{EF}) were calculated using the peak area of ^{27}Al NMR spectra.

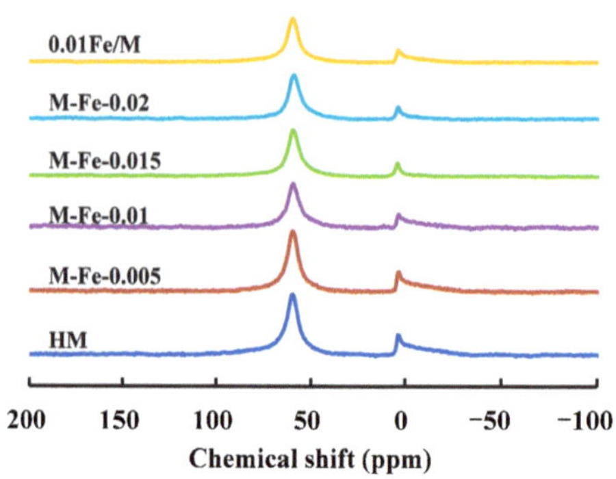

Figure 2. ^{27}Al NMR spectra of the samples.

As shown in the SEM images (Figure S2), all samples are micro-sized bulk irregular crystals with a diameter of ca. 6 μm, which are assemblies of nanoscale particles. Fe addition does not affect the morphology of the zeolites apparently. The crystal of 0.01Fe/M is larger than that of M-Fe-x, which exhibits a diameter of ca. 8 μm. Additionally, for 0.01Fe/M, the edges of the primary particles and the voids between the nanoparticles become less clear (Figure S3).

The N_2 adsorption–desorption isotherms of all samples are given in Figure S4. HM and M-Fe-x samples exhibit type I + IV isotherms. The hysteresis loop is in the region of $0.4 < P/P_0 < 1$, which indicates the existence of intercrystalline mesopores. As mentioned above, HM and M-Fe-x samples are bulk crystallites consisting of nano-structures; thus, a small number of mesopores are formed by the aggregation of the primary particles [23].

However, 0.01Fe/M exhibits type I isotherm, typical of microporous materials. The lack of a mesopore structure in 0.01Fe/M might be attributed to the dense aggregation of the nanoparticles. The pore size distribution of the samples is shown in Figure S5. The textural properties of the samples are summarized in Table 2. At relatively low Fe/Si ratios (0.005 and 0.01), the micropore surface area and micropore volume of the samples increase compared to M-Fe-0. A similar phenomenon has also been reported by Zhang [19]. The Fe–O bond is longer than the Al–O bond, which might result in an increase in S_{micro}. With the Fe/Si ratio further increasing to 0.015, the S_{micro} and V_{micro} slightly decrease. When the Fe/Si ratio increases to 0.02, the S_{micro} and V_{micro} decrease sharply; the external surface area and mesopore volume also decrease. As the Fe/Si ratio increases to 0.015 and 0.02, the extraframework iron species might also increase, which results in the blockage of micropores and even mesopores. The S_{micro} and V_{micro} of 0.01Fe/M are similar to those of M-Fe-0, but the S_{ext} and V_{meso} are lower.

Table 2. Textural properties of synthesized MOR zeolites.

Sample	Surface Area (m^2/g)			Pore Volume (cm^3/g)		
	S_{BET}	S_{micro}	S_{ext}	V_{total}	V_{micro}	V_{meso}
HM	449	412	37	0.194	0.163	0.031
M-Fe-0.005	489	449	40	0.214	0.179	0.035
M-Fe-0.01	488	450	38	0.211	0.179	0.032
M-Fe-0.015	459	424	35	0.207	0.169	0.038
M-Fe-0.02	364	340	24	0.159	0.136	0.023
0.01Fe/M	445	426	19	0.180	0.169	0.011

3.2. Framework Incorporation of Fe Atoms

The FTIR spectra in the region of 500–1500 cm^{-1} are shown in Figure S6. The bands located at approximately 570 cm^{-1} are attributed to T-O (T = Si, Al, Fe) vibrations, which is typical of zeolites containing 5-membered structures, including mordenite [24]. This band is sensitive to the crystallinity of the zeolite [25]. The band intensities of Fe-modified samples are comparable to HM, indicating that all samples are well-crystallized mordenite. The bands of symmetric and asymmetric stretching of T-O are located at 800 cm^{-1} and 1089 cm^{-1}, respectively [26]. The bands at around 1089 cm^{-1} for Fe-modified samples gradually become blunt and shift to lower wavenumbers compared to HM, which is due to the longer bond length of Fe–O than Al–O. The results of FTIR are consistent with XRD patterns, further confirming that the Fe atoms have been incorporated into the MOR framework.

The chemical states of the iron species were detected by XPS analysis. The O 1s XPS spectra are shown in Figure S7. Only one strong peak centered at 532.4 eV is observed, which is accredited to the oxygen in T–O–T bonds. The peak of oxygen in the iron oxides at 530 eV is not observed [17], which reveals that there are no bulk iron oxide particles on the zeolite surfaces. As shown in Figure 3, the Fe 2p spectra of the Fe-HMOR samples exhibit Fe $2p_{3/2}$ and Fe $2p_{1/2}$ signals at ca. 711.9 and 725.0 eV, respectively. There are also shake-up satellite peaks at approximately 717.7 and 731.2 eV. The binding energy for Fe $2p_{3/2}$ and Fe $2p_{1/2}$ of pure Fe_2O_3 is 710.8 and 724.2 eV, respectively [27]. The signals of Fe^{2+} $2p_{3/2}$ and $2p_{1/2}$ are located at lower values (710 and 723 eV) [28]. Therefore, the Fe atoms in the zeolites are Fe (III) species. The peaks of the Fe-modified samples shift to higher values by 0.8–1.1 eV compared with Fe_2O_3, which should be due to the higher electronegativity of Si compared to Fe. This further confirms that the Fe atoms have been incorporated into the zeolite framework. For M-Fe-0.015 and M-Fe-0.02, the Fe $2p_{3/2}$ and Fe $2p_{1/2}$ signals shift to lower values of 711.4 and 724.7 eV, which might be attributed to the increased extraframework iron species.

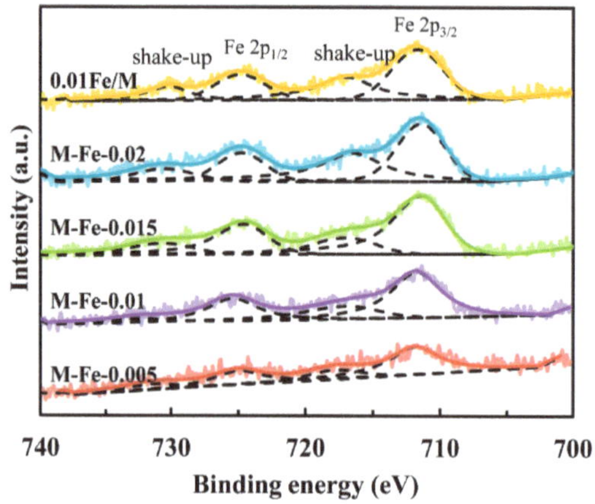

Figure 3. XPS spectra of Fe 2p for Fe-modified samples.

The UV–Vis spectrometer was employed to investigate the framework and extraframework Fe species. The spectra are shown in Figure 4. The bands at 200 nm are observed for all samples, which belong to the Al–O charge transfer transition of four-coordinated framework aluminum [29]. The bands at ca. 216 and 274 nm are ascribed to the isolated Fe^{3+} ions incorporated into the zeolite matrix in tetrahedral coordination [30]. With the increase in the Fe/Si ratio, the band at 200 nm is gradually overlapped by the band at 216 nm, and the band at 274 nm becomes more intense, indicating the increase in incorporated Fe atoms. The weak and broad bands in the range of 300–450 nm are ascribed to Fe^{3+} ions forming oligonuclear clusters and iron oxide nanoparticles, while the adsorption bands above 450 nm are due to large iron oxides [31,32]. The bands ascribed to iron oxides become more intense for the samples with higher Fe/Si ratios, indicating the increase in extraframework iron species. Comparing the spectra of M-Fe-0.01 and 0.01Fe/M, the peak areas of the bands at 300–600 nm of 0.01Fe/M are smaller than those of M-Fe-0.01, suggesting there are less extraframework iron species in 0.01Fe/M.

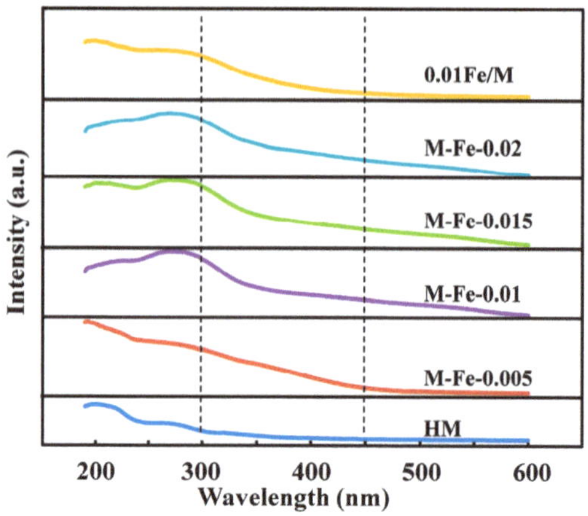

Figure 4. UV–Vis spectra of Fe-modified samples.

Raman spectroscopy is applied to characterize the structure of zeolites [33]. The framework structure of MOR consists of 5- and 4-membered T-O-T rings, corresponding to bands at around 402 cm^{-1} and 465 cm^{-1} in the Raman spectra [34]. As shown in Scheme 1, mordenite has four non-equivalent crystallographic tetrahedral sites: T1, T2 and T4 in the 12-MR main channels and T3 in the side pockets [35]. In other words, Fe atoms have four possible different locations in the MOR framework. The 5-membered rings contain T1 and T2, while the 4-membered rings contain T3 and T4 [36]. To distinguish the location of framework Fe atoms, Raman spectra were taken and are shown in Figure 5. For HM, the bands corresponding to 5- and 4-membered rings are located at 390 and 456 cm^{-1}, respectively. Regarding the Fe-HMOR modified by adding ferric complex, the band corresponding to 5-membered rings shifts from 390 to 378 cm^{-1}, whereas the other band remains at 456 cm^{-1}. This indicates that the Fe atoms preferably occupy the T1 and T2 sites in the 12-MR main channels [17]. These two bands of 0.01Fe/M are located at 392 and 450 cm^{-1}, respectively, suggesting that the Fe atoms in this sample prefer to occupy the T3 and T4 sites [19]. The large size of the EDTA–Fe complex might cause difficulties for the complexed iron (III) to enter the 8-MR channels; thus, the Fe atoms are located in the main channels for M-Fe-x (x = 0.005, 0.01, 0.015, 0.02). However, when the Fe source is $Fe(NO_3)_3$, Fe^{3+} species can enter both the 12-MR and 8-MR channels; therefore, Al atoms in the T3 sites and T4 sites can both be substituted by Fe atoms.

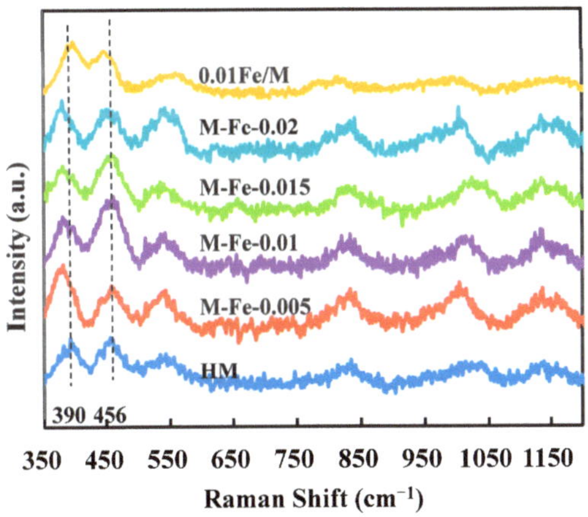

Figure 5. Raman spectra of the samples.

3.3. Acid Properties of the Catalysts

Mordenite is a kind of solid acid catalyst containing Lewis acid sites and Brønsted acid sites. The Lewis acid sites are related to the tri- or penta-coordinated Al sites, while the Brønsted acid sites correspond to the bridging OH sites (Si-OH-Al) [37]. It is of great importance to distinguish between the Brønsted acid sites in the 12-MR channels and the side pockets, as the latter are the active sites in the DME carbonylation reaction. The acid strength of Brønsted acid sites in the 8-MR is higher than those in the 12-MR [38]. NH_3 can reach all acid sites because of its strong alkalinity and small molecular diameter; thus, it has been intensively used as a probe molecule [39]. The NH_3-TPD profiles of all HMOR catalysts are shown in Figure 6. The NH_3 desorption on the H-form mordenite occurs in two temperature regions, corresponding to weak acid sites formed by amorphous Al species and strong Brønsted and/or Lewis acid sites [40]. The TPD profiles can be further deconvoluted into weak, moderate and strong peaks. The moderate peak is related to the Lewis acid sites, which are usually located below 300 °C [20,40,41]. The moderate peak temperature of the Fe-HMOR shifts to a higher region compared to that of HM (Table S2). It was reported that the NH_3 desorption peak of iron oxide was positioned at 368 °C [42]. Therefore, the shift of moderate peak temperature should be due to the higher acid strength of iron oxides compared to extraframework Al species. This phenomenon is not observed for 0.01Fe/M, indicating that most of the iron species exist as framework Fe atoms. The strong peak at the highest temperature corresponds to the framework Brønsted acid sites, which was verified by various researchers [41,43,44]. Therefore, the total amount of Brønsted acid sites in both the 12-MR ($B_{12\text{-}MR}$) and 8-MR side pockets ($B_{8\text{-}MR}$) is determined by the integration of the deconvoluted strong peak area. The amounts of weak, moderate and strong acid sites in the samples are calculated and summarized in Table 3. The total Brønsted acid amount (B_{total}) of Fe-HMOR increased compared to that of HM. As the Fe/Si ratio increases, the B_{total} of the Fe-modified samples also increases. As described above, with the Fe/Si ratio increasing, the Si/Al ratios increase, but the Si/(Al+Fe) ratios slightly decrease. Therefore, the increased B_{total} could result from the formed bridging hydroxyl groups of Si-OH-Fe. The B_{total} of 0.01Fe/M is 0.573 mmol/g, slightly higher than that of M-Fe-0.01 (0.555 mmol/g), which is consistent with the higher framework Fe fraction of 0.01Fe/M. The peak temperature of M-Fe-0 for strong acid sites is 503 °C. When ferric complex is added, the peak temperature increases to around 527 °C. The peak temperature of M-Fe-0.01 is the highest (536 °C), indicating that the strength of total Brønsted acid sites is enhanced through Fe incorporation. The acid strength of Si-OH-Fe is weaker than that of Si-OH-Al because the electron-acceptor property of Fe^{3+} is weaker compared to Al^{3+} [45–47]. According to the Raman spectra, for M-Fe-x (x = 0.005, 0.01, 0.015, 0.02), the Fe atoms mainly enter the 12-MR channels. Thus, it is speculated that the strong acid site strength of these Fe-incorporated samples is enhanced as a result of the increased Si-OH-Al in the 8-MR. On the other hand, the peak temperature of 0.01Fe/M decreased to 490 °C compared with HM, suggesting a decrease in the acidic strength.

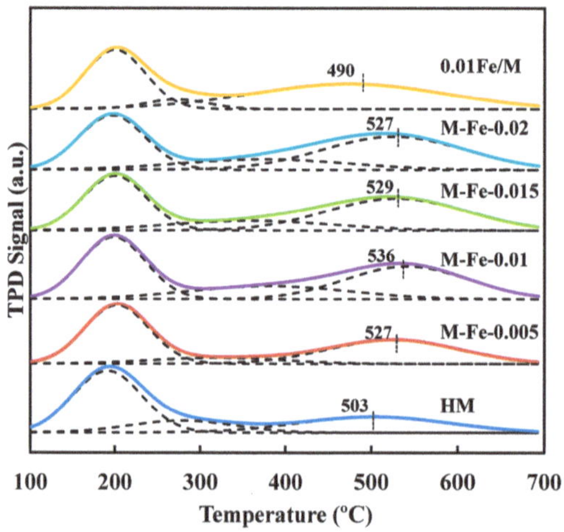

Figure 6. NH_3-TPD profiles of the samples.

Table 3. Acid amounts and distributions of H-MOR samples.

Sample	NH_3-TPD (mmol/g) [a]			Py-IR (mmol/g)		NH_3-IR (mmol/g) [c]	
	Weak	Moderate	Strong	$B_{8\text{-}MR}$	$B_{12\text{-}MR}$ [b]	B	L
HM	0.550	0.164	0.307	0.157	0.150	0.304	0.157
M-Fe-0.005	0.492	0.101	0.407	0.246	0.161	0.399	0.127
M-Fe-0.01	0.519	0.223	0.555	0.395	0.160	0.559	0.214
M-Fe-0.015	0.451	0.207	0.568	0.336	0.232	0.566	0.208
M-Fe-0.02	0.464	0.227	0.586	0.290	0.296	0.604	0.228
0.01Fe/M	0.479	0.155	0.573	0.293	0.280	0.552	0.161

[a] Calculated using the integrated peak areas of the weak, moderate and strong peaks from NH_3-TPD profiles. [b] Based on the peak integration of 1545 cm^{-1} peaks in py-IR spectra. [c] Calculated based on the extinction coefficients of 0.147 and 0.022 $cm^2/\mu mol$ for the bands at 1450 and 1620 cm^{-1}, respectively [37].

NH_3-adsorption FTIR (NH_3-IR) was also employed to quantify the Brønsted (B) and Lewis (L) acid sites. As shown in Figure 7, the bands at 1450 and 1620 cm^{-1} are assigned to NH_3 interacted with B and L acid sites, respectively [17]. The B and L acid amounts calculated based on the NH_3-IR spectra are summarized in Table 3. The values of the B acid amount are similar to those obtained for NH_3-TPD. The L acid amounts show a similar variation trend to the moderate acid amounts.

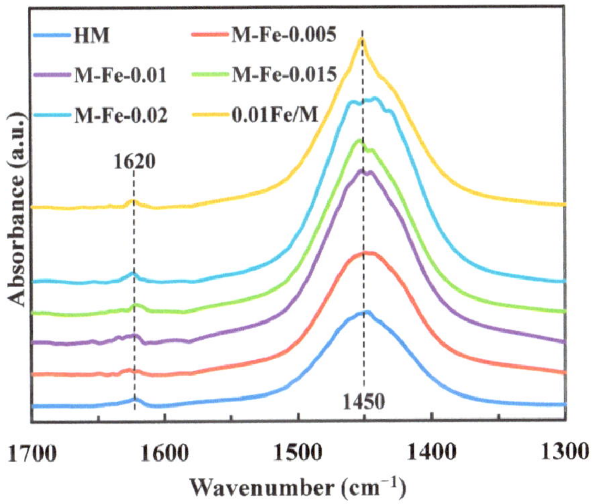

Figure 7. NH_3-IR spectra of the samples.

Pyridine is widely employed as a probe molecule to assess both the Brønsted and Lewis acidity of zeolites. The FTIR spectra of pyridine-adsorbed HMOR (Py-IR) are shown in Figure 8. The adsorption bands at 1545 and 1446 cm^{-1} are ascribed to PyH^+ and Py-L: pyridine adsorbed on protonic acid sites or pyridine-forming complexes with Lewis acid sites, respectively [48]. The band at 1490 cm^{-1} is due to pyridine interacted with both Brønsted and Lewis acid sites. The kinetic diameter of the pyridine molecule is 0.57 nm [49], which makes it difficult to adsorb on the acid sites in the side pockets. Thus, Py-IR spectra are suitable for measuring the acidity in the 12-MR. The quantities of $B_{12\text{-}MR}$ are calculated according to the integrated areas of the peak at 1545 cm^{-1} and the relative molar extinction coefficient reported by Emeis [50]. The $B_{12\text{-}MR}$ of the samples are summarized in Table 3. The $B_{12\text{-}MR}$ increases from 0.150 mmol/g for HM to 0.296 mmol/g for M-Fe-0.02 as the Fe/Si ratio increases. The $B_{12\text{-}MR}$ of 0.01Fe/M is 0.280 mmol/g, which is higher than that of M-Fe-0.01 (0.160 mmol/g). The $B_{8\text{-}MR}$ is calculated as B_{total} minus $B_{12\text{-}MR}$ and is also given in Table 3. With the Fe/Si ratio increasing from 0 to 0.02, the $B_{8\text{-}MR}$ increases first from 0.157 mmol/g (HM) to 0.395 mmol/g (M-Fe-0.01) and then decreases to 0.290 mmol/g (M-Fe-0.02). The $B_{8\text{-}MR}$ of the Fe-HMOR samples is higher than that of HM, which verifies that the improved strong acid strength of Fe-HMOR should be ascribed to the increased Si-OH-Al in the 8-MR. The $B_{8\text{-}MR}$ of 0.01Fe/M (0.293 mmol/g) is apparently lower than that of M-Fe-0.01, while the B_{total} of 0.01Fe/M is slightly higher. As discussed in Section 3.2, for 0.01Fe/M, the Fe atoms substitute the Al atoms both in the 8-MR and 12-MR. Therefore, the Si–OH–Fe bonds of 0.01Fe/M are formed both in the 8-MR and 12-MR channels, which does not affect the Si-OH-Al distribution.

Combining the results of Raman spectra and the acid amounts of the samples, by adding ferric complex into the synthetic gel, the Fe atoms selectively located in the 12-MR channels, thus greatly affecting the acid distribution of Brønsted acid sites and facilitating the formation of Si-OH-Al in the 8-MR.

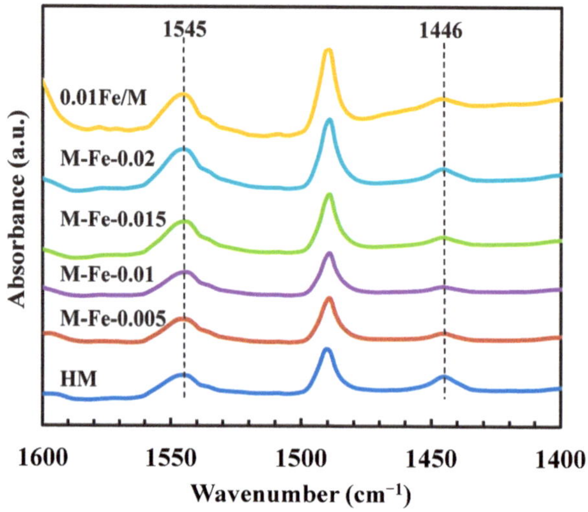

Figure 8. Py-IR spectra of the samples.

3.4. Catalytic Performance in DME Carbonylation

The catalytic performances in the DME carbonylation to MA reaction of all the samples were investigated. The changes in the DME conversion, MA selectivity, methanol (MeOH) selectivity and hydrocarbon (HCs) selectivity with the reaction time are shown in Figure 9a–d. All the catalysts exhibit a relatively low DME conversion at the beginning of the reaction. Then, the DME conversion gradually increases and reaches the maximum value. Then, the DME conversion decreases over time. The maximum values of DME conversions decrease in the following order: M-Fe-0.01 (30.9%) > M-Fe-0.015 (21.2%) > M-Fe-0.005 (19.2%) > 0.01Fe/M (16.4%) > HM (11.7%) > M-Fe-0.02 (9.0%). This is consistent with the first increasing and then decreasing trend of $B_{8\text{-}MR}$. The maximum space-time yield of MA (STY_{MA}) in each catalyst is calculated and shown in Figure 10a. The STY_{MA} of M-Fe-0.01 is 0.11 g/(g_{cat}·h), which is nearly three times that of HM (0.04 g/(g_{cat}·h)). The relations between STY_{MA} and $B_{8\text{-}MR}$ are shown in Figure 10b. It was reported that STY_{MA} had a positive correlation with $B_{8\text{-}MR}$ because the Brønsted acid sites in the 8-MR were the active center for the DME carbonylation reaction [51,52]. In Figure S8b, a positive correlation is observed for HM, M-Fe-0.005, M-Fe-0.01 and M-Fe-0.015. The STY_{MA} of M-Fe-0.02 is relatively low, considering that the $B_{8\text{-}MR}$ of M-Fe-0.02 is higher than that of HM. Zhou [18] ascribed the low catalytic activity of the catalyst with a high iron content to the decreased BET surface area and pore volume, which decrease the accessibility of the acid sites. The decrease in the BET surface area and pore volume of M-Fe-0.02 is also observed, which might be attributed to the increased extraframework iron species. The STY_{MA} of 0.01Fe/M is also low, which should be due to the substitution of Al by Fe. The Si–OH–Fe bonds formed in the 8-MR contribute to the $B_{8\text{-}MR}$, but they are less active in the DME carbonylation reaction.

The MA selectivity of all catalysts is high in the first 5 h (>95%) and then decreases as the reaction proceeds. The selectivity of methanol is relatively high at the beginning of the reaction, which should be ascribed to the methanol formed via DME dissociation on the acid sites [12]. When the acid sites are saturated with methoxy species, methanol selectivity reaches the lowest level. Then, methanol selectivity increases continuously with the reaction time. The variation trend of hydrocarbon selectivity is similar to that of methanol. The relatively high initial hydrocarbon selectivity might be due to the decomposition of DME and methanol on acid sites [53]. Liu [54] reported that methanol-to-hydrocarbon (MTH) reactions in the 12-MR were responsible for the decrease in DME conversion. The water molecules formed in MTH reactions competitively adsorbed on the acid sites in the 8-MR, thus leading to the decrease in MA selectivity and the increase in byproducts [55]. The MA selectivity of HM, M-Fe-0.005 and 0.01Fe/M decreases fast and drops down to less than 50% at 24 h, while the MA selectivity of M-Fe-0.01, M-Fe-0.015 and M-Fe-0.02 remains high during the reaction process. The byproduct selectivities of these samples are also lower. As discussed in Sections 3.2 and 3.3, for M-Fe-x (x = 0.005, 0.01, 0.015, 0.02), the Fe species exist in the zeolite framework as Si-OH-Fe in the 12-MR. Additionally, the framework Al species of Fe-HMOR decrease compared to those of HM. Therefore, the Si–OH–Al bonds in the 12-MR of M-Fe-x (x = 0.005, 0.01, 0.015, 0.02) are decreased. Meng [56] reported that Fe-substituted MFI zeolite exhibited very low activity in the MTH reaction due to the low acid strength of Si-OH-Fe. It is speculated that the activity of Si-OH-Fe is also low in the side reactions of DME carbonylation; thus, fewer byproducts are formed when Al is substituted by Fe in the 12-MR channels of MOR. The contents of extraframework Al species, which would decrease the accessibility of acid sites in the 8-MR and favor methanol formation [57], are relatively low in the Fe-substituted samples. This might also contribute to the lower byproduct selectivity of M-Fe-x (x = 0.01, 0.015, 0.02). The amount of framework Fe in M-Fe-0.005 is relatively low; thus, there is no significant difference in the MA and byproduct selectivity compared to HMOR. The Si-OH-Al distribution in the 12-MR and 8-MR of 0.01Fe/M is not apparently affected, which results in the lower MA selectivity compared to M-Fe-0.01.

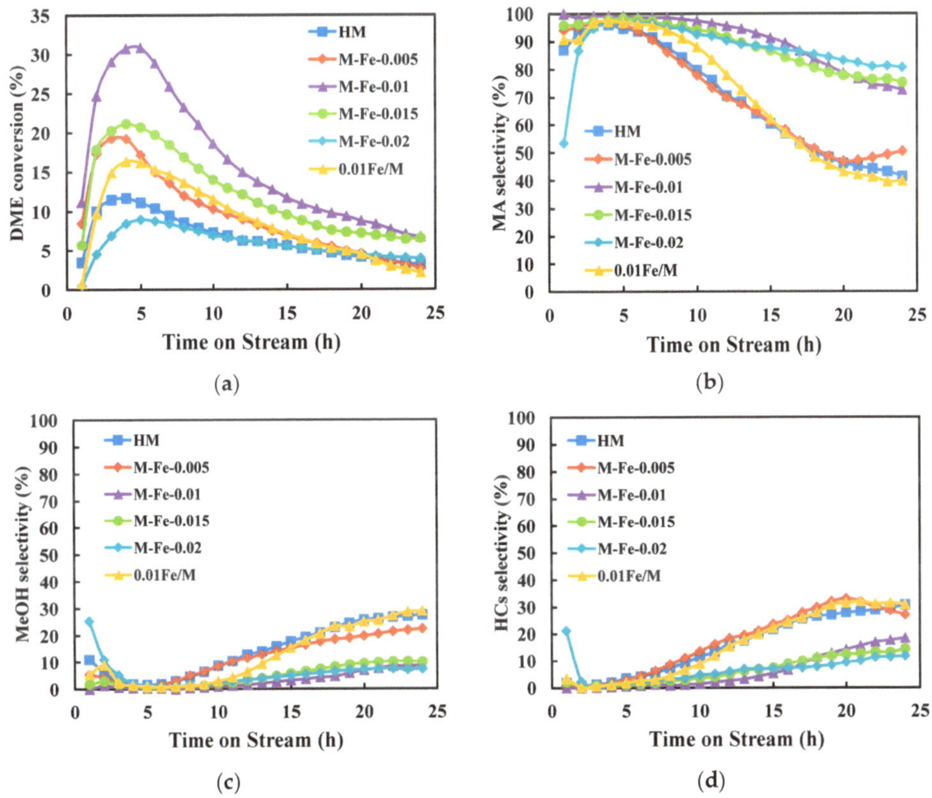

Figure 9. (**a**) DME conversion; (**b–d**) MA, methanol and hydrocarbon selectivities of DME carbonylation over the catalysts. Reaction conditions: T = 200 °C, P = 2 MPa, DME/CO/Ar = 3/30/67, GHSV = 3600 mL/(g·h).

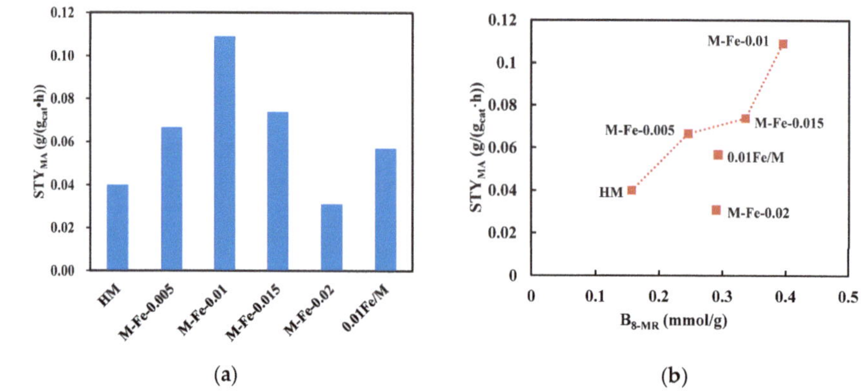

Figure 10. (**a**) STY_{MA} of the samples; (**b**) The relation between STY_{MA} and $B_{8\text{-}MR}$ of the samples.

The TG and DTG profiles of the spent catalysts are presented in Figure 11a,b. The weight loss ratio of HM is the highest. With the increase in the Fe/Si ratio, the carbon deposition gradually decreases from 9.9 wt% in HM to 5.6 wt% in M-Fe-0.02. The weight loss ratio of 0.01Fe/M is 9.6 wt%, which is comparable to that of HM. The Brønsted acid sites in the 12-MR are thought to be favorable for side reactions. As discussed above, Si-OH-Fe might exhibit low activity in the side reactions. Therefore, carbon deposition could be inhibited only when Fe atoms are incorporated into the main channels of MOR. The weight loss in the region of 250–350 °C is due to the oxidation of soft coke composed of reaction intermediates, such as surface methyl and acetyl species associated with the formation of MA [7,58]. The heavy coke composed of large hydrocarbons, such as hydrogen-deficient aromatic species, oxidizes above 350 °C [7,11]. The DTG curves of all spent catalysts show two peaks in the temperature range of 250–350 °C and above 350 °C, respectively. The peak temperature related to heavy coke of the spent catalysts varies. The peak temperature is 600 °C for M-Fe-0 and M-Fe-0.005, and it then decreases to 450 °C for M-Fe-0.01. With the Fe/Si ratio further increasing, the peak temperature of M-Fe-0.015 and M-Fe-0.02 decreases to 425 °C and 400 °C, respectively. The peak temperature of 0.01Fe/M is 420 °C. The results of DTG analysis indicate that the incorporation of Fe into the zeolite framework is an effective way to suppress the formation of heavy coke species with lower H/C ratios [59].

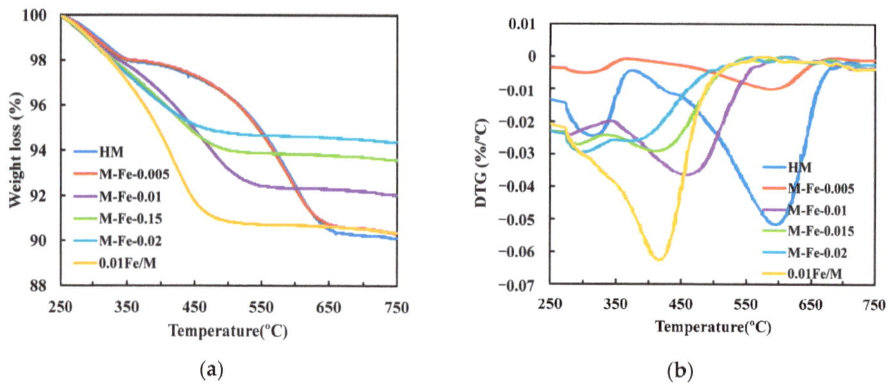

Figure 11. (**a**) TG and (**b**) DTG profiles of the spent catalysts.

The morphology and acidity of the spent and regenerated catalysts were investigated. The spent HM, M-Fe-0.01 and 0.01Fe/M are referred to as HM-spent, M-Fe-0.01-spent and 0.01Fe/M-spent, respectively. The spent catalysts were calcined in air at 500 °C for 6 h to obtain regenerated catalysts. The regenerated catalysts are referred to as HM-R,

M-Fe-0.01-R and 0.01Fe/M-R, respectively. SEM images of the spent catalysts are shown in Figure S8a–c. The morphology of the samples remains largely steady after 24 h of reaction, except for some small fragments or debris on the surface of the assemblies. This suggests that the carbon deposit has little influence on the morphology of the catalysts. As shown in Figure S8d–f, more fragments and debris appear on the zeolite surface after regeneration, indicating that the calcination process has a negative effect on the morphology. The NH_3-TPD profiles of the spent and regenerated catalysts are shown in Figure S9. The weak, moderate and strong acid amounts calculated based on the deconvoluted peak areas of the samples are listed in Table S3. The weak, moderate and strong acid amounts of the spent catalysts decrease compared with fresh ones, which might be due to the coverage of carbon deposit [7]. After the regeneration, the acid amounts of the samples increase compared to spent catalysts, which should be attributed to the combustion of carbon deposit. The moderate acid amounts of HM-R, M-Fe-0.01-R and 0.01Fe/M-R are higher than those of the fresh ones, indicating there is an increase in the Lewis acid sites of the regenerated catalysts. The strong acid amounts (i.e., total Brønsted acid amounts) of regenerated catalysts decrease compared with the fresh catalysts, suggesting that part of the framework Al or Fe atoms are removed during the reaction and the regeneration process. The Py-IR spectra of the spent and regenerated catalysts are shown in Figure S10. The calculated $B_{12\text{-}MR}$ and $B_{8\text{-}MR}$ are listed in Table S3. For the spent catalysts, both Brønsted acid sites in the 12-MR and 8-MR are largely reduced compared to the fresh ones due to the formation of a coke precursor or carbon deposit [7]. After the regeneration, the Brønsted acid sites in both channels are recovered, but the $B_{12\text{-}MR}/B_{8\text{-}MR}$ are still lower than those of the fresh catalysts.

4. Conclusions

Fe-modified mordenite catalysts with different Fe/Si ratios from 0.005 to 0.02 were synthesized by adding Fe complex as Fe source through one-step hydrothermal conversion. As a reference, 0.01Fe/M with a Fe/Si ratio of 0.01 was synthesized by adding $Fe(NO_3)_3$ as Fe source. The Si/Al ratios of Fe-modified samples increased with the increasing Fe/Si ratio, while the Si/(Al+Fe) ratios slightly decreased. The influence of Fe source on the constitution of Fe species was investigated. It was found that most of the Fe atoms successfully substituted for the Al atoms in the zeolite framework. The framework Fe atoms were preferentially located in the 12-MR channels of the Fe complex modified samples, while the framework Fe atoms of 0.01Fe/M were located both in the 12-MR and 8-MR. Except for 0.01Fe/M, the Brønsted acid strength of the Fe-HMOR was improved compared to the HMOR sample without Fe modification. By increasing the Fe/Si ratio, the total Brønsted acid amount increased due to the decrease in Si/(Al+Fe) ratio. The Brønsted acid amounts in the 8-MR ($B_{8\text{-}MR}$) first increased and then decreased with the Fe/Si ratio increasing from 0 to 0.02. M-Fe-0.01 with the Fe/Si ratio of 0.01 exhibited the highest $B_{8\text{-}MR}$. The results of acid strength and Brønsted acid distribution suggested that the selective incorporation of Fe atoms into the 12-MR might facilitate the formation of Si–OH–Al bonds in the side pockets. However, the Si–OH–Al bond distribution of 0.01Fe/M was not affected because Si–OH–Fe bonds were formed in both channels. By incorporating a certain amount of Fe through the Fe complex into the zeolite framework—for instance, M-Fe-0.01 in this work—the Brønsted acid sites derived from framework Al in the side pockets could be largely enhanced, while those in the 12-MR decreased. In the DME carbonylation reaction, M-Fe-0.01 showed the highest DME conversion, which was nearly three times that of HMOR and two times that of 0.01Fe/M, respectively. The MA selectivity of M-Fe-0.01 also remained high during the reaction process due to the improved Brønsted acid distribution. By adding Fe complex into the synthetic gel, the Brønsted acid strength, amount and distribution of the synthesized Fe-HMOR were effectively regulated, thus affecting the catalytic performance.

Supplementary Materials: The following supporting information can be downloaded at: https://www.mdpi.com/article/10.3390/ma17102417/s1, Figure S1: Pictures of experimental apparatus; Figure S2: SEM images of the samples; Figure S3: SEM images in higher magnifications; Figure S4: N_2 adsorption–desorption isotherms of the samples; Figure S5: Pore size distribution of the samples; Figure S6: FTIR spectra in 500–1500 cm^{-1}; Figure S7: XPS spectra of O 1s for Fe-modified samples; Figure S8: SEM images of the spent catalysts and the regenerated catalysts; Figure S9: NH_3-TPD profiles of the spent and regenerated catalysts; Figure S10: Py-IR spectra of the spent and regenerated catalysts; Table S1: Crystal sizes calculated from XRD patterns using the Scherrer equation and average particle sizes obtained from SEM images; Table S2: Weak, moderate and strong peak temperatures of the NH_3-TPD profiles; Table S3: Acid amounts and distributions of the spent and regenerated catalysts. Reference [60] is cited in the Supplementary Materials.

Author Contributions: Conceptualization, Y.W. and W.L.; validation, L.B., K.C. and Y.H.; formal analysis, K.C. and L.Q.; investigation, W.L., N.G., X.S. and Y.L.; resources, J.S.; writing—original draft preparation, W.L.; writing—review and editing, Y.W. and L.B.; visualization, W.L. and X.Z.; supervision, Y.W.; project administration, Y.W.; funding acquisition, Y.W. All authors have read and agreed to the published version of the manuscript.

Funding: This work is partially supported by the National Natural Science Foundation of China (21276183).

Data Availability Statement: Data are contained within the article.

Conflicts of Interest: The authors declare no conflicts of interest.

References

1. Feng, X.; He, Z.; Zhang, L.; Zhao, X.; Cao, J. Facile designing a nanosheet HMOR zeolite for enhancing the efficiency of ethanol synthesis from dimethyl ether and syngas. *Int. J. Hydrogen Energ.* **2022**, *47*, 9273–9282. [CrossRef]
2. Balat, M.; Balat, H. Recent trends in global production and utilization of bio-ethanol fuel. *Appl. Energ.* **2009**, *86*, 2273–2282. [CrossRef]
3. Zhan, E.; Xiong, Z.; Shen, W. Dimethyl ether carbonylation over zeolites. *J. Energy Chem.* **2019**, *36*, 51–63. [CrossRef]
4. Chen, F.; Feng, X.; Zhang, L.; Zhao, J.; He, Z.; Yi, F.; Zhao, X.; Cao, J. Selective enrichment of Brønsted acid site in 8-membered ring channels of MOR zeolite to enhance the catalytic reactivity of dimethyl ether carbonylation. *Chem. Eng. Sci.* **2022**, *263*, 118110. [CrossRef]
5. Yao, J.; Wu, Q.; Fan, J.; Komiyama, S.; Yong, X.; Zhang, W.; Zhao, T.; Guo, Z.; Yang, G.; Tsubaki, N. A carbonylation zeolite with specific nanosheet structure for efficient catalysis. *ACS Nano* **2021**, *15*, 13568–13578. [CrossRef] [PubMed]
6. Zhao, J.; Qian, W.; Ma, H.; Ying, W.; Yuan, P.; Zhang, H. Theoretical study for adsorption-diffusion on H-MOR and pyridine pre-adsorbed H-MOR of dimethyl ether carbonylation. *ACS Omega* **2023**, *8*, 22067–22076. [CrossRef] [PubMed]
7. Wang, X.; Li, R.; Yu, C.; Liu, Y.; Xu, C.; Lu, C. Study on the deactivation process of dimethyl ether carbonylation reaction over mordenite catalyst. *Fuel* **2021**, *286*, 119480. [CrossRef]
8. Chu, Y.; Lo, A.; Wang, C.; Deng, F. Origin of high selectivity of dimethyl ether carbonylation in the 8-membered ring channel of mordenite zeolite. *J. Phys. Chem. C* **2019**, *123*, 15503–15512. [CrossRef]
9. Rasmussen, D.B.; Christensen, J.M.; Temel, B.; Studt, F.; Moses, P.G.; Rossmeisl, J.; Riisager, A.; Jensen, A.D. Ketene as a reaction intermediate in the carbonylation of dimethyl ether to methyl acetate over mordenite. *Angew. Chem. Int. Ed.* **2015**, *54*, 7261–7264. [CrossRef]
10. Chen, W.; Li, G.; Yi, X.; Day, S.J.; Tarach, K.A.; Liu, Z.; Liu, S.; Tsang, S.C.E.; Góra-Marek, K.; Zheng, A. Molecular understanding of the catalytic consequence of ketene intermediates under confinement. *J. Am. Chem. Soc.* **2021**, *143*, 15440–15452. [CrossRef]
11. Cheng, Z.; Huang, S.; Li, Y.; Cai, K.; Wang, Y.; Wang, M.; Lv, J.; Ma, X. Role of Brønsted acid sites within 8-MR of mordenite in the deactivation roadmap for dimethyl ether carbonylation. *ACS Catal.* **2021**, *11*, 5647–5657. [CrossRef]
12. He, T.; Liu, X.; Xu, S.; Han, X.; Pan, X.; Hou, G.; Bao, X. Role of 12-ring channels of mordenite in DME carbonylation investigated by solid-state NMR. *J. Phys. Chem. C* **2016**, *120*, 22526–22531. [CrossRef]
13. Bhan, A.; Allian, A.D.; Sunley, G.J.; Law, D.J.; Iglesia, E. Specificity of sites within eight-membered ring zeolite channels for carbonylation of methyls to acetyls. *J. Am. Chem. Soc.* **2007**, *129*, 4919–4924. [CrossRef] [PubMed]
14. Issa, H.; Chaouati, N.; Toufaily, J.; Hamieh, T.; Sachse, A.; Pinard, L. Toolbox of post-synthetic mordenite modification strategies: Impact on textural, acidic and catalytic properties. *ChemCatChem* **2019**, *18*, 4581–4592. [CrossRef]
15. Liu, Y.; Zheng, D.; Li, B.; Lyu, Y.; Wang, X.; Liu, X.; Li, L.; Yu, S.; Liu, X.; Yan, Z. Isomerization of α-pinene with a hierarchical mordenite molecular sieve prepared by the microwave assisted alkaline treatment. *Micropor. Mesopor. Mater.* **2020**, *299*, 110117. [CrossRef]
16. Xu, F.; Lv, J.; Chen, C.; Hong, Z.; Zhao, G.; Miao, L.; Yang, W.; Zhu, Z. Effect of steam treatment on the properties of mordenite and its catalytic performance in a DME carbonylation reaction. *Ind. Eng. Chem. Res.* **2022**, *61*, 1258–1266. [CrossRef]

17. Li, Y.; Huang, S.; Cheng, Z.; Cai, K.; Li, L.; Milan, E.; Lv, J.; Wang, Y.; Sun, Q.; Ma, X. Promoting the activity of Ce-incorporated MOR in dimethyl ether carbonylation through tailoring the distribution of Brønsted acids. *Appl. Catal. B Environ.* **2019**, *256*, 117777. [CrossRef]
18. Zhou, H.; Zhu, W.; Shi, L.; Liu, H.; Liu, S.; Xu, S.; Ni, Y.; Liu, Y.; Li, L.; Liu, Z. Promotion effect of Fe in mordenite zeolite on carbonylation of dimethyl ether to methyl acetate. *Catal. Sci. Technol.* **2015**, *5*, 1961–1968. [CrossRef]
19. Zhang, L.; Feng, X.; He, Z.; Chen, F.; Su, C.; Zhao, X.; Cao, J.; He, Y. Enhancing the stability of dimethyl ether carbonylation over Fe-doped MOR zeolites with tunable 8-MR acidity, Chem. *Eng. Sci.* **2022**, *256*, 117671. [CrossRef]
20. Liu, W.; Wang, Y.; Bu, L.; Zhi, Y.; Wang, Z.; Yang, M.; Chu, K.; Huang, Y.; Guo, N.; Qu, L.; et al. Wheel-like mordenite nanoassemblies with shortened channel lengths for improved catalytic performance in dimethyl ether carbonylation. *ACS Appl. Nano Mater.* **2023**, *6*, 18005–18015. [CrossRef]
21. Wu, P.; Komatsu, T.; Yashima, T. Isomorphous substitution of Fe^{3+} in the framework of aluminosilicate mordenite by hydrothermal synthesis. *Micropor. Mesopor. Mater.* **1998**, *20*, 139–147. [CrossRef]
22. Xiong, Z.; Qi, G.; Zhan, E.; Chu, Y.; Xu, J.; Wei, J.; Ta, N.; Hao, A.; Zhou, Y.; Deng, F.; et al. Experimental identification of the active sites over a plate-like mordenite for the carbonylation of dimethyl ether. *Chem* **2023**, *9*, 76–92. [CrossRef]
23. Zhang, W.; Xie, J.; Hou, W.; Liu, Y.; Zhou, Y.; Wang, J. One-pot template-free synthesis of Cu-MOR zeolite toward efficient catalyst support for aerobic oxidation of 5-hydroxymethylfurfural under ambient pressure. *ACS Appl. Mater. Inter.* **2016**, *8*, 23122–23132. [CrossRef] [PubMed]
24. Lu, J.; Wang, Y.; Sun, C.; Zhao, T.; Zhao, J.; Wang, Z.; Liu, W.; Wu, S.; Shi, M.; Bu, L. Novel synthesis and catalytic performance of hierarchical MOR. *New J. Chem.* **2021**, *45*, 8629–8638. [CrossRef]
25. Jacobs, P.A.; Beyer, H.K.; Valyon, J. Properties of the end members in the Pentasil-family of zeolites: Characterization as adsorbents. *Zeolites* **1981**, *1*, 161–168. [CrossRef]
26. Liu, C.; Gu, W.; Kong, D.; Guo, H. The significant effects of the alkali-metal cations on ZSM-5 zeolite synthesis: From mechanism to morphology. *Micropor. Mesopor. Mater.* **2014**, *183*, 30–36. [CrossRef]
27. Zhang, J.; Tang, X.; Yi, H.; Yu, Q.; Zhang, Y.; Wei, J.; Yuan, Y. Synthesis, characterization and application of Fe-zeolite: A review. *Appl. Catal. A Gen.* **2022**, *630*, 118467. [CrossRef]
28. Hu, X.; Chen, J.; He, S.; Liang, T.; Zheng, S.; Lu, L.; Hao, C.; Chen, K.; Li, T.; Yi, L.; et al. Study on the catalytic activity and hydrothermal stability of one-pot synthesized Fe-based FER zeolites for NH_3-SCR. *Catal. Sci. Technol.* **2023**, *13*, 5435–5448. [CrossRef]
29. Gomez, S.; Lerici, L.; Saux, C.; Perez, A.L.; Brondino, C.D.; Pierella, L.; Pizzio, L. Fe/ZSM-11 as a novel and efficient photocatalyst to degrade dichlorvos on water solutions. *Appl. Catal. B Environ.* **2017**, *202*, 580–586. [CrossRef]
30. Kotolevich, Y.; Khramov, E.; Sánchez-López, P.; Pestryakov, A.; Zubavichus, Y.; Antúnez-Garcia, J.; Petranovskii, V. Formation of Ag-Fe bimetallic nano-species on mordenite depending on the initial ratio of components. *Materials* **2023**, *16*, 3026. [CrossRef]
31. Yu, D.; Wang, P.; Li, X.; Zhao, H.; Lv, X. Study on the role of Fe species and acid sites in NH_3-SCR over the Fe-based zeolites. *Fuel* **2023**, *336*, 126759. [CrossRef]
32. Deng, Y.; Veser, G. Toward accelerated activation of Fe-ZSM-5 in methane dehydroaromatization. *Energ. Fuel* **2023**, *37*, 13282–13295. [CrossRef]
33. Yu, Y.; Xiong, G.; Li, C.; Xiao, F. Characterization of aluminosilicate zeolites by UV Raman spectroscopy. *Micropor. Mesopor. Mat.* **2001**, *46*, 23–34. [CrossRef]
34. Twu, J.; Dutta, P.K.; Kresge, C.T. Vibrational spectroscopic examination of the formation of mordenite crystals. *J. Phys. Chem.* **1991**, *13*, 5267–5271. [CrossRef]
35. Boronat, M.; Martínez-Sánchez, C.; Law, D.; Corma, A. Enzyme-like specificity in zeolites: A unique site position in mordenite for selective carbonylation of methanol and dimethyl ether with CO. *J. Am. Chem. Soc.* **2008**, *130*, 16316–16323. [CrossRef] [PubMed]
36. Reule, A.A.C.; Sawada, J.A.; Semagina, N. Effect of selective 4-membered ring dealumination on mordenite-catalyzed dimethyl ether carbonylation. *J. Catal.* **2017**, *349*, 98–109. [CrossRef]
37. Yakimov, A.V.; Ravi, M.; Verel, R.; Sushkevich, V.L.; van Bokhoven, J.A.; Copéret, C. Structure and framework association of Lewis acid sites in MOR zeolite. *J. Am. Chem. Soc.* **2022**, *144*, 10377–10385. [CrossRef]
38. Lukyanov, D.B.; Vazhnova, T.; Cherkasov, N.; Casci, J.L.; Birtill, J.J. Insights into Brønsted acid sites in the zeolite mordenite. *J. Phys. Chem. C* **2014**, *118*, 23918–23929. [CrossRef]
39. Datka, J.; Gil, B.; Kubacka, A. Acid properties of NaH-mordenites: Infrared spectroscopic studies of ammonia sorption. *Zeolites* **1995**, *15*, 501–506. [CrossRef]
40. Sheng, H.; Ma, H.; Qian, W.; Fei, N.; Zhang, H.; Ying, W. Platinum-copper bimetallic-modified nanoprism mordenite for carbonylation of dimethyl ether. *Energ. Fuel.* **2019**, *33*, 10159–10166. [CrossRef]
41. Lónyi, F.; Valyon, J. On the interpretation of the NH_3-TPD patterns of H-ZSM-5 and H-mordenite. *Micropor. Mesopor. Mat.* **2001**, *47*, 293–301. [CrossRef]
42. Wang, J.; Xu, Z.; Zhao, W.; Li, X. Selective catalytic reduction of NO by NH_3 over MoO_3 promoted Fe_2O_3 catalyst. *Kinet. Catal.* **2018**, *59*, 628–634. [CrossRef]
43. Wang, M.; Huang, S.; Lü, J.; Cheng, Z.; Li, Y.; Wang, S.; Ma, X. Modifying the acidity of H-MOR and its catalytic carbonylation of dimethyl ether. *Chin. J. Catal.* **2016**, *9*, 1530–1537. [CrossRef]

44. Liu, Y.; Zhao, N.; Xian, H.; Cheng, Q.; Tan, Y.; Tsubaki, N.; Li, X. Facilely synthesized H-mordenite nanosheet assembly for carbonylation of dimethyl ether. *ACS Appl. Mater. Inter.* **2015**, *7*, 8398–8403. [CrossRef]
45. Lee, K.; Lee, S.; Ihm, S. Acid strength control in MFI zeolite for the methanol-to-hydrocarbons (MTH) reaction. *Ind. Eng. Chem. Res.* **2014**, *53*, 10072–10079. [CrossRef]
46. Lew, C.M.; Chen, C.; Long, G.J.; Grandjean, F.; Ichimura, A.S.; Xie, D.; Grosso-Giordano, N.A.; Chakarawet, K.; Lacheen, H.S.; Jensen, K.O.; et al. Synthesis, physicochemical characterization, and catalytic evaluation of Fe^{3+}-containing SSZ-70 zeolite. *ACS Catal.* **2022**, *12*, 6464–6477. [CrossRef]
47. Tsubota, S.; Kokuryo, S.; Tamura, K.; Miyake, K.; Uchida, Y.; Mizusawa, A.; Kubo, T.; Nishiyama, N. Exploring the Effect of Brønsted Acidity of MFI-type Zeolites on Catalytic Cracking Temperature of Low Density Polyethylene. *Catal. Sci. Technol.* **2024**, *14*, 1369–1374. [CrossRef]
48. Barzetti, T.; Selli, E.; Moscotti, D.; Forni, L. Pyridine and ammonia as probes for FTIR analysis of solid acid catalysts. *J. Chem. Soc. Faraday Trans.* **1996**, *92*, 1401. [CrossRef]
49. Palčić, A.; Valtchev, V. Analysis and control of acid sites in zeolites. *Appl. Catal. A Gen.* **2020**, *606*, 117795. [CrossRef]
50. Emeis, C.A. Determination of integrated molar extinction coefficients for infrared adsorption bands of pyridine adsorbed on solid acid catalysts. *J. Catal.* **1993**, *141*, 347–354. [CrossRef]
51. Chen, F.; Feng, X.; Zhao, J.; He, Z.; Zhang, L.; Wang, Y.; Deng, P.; Gao, X.; Zhao, X.; Cao, J. Designing Mordenite Zeolites with Tunable Distribution of Acid Sites in Channels and Nano-Morphology to Boost the Catalytic Behavior Performance of Dimethyl Ether Carbonylation. *Chem. Eng. Sci.* **2023**, *282*, 119250. [CrossRef]
52. Chen, N.; Zhang, J.; Gu, Y.; Zhang, W.; Cao, K.; Cui, W.; Xu, S.; Fan, D.; Tian, P.; Liu, Z. Designed Synthesis of MOR Zeolites Using Gemini-Type Bis(Methylpyrrolidinium) Dications as Structure Directing Agents and their DME Carbonylation Performance. *J. Mater. Chem. A* **2022**, *10*, 8334–8343. [CrossRef]
53. Lu, P.; Yang, G.; Tanaka, Y.; Tsubaki, N. Ethanol direct synthesis from dimethyl ether and syngas on the combination of noble metal impregnated zeolite with Cu/ZnO catalyst. *Catal. Today* **2014**, *232*, 22–26. [CrossRef]
54. Liu, S.; Liu, H.; Ma, X.; Liu, Y.; Zhu, W.; Liu, Z. Identifying and controlling the acid site distributions in mordenite zeolite for dimethyl ether carbonylation reaction by means of selective ion-exchange. *Catal. Sci. Technol.* **2020**, *1*, 4663–4672. [CrossRef]
55. Liu, S.; Fang, X.; Liu, Y.; Liu, H.; Ma, X.; Zhu, W.; Liu, Z. Dimethyl ether carbonylation over mordenite zeolite modified by alkyimidazolium ions. *Catal. Commun.* **2020**, *147*, 106161. [CrossRef]
56. Meng, L.; Zhu, X.; Mezari, B.; Pestman, R.; Wannapakdee, W.; Hensen, E.J.M. On the role of acidity in bulk and nanosheet [T]MFI (T=Al^{3+}, Ga^{3+}, Fe^{3+}, B^{3+}) zeolites in the methanol-to-hydrocarbons reaction. *ChemCatChem* **2017**, *9*, 3942–3954. [CrossRef] [PubMed]
57. Ham, H.; Jung, H.S.; Kim, H.S.; Kim, J.; Cho, S.J.; Lee, W.B.; Park, M.; Bae, J.W. Gas-Phase Carbonylation of Dimethyl Ether on the Stable Seed-Derived Ferrierite. *ACS Catal.* **2020**, *10*, 5135–5146. [CrossRef]
58. Liu, J.; Xue, H.; Huang, X.; Wu, P.; Huang, S.; Liu, S.; Shen, W. Stability Enhancement of H-Mordenite in Dimethyl Ether Carbonylation to Methyl Acetate by Pre-Adsorption of Pyridine. *Chin. J. Catal.* **2010**, *31*, 729–738. [CrossRef]
59. Wang, X.; Li, R.; Yu, C.; Zhang, L.; Xu, C.; Zhou, H. Dimethyl ether carbonylation over nanosheet-assembled hierarchical mordenite. *Micropor. Mesopor. Mater.* **2019**, *274*, 227–235. [CrossRef]
60. Vinila, V.S.; Isac, J. Synthesis and structural studies of superconducting perovskite GdBa2Ca3Cu4O10.5þd nanosystems. In *Design, Fabrication, and Characterization of Multifunctional Nanomaterials*, 1st ed.; Thomas, S., Kalarikkal, N., Abraham, A.R., Eds.; Elsevier: Amsterdam, The Netherlands, 2022; Volume 2, pp. 319–338.

nanomaterials

Article

Novel High-Entropy FeCoNiMoZn-Layered Hydroxide as an Efficient Electrocatalyst for the Oxygen Evolution Reaction

Zhihao Cheng [1], Xin Han [1], Liying Han [1], Jinfeng Zhang [1], Jie Liu [1,*], Zhong Wu [1] and Cheng Zhong [1,2]

1 Key Laboratory of Advanced Ceramics and Machining Technology (Ministry of Education), and Tianjin Key Laboratory of Composite and Functional Materials, School of Materials Science and Engineering, Tianjin University, Tianjin 300072, China; 18002171015@163.com (Z.C.); 18822144032@163.com (X.H.); 18822029109@163.com (L.H.); jinfeng@tju.edu.cn (J.Z.); zhong.wu@tju.edu.cn (Z.W.); cheng.zhong@tju.edu.cn (C.Z.)

2 Joint School of National University of Singapore and Tianjin University, International Campus of Tianjin University, Binhai New City, Fuzhou 350207, China

* Correspondence: jieliu0109@tju.edu.cn

Abstract: The exploration of catalysts for the oxygen evolution reaction (OER) with high activity and acceptable price is essential for water splitting to hydrogen generation. High-entropy materials (HEMs) have aroused increasing interest in the field of electrocatalysis due to their unusual physicochemical properties. In this work, we reported a novel FeCoNiMoZn-OH high entropy hydroxide (HEH)/nickel foam (NF) synthesized by a facile pulsed electrochemical deposition method at room temperature. The FeCoNiMoZn-OH HEH displays a 3D porous nanosheet morphology and polycrystalline structure, which exhibits extraordinary OER activity in alkaline media, including much lower overpotential (248 mV at 10 mA cm^{-2}) and Tafel slope (30 mV dec^{-1}). Furthermore, FeCoNiMoZn-OH HEH demonstrates excellent OER catalytic stability. The enhanced catalytic performance of the FeCoNiMoZn-OH HEH primarily contributed to the porous morphology and the positive synergistic effect between Mo and Zn. This work provides a novel insight into the design of HEMs in catalytic application.

Keywords: high entropy; water splitting; oxygen evolution reaction; electrocatalytic

Citation: Cheng, Z.; Han, X.; Han, L.; Zhang, J.; Liu, J.; Wu, Z.; Zhong, C. Novel High-Entropy FeCoNiMoZn-Layered Hydroxide as an Efficient Electrocatalyst for the Oxygen Evolution Reaction. *Nanomaterials* **2024**, *14*, 889. https://doi.org/10.3390/nano14100889

Academic Editor: Francesc Viñes Solana

Received: 13 April 2024
Revised: 12 May 2024
Accepted: 13 May 2024
Published: 20 May 2024

1. Introduction

Hydrogen with high combustion efficiency and green renewability has been widely recognized as a promising and achievable new energy source to replace fossil fuel [1,2]. Electrochemical water splitting is a highly efficient hydrogen production technology without polluting byproducts [3,4]. Nevertheless, as one of the key reactions of water electrolysis, the oxygen evolution reaction (OER) suffers from high overpotential and sluggish reaction kinetics as a result of the complicated four-electron transfer process [5–7]. Currently, noble metal-based electrocatalysts, including RuO_2 and IrO_2, are considered excellent electrocatalysts for OER [3,5,8]. However, their scarcity, high price, poor durability, and low selectivity restricted their large-scale commercial application [9–11]. Consequently, it is of great importance to develop non-noble metal electrocatalysts with acceptable cost, high efficiency, and satisfactory durability for OER.

In this scenario, catalysts composed of transition metal have been widely studied as promising OER catalysts because of their abundant resources and low price. To date, various electrocatalysts consisting of the transition metal have been extensively developed, including NiFe [12], NiFe hydroxide [13], FeCoW hydroxide [14], and NiFeMo oxide [15]. However, the binary or ternary catalysts lack broad compositional tunability [16], resulting in a limited number and types of active sites [17].

Recently, high-entropy materials (HEMs), comprised at least five principal elements, have aroused extensive interest in the field of electrocatalysis as a result of their unusual physicochemical properties. The large lattice distortion in HEMs not only induces plenty of lattice defects on the surface, which allows the introduction of plentiful active sites but also limits the atomic diffusion to enhance the structural stability. Furthermore, the synergistic interaction among multiple principle components improves the tunability of the electronic structure of HEMs [6,18–20]. Therefore, HEM electrocatalysts have been extensively studied in the past few years. Recently, Huang et al. [21] synthesized FeCoNiMnCu high entropy alloy (HEA) using a solvothermal method; the synthesized HEA catalyst exhibited outstanding catalytic activity and stability for catalyzing OER. Strotkoetter et al. [22] fabricated the FeCoNiCrMn high entropy oxide via the codeposition method and this catalyst showed an outstanding OER electrocatalytic performance in an alkaline medium. So far, various methods have been developed to synthesize HEM catalysts, including plasma sintering [23], arc-melting [24], carbothermal shock [25], and laser ablation [26]. Nevertheless, most of these methods usually involve harsh high-temperature conditions [27,28]. Consequently, it is highly desirable to explore a facile technique under mild atmospheres for the synthesis of HEM catalysts with high activity and excellent durability [29]. Electrodeposition, as a conventional method, has been extensively used to fabricate electrocatalysts owing to its low cost and simple process. For example, He et al. [30] synthesized FeCoNiMn (oxy) hydroxide by electrodeposition for OER catalysis. Bian et al. [31] prepared FeMnCuCo HEA as an efficient OER catalyst via one-step electrodeposition.

It is noteworthy that the performance of high entropy catalysts is strongly dependent on component selection. The adjustable elemental components and ratios of high entropy catalysts offer a possibility to develop catalysts with higher catalytic activity [19]. Especially, the catalysts based on 3d-transition metals including Fe, Ni, Co, etc., demonstrated excellent electrocatalytic performance in alkaline media [19,32]. Furthermore, Zn could be combined with active metal Ni to modulate the electronic property of the catalyst [6,33]. For instance, Huang et al. [34] synthesized the $Fe_{0.5}CoNiCuZn_x$ catalyst by electro-deoxidizing molten salts at high temperature, this electrode shows a low overpotential (340 mV at 10 mA cm^{-2}) and superior durability for 24 h. The introduction of Zn alters the HEA bond with the element and thereby increases the adsorption energies between the surface of the HEA catalyst and OH^-/H^+. Moreover, the introduction of Zn also increases the interplanar spacing of HEA and thus enhances the OER activity [34]. Additionally, as a 4d high-valence metal, Mo has a larger spatial extent, being endowed with strong hybridization with adjacent ligand orbitals. The large electronic bandwidth of Mo allows for a wider tunable range of electronic structures [35]. Furthermore, Mo metals have additional orbital degrees of freedom, which facilitates the tuning of the electronic band structure and the adsorption/desorption energy of the intermediates [32]. For instance, Qiu et al. [36] developed the AlNiCoRuMo HEA by a melting and de-alloying process. The catalyst exhibited an outstanding OER activity. The addition of Mo makes the e_g-orbital occupation of both Co and Ru close to unity, which results in a high OER of the catalyst. Li et al. [37] designed a FeCoNiMnMo HEA by molten salt electrolysis of oxides and this catalyst showed lower overpotentials (279 mV at 10 mA cm^{-2}) and smaller tafel slopes (56.1 mV dec^{-1}). However, the combination of Fe, Co, Ni, Mo, and Zn in OER catalysis has not already been reported and its interactions are not clear. Considering that both Mo and Zn have promotive effects on OER electrocatalysis, it is therefore possible that the combination of Mo and Zn may have positive synergistic effects on FeCoNiMoZn catalysts.

Herein, we synthesized the FeCoNiMoZn-OH-layered high entropy hydroxide (HEH) on the surface of nickel foam (NF) via the pulse electrodeposition method under a mild atmosphere. The obtained FeCoNiMoZn-OH HEH exhibits a three-dimensional (3D) porous morphology, which consists of a large number of nanosheets. This unique structure ensures an ample specific surface area and facilitates the ions transfer. The obtained FeCoNiMoZn-OH catalyst afford outstanding OER catalytic performance under alkaline condition, including a lower overpotential of 248 mV at 10 mA cm^{-2}, a smaller tafel slope of 30 mV dec^{-1}, and a long-term durability of 55 h compared with the commercial RuO_2.

2. Experimental Section

2.1. Reagents and Materials

Ferrous (II) sulfate heptahydrate ($FeSO_4 \cdot 7H_2O$, 99%), sodium dodecylsulfate (SDS), hydrochloric acid (HCl, 37%), and anhydrous athanol (C_2H_5OH, 99.7%) were purchased from Feng Chuan Chemical Reagent Co., Ltd. (Tianjin, China). Cobalt sulfate heptahydrate ($CoSO_4 \cdot 7H_2O$, 99.5%) was bought from Macklin Biochemical Co., Ltd. (Shanghai, China). Isopropyl alcohol (C_3H_8O, >99%) and nickel sulfate hexahydrate ($NiSO_4 \cdot 6H_2O$, 99%) was obtained from Aladdin Chemical Technology Co., Ltd. (Shanghai, China). Sodium molybdate dihydrate ($Na_2MoO_4 \cdot 2H_2O$, $\geq$99.5%) was obtained from Dibo Chemicals Technology Co., Ltd. (Shanghai, China) Zinc(II) sulfate heptahydrate ($ZnSO_4 \cdot 7H_2O$), trisodium citrate dihydrate ($Na_3C_6H_5O_7 \cdot 2H_2O$, $\geq$99.5%), and potassium hydroxide (KOH, 85%) were supplied by Kermel Chemical Reagent Co., Ltd. (Tianjin, China). Ruthenium (IV) oxide (RuO_2, 97%) was bought from Haohong Scientific Co., Ltd. (Shanghai, China). Nickel foam (NF) was obtained from Leviathan Technology Co., Ltd. (Tianjin, China).

2.2. Synthesis of HEH Catalyst

FeCoNiMoZn-OH HEH was synthesized on the surface of NF by a pulse electrodeposition method. Firstly, a NF (1mm thick, 110 ppi, exposed area of 1 cm^2) was ultrasonically cleaned in hydrochloric acid, ethanol, and deionized water (DI). Secondly, a conventional three-electrode system was used to prepare the FeCoNiMoZn-OH HEH catalyst. The working electrode was the cleaned NF. The reference electrode and counter electrode were the Ag/AgCl electrode and graphite rod, separately. The distance between the working electrode and the counter electrode was 2.50 cm. In total, 0.04 M $FeSO_4$, 0.04 M $CoSO_4$, 0.05 M $NiSO_4$, 0.025 M Na_2MoO_4, 0.03 M $ZnSO_4$, 0.1 M $Na_3C_6H_5O_7$, and 0.15 g L^{-1} SDS were dissolved sequentially into an aqueous solution to obtain the electrolyte. The FeCoNiMoZn HEH was electrodeposited by square pulse potential mode at 25 °C using an electrochemical workstation (CHI 760E, Chenhua Instruments, Inc., Shanghai, China). The upper potential limit and pulse duration were 0 V and 0.1 s; the lower potential limit and pulse duration were set as −2 V and 0.1 s, respectively. The entire electrode position process lasted for 3000 cycles. Thirdly, the obtained electrode was cleaned with DI water repeatedly. FeCoNi-OH, FeCoNiMo-OH, and FeCoNiZn-OH catalysts were prepared following a similar process to FeCoNiMoZn-OH HEH. The concentrations of the corresponding metal salts were kept consistent throughout the synthesis.

2.3. Materials Characterization

The morphology of FeCoNiMoZn-OH HEH was characterized by a field-emission scanning electron microscope (FE-SEM, JSM-7800F, JEOL, Tokyo, Japan) and a high-resolution transmission electron microscope (HRTEM, JEM-F200, JEOL, Tokyo, Japan) equipped with energy-dispersive X-ray spectroscopy (EDX). The phase characteristics of the sample were investigated via the X-ray powder diffractometer (XRD, Bruker D8 Advanced, Ettlingen, Germany). The chemical states of the obtained catalyst elements were detected via X-ray photoelectron spectroscopy (XPS, AXIS SUPRA, Kratos, Manchester, UK) using an Al Kα X-ray as the exciting source. The coordination environment as well as functional group within the material were characterized by Raman spectroscopy (Thermo Fischer DXR, Waltham, MA, USA) and Fourier transform infrared spectroscopy (FTIR, Thermo Fisher is 10, Waltham, MA, USA).

2.4. Electrochemical Tests

In this work, OER catalytic performance tests were conducted with a standard three-electrode system (the electrochemical workstation, CHI 760E). The working electrode was NF-loaded with catalysts and the counter electrode and reference electrode were Pt foil and Ag/AgCl, separately. The electrolyte was 1.0 M KOH aqueous solution. All potentials (vs. Ag/AgCl) were converted into potentials against the reversible hydrogen electrode (RHE) ($E_{RHE} = E_{Ag/AgCl} + 0.059\ pH + 0.197$ V). Before all electrochemical testing, cyclic voltammetry (CV) was conducted for 20 cycles at 40 mV s^{-1} to energize the catalyst activity. Linear sweep voltammetry (LSV) was employed to measure the catalytic efficiency for the OER of the samples. The LSV polarization curves ranged from 1.2 to 1.7 V (vs. RHE) and were collected at 5 mV s^{-1}. Tafel slopes of these samples were determined from the LSV curves (the linear region). Tafel slopes were estimated according to equation η = b log j + a, where η is the overpotential of the electrode, j is the current density, b is the tafel slope, and a is the intercept of the fitted curve. All the polarization potentials obtained in the LSV test were implemented with iR compensation (90%). Electrochemical impedance spectroscopy (EIS) tests were performed in a KOH solution at the overpotential corresponding to the current density of 10 mA cm^{-2}. The lower- and upper-frequency limits during the test were 0.01 Hz and 100 kHz, separately. The obtained catalysts were subjected to CV tests in 1.0 M KOH solution at different scan rates and CV curves ranging from 0.72 to 0.82 V (vs. RHE) were collected to evaluate the electrochemical active surface area (ECSA). The ECSA of the samples were evaluated by the equation ECSA = C_{dl}/C_s, where C_{dl} (double-layer capacitance) was obtained by plotting $\Delta j = (j_a - j_c)$ against the scan rates, which is half of the value of the linear slope. C_s (specific capacitance) was set as 40 μF cm^{-2} [38]. The long-term durability was evaluated by chronopotentiometry tests. For comparison, the benchmark RuO_2 electrodes (2.5 mg cm^{-2}) were fabricated. Overall, 10 mg of commercial RuO_2 powder, 35 μL Nafion (5 wt. % in ethanol), and 965 μL of isopropanol were mixed and then ultrasounded for 30 min, followed by the formation of a homogeneous ink suspension. Then, 250 μL of this suspension was evenly drop-coated on the surface of NF and the sample was finally dried. In all testing, each set of tests is performed more than three times to ensure that the obtained data are within the permissible range of error.

3. Results and Discussion

Compared with the smooth morphology of bare nickel foam (Figure S1), the surface of the deposited NF is entirely covered by the FeCoNiMoZn-OH HEH catalyst layer (Figure 1a,b). The HEH layer exhibits the 3D porous morphology, which consists of a large number of nanosheets (Figure 1b). This unique 3D porous structure not only favors the provision of ample specific surface area but also promotes ions/mass transportation. Figure 1c displays the XRD spectrum of the FeCoNiMoZn-OH HEH. The main diffraction peaks appear at 19.3°, 33.1°, 38.4°, 51.8°, 60.2°, 70.8°, and 81.2°, indexing to (001), (100), (002), (012), (111), (201), and (202) of FeCoNiMoZn-OH HEH (JCPDS 74-1057). This indicates that the FeCoNiMoZn-OH HEH has a polycrystalline structure. To further investigate the microstructure of FeCoNiMoZn-OH HEH, Figure 1d shows a TEM image of FeCoNiMoZn-OH HEH. The HEH possesses the ultrathin nanosheet structure, which is in accordance with the morphology shown in SEM (Figure 1b). The HRTEM of FeCoNiMoZn-OH HEH (Figure 1e) displays the interplanar spacing of 0.15, 0.20, and 0.23 nm (Figure 1e), indicating (111), (202), and (002) planes of FeCoNiMoZn-OH HEH, respectively. Also, FeCoNiMoZn-OH HEH displays a large number of lattice defects, demonstrating the lattice distortion effect of high entropy materials. The red and green symbols indicate the dislocation and stacking faults, separately. The selected area electron diffraction (SAED) pattern exhibits multiple sets of bright diffraction concentric rings (Figure 1f), further confirming the polycrystalline structure of FeCoNiMoZn-OH HEH. Furthermore, the EDS line and mapping spectra of FeCoNiMoZn-OH HEH show the existence of Fe, Co, Ni, Mo, Zn, and O elements (Figures S2 and S3, Supplementary Materials; Figure 1g) and these elements are uniformly distributed on the whole sample (Figure 1g). The atomic weight percentage of the corresponding elements is also listed in Table S1 (Supplementary Materials).

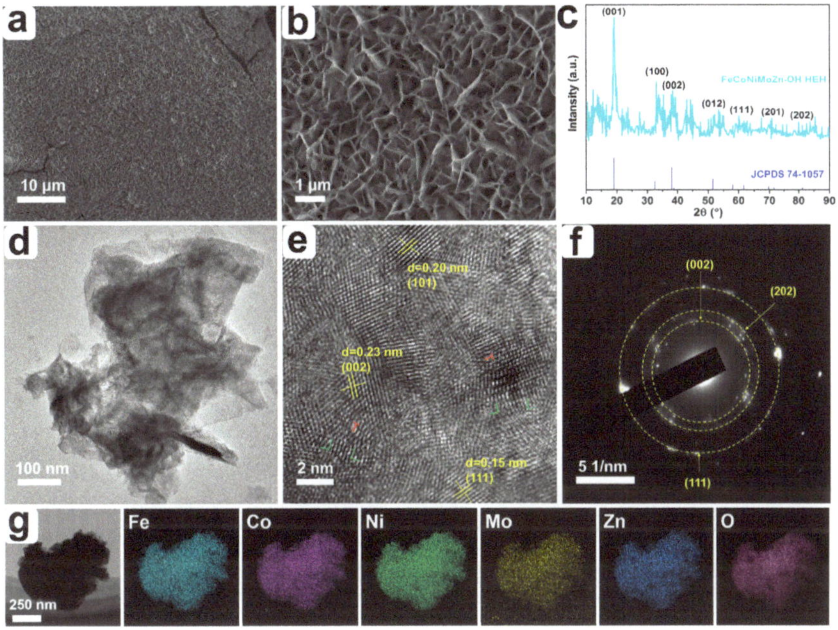

Figure 1. (**a**,**b**) SEM images, (**c**) XRD spectrum, (**d**) TEM image, (**e**) HRTEM image, (**f**) the corresponding SAED pattern, and (**g**) elemental mappings of FeCoNiMoZn-OH HEH.

An XPS test was conducted to characterize the chemical states of FeCoNiMoZn-OH HEH. The XPS survey spectra of FeCoNiMoZn-OH HEH present the peaks of Fe 2p, Co 2p, Ni 2p, Mo 3d, Zn 2p, and O 1s (Figure S4, Supplementary Materials). The high-resolution spectrum of Fe 2p (Figure 2a) shows peaks at 706.2 eV, 708.8 eV, and 711.9 eV, corresponding to Fe^0 $2p_{3/2}$, Fe^{2+} $2p_{3/2}$, and Fe^{3+} $2p_{3/2}$. The peaks appear at 718.2 eV, 720.6 eV, and 722.5 eV, indexing Fe^0 $2p_{1/2}$, Fe^{2+} $2p_{1/2}$, and Fe^{3+} $2p_{1/2}$ [39]. Two typical satellite peaks of Fe at 715 and 724.2 eV were also observed [3]. These suggest the presence of metallic Fe and iron hydroxides/oxides. For the Co 2p spectrum (Figure 2b), the peaks appear at 777.4 eV, 780.4 eV, 793.2 eV, and 794.9 eV, which index to Co^0 $2p_{3/2}$, Co^{2+} $2p_{3/2}$, Co^0 $2p_{1/2}$, and Co^{2+} $2p_{1/2}$ separately [39]. Two typical satellite peaks are situated at 784.1 and 799.5 eV. Obviously, Co has only a zero and divalent valence state. In the high-resolution spectrum of Ni 2p (Figure 2c), two typical satellite peaks of Ni are found at 858.3 and 875.9 eV. Moreover, the peaks at 852.0 eV, 853.2 eV, 869.6 eV, and 871.9 eV conform to Ni^0 $2p_{3/2}$, Ni^{2+} $2p_{3/2}$, Ni^0 $2p_{1/2}$, and Ni^{2+} $2p_{1/2}$, separately, which demonstrates the presence of metallic Ni and divalent Ni compounds [18,21]. The high-resolution spectrum of Mo 3d (Figure 2d) displays the peaks located at 230.9 eV, 232.0 eV, 233.7 eV, and 235.2 eV, assigning to Mo^{4+} $3d_{5/2}$, Mo^{6+} $3d_{5/2}$, Mo^{4+} $3d_{3/2}$, and Mo^{6+} $3d_{3/2}$ separately [3]. The two peaks of the Zn 2p spectrum at 1020.8 and 1043.9 eV belong to Zn^{2+} $2p_{3/2}$ and Zn^{2+} $2p_{1/2}$ separately (Figure 2e) [40]. In the O 1s spectrum, the peaks at 531.8 and 533.1 eV indicate M-OH and H_2O (Figure 2f) [41]. The intense signal of M-OH implies that the synthesized sample is mainly composed of hydroxides, which is also confirmed by the conclusion of XRD analysis (Figure 1c).

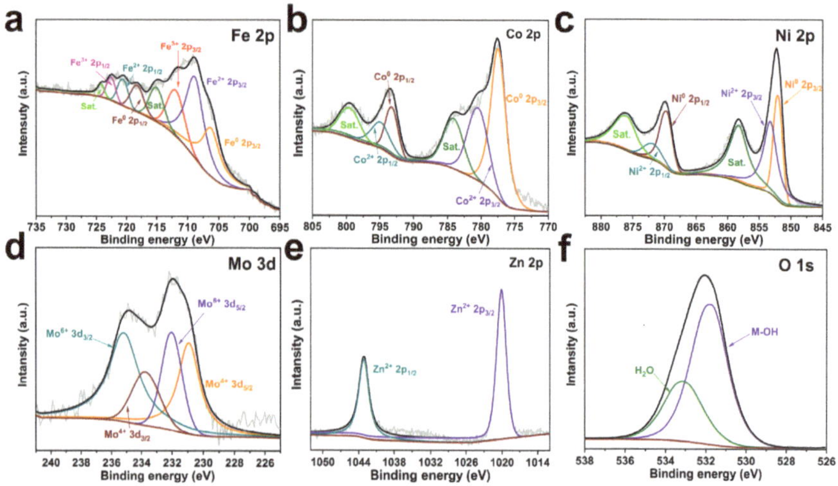

Figure 2. XPS high-resolution spectra of FeCoNiMoZn-OH HEH. (**a**) Fe 2p, (**b**) Co 2p, (**c**) Ni 2p, (**d**) Mo 3d, (**e**) Zn 2p, and (**f**) O 1s spectra of FeCoNiMoZn-OH HEH.

To understand the coordination environment of FeCoNiMoZn-OH HEH, Raman spectroscopy measurement was performed (Figure S5, Supplementary Materials). The prominent Raman signals at about 460 and 530 cm^{-1} are derived from the existence of metal-hydroxide (M-OH) bonds [42]. In addition, the spectral features of FeCoNiMoZn-OH HEH are relatively broad and diffuse. This demonstrates the lower crystallinity of the obtained catalysts and the higher density of lattice defects [43]. FTIR spectrum revealed the presence of oxygen-containing groups in FeCoNiMoZn-OH HEH (Figure S6, Supplementary Materials). The peaks at 3380 cm^{-1} and 1650 cm^{-1} indicate the hydroxyl radicals and the interlayer water, separately [44]. These are in accordance with the findings of XRD and XPS.

The OER catalytic activity of the obtained samples was revealed by LSV curves tested in O_2-saturated 1.0 M KOH aqueous solution. It is found that FeCoNiMoZn-OH HEH demonstrates the most excellent OER catalytic activity compared to FeCoNiMo-OH, FeCoNiZn-OH, FeCoNi-OH, RuO_2, and NF samples (Figure 3a). Figure 3b shows the histogram of overpotential at different current densities. FeCoNiMoZn-OH HEH possesses the lowest overpotential of 248 mV at the current density of 10 mA cm^{-2} in comparison to FeCoNiMo-OH (267 mV), FeCoNiZn-OH (287 mV), and FeCoNi-OH (280 mV) catalysts and commercial RuO_2 (261 mV). Moreover, FeCoNiMoZn-OH HEH also has the lowest overpotential of only 269 mV at 50 mA cm^{-2} among the obtained samples, followed by FeCoNiMo-OH (310 mV), FeCoNi-OH (328 mV), and FeCoNiZn-OH (332 mV) catalysts and commercial RuO_2 (340 mV). Obviously, the introduction of Zn almost has no effect on the overpotential of FeCoNiZn-OH catalyst compared to FeCoNi-OH. In contrast, the introduction of Mo significantly lowers the overpotential of FeCoNiMo-OH, suggesting that Mo can promote the catalytic efficiency of OER. Although Zn has little effect on lowering the overpotential of FeCoNiZn-OH, its introduction further reduces the overpotential of FeCoNiMoZn-OH. This demonstrates the positive synergistic effect of Mo and Zn in FeCoNiMoZn-OH HEH. The Tafel slope is a significant indicator to reflect the kinetic characteristics of the OER-catalyzed process. Figure 3c shows that FeCoNiMoZn-OH HEH has the smallest tafel slope of 30 mV dec^{-1} in comparison to FeCoNi-OH (53 mV dec^{-1}), FeCoNiMo-OH (42 mV dec^{-1}), FeCoNiZn-OH (45 mV dec^{-1}), and RuO_2 (108 mV dec^{-1}), demonstrating the superior OER kinetics of FeCoNiMoZn-OH HEH.

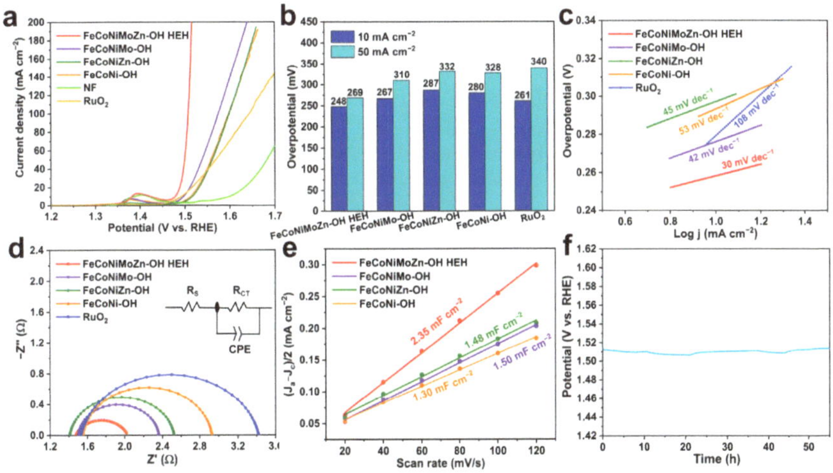

Figure 3. Electrocatalytic performance of FeCoNiMoZn-OH electrocatalysts. (**a**) LSV, (**b**) histogram of overpotential, (**c**) Tafel slope, (**d**) EIS, (**e**) C_{dl} values, and (**f**) chronopotentiometry curve.

Figure 3d displays the typical Nyquist plots of the obtained catalysts and the corresponding equivalent electrical circuit. R_{ct} and R_s represent charge transfer resistance and electrolyte solution resistance, separately. Among all obtained electrocatalysts, FeCoNiMoZn-OH HEH displays the smallest diameter of the semi-arc of R_{ct}, demonstrating its superior properties in charge transfer. This is conducive to the acceleration of OER and therefore improves the catalytic activity. Additionally, the CV curves of the catalysts tested at various scan rates were collected (Figure S7, Supplementary Materials) to estimate the ECSA of the obtained catalysts. ECSA is determined by C_{dl} (ECSA = C_{dl}/C_s, the C_s was set as 40 μF cm^{-2}) [21]. As shown in Figure 3e, FeCoNiMoZn-OH HEH electrocatalyst displays the largest C_{dl} value (2.35 mF cm^{-2}) among the obtained catalysts, almost two times higher than the FeCoNi-OH catalyst (1.30 mF cm^{-2}). Correspondingly, the FeCoNiMoZn-OH HEH has the largest ECSA of 58.8 cm^2 compared with FeCoNiMo-OH (37.5 cm^2), FeCoNiZn-OH (37.0 cm^2), and FeCoNi-OH (32.5 cm^2) catalysts (Figure S8, Supplementary Materials), suggesting the most potential active sites of FeCoNiMoZn-OH HEH. This is ascribed to the abundant structural defects of FeCoNiMoZn-OH HEH.

The LSV curves normalized by ECSA are employed to investigate the intrinsic activity of the obtained electrocatalysts. FeCoNiMoZn-OH HEH has the highest intrinsic activity among the obtained samples, followed by FeCoNiMo-OH, FeCoNi-OH, and FeCoNiZn-OH catalysts (Figure S9, Supplementary Materials). The introduction of Zn results in a slight reduction in the intrinsic activity of FeCoNiZn-OH eletrocatalyst compared to FeCoNi-OH. This may be attributed to the fact that Zn itself is catalytically inactive and occupies some active sites or covers the active sites. On the contrary, the introduction of Mo plays a vital role in improving the intrinsic activity of the FeCoNiMo-OH HEH catalyst. As a high valence metal, Mo allows fast multi-electron transfer between the species with multiple oxidation valence states, favoring the improvements in catalytic intrinsic activity [45]. Interestingly, the combination of inactive Zn and Mo in FeCoNiMoZn-OH HEH produces a positive synergistic effect between Zn and Mo, which results in the further improvement in the intrinsic activity of FeCoNiMoZn-OH HEH. It is possible that the Zn can effectively tune the amounts of Mo^{4+} and Mo^{6+} and improve the distribution of high valence ions, thus regulating the adsorption process of the OER intermediates.

In addition, the mass of the prepared samples before and after electrodeposition were measured and the loading of the obtained catalysts was compared (Table 1). There was little difference in the loading (ΔM) of the prepared catalysts. Moreover, the turnover frequency (TOF) of FeCoNiMoZn-OH HEH is calculated assuming that all materials obtained by deposition are active sites. Notably, the TOF of the FeCoNiMoZn-OH HEH is as high as 1.81 s^{-1} at 10 mA cm^{-2} (Figure S10), suggesting an enhanced electrocatalytic oxygen production capability. Compared with other reported advanced catalysts, the obtained FeCoNiMoZn-OH HEH exhibits a high TOF (Table S2, Supplementary Materials).

Table 1. Comparison of the mass of the prepared samples before and after electrodeposition.

Materials	M_1 (mg)	M_2 (mg)	ΔM (mg)
FeCoNiMoZn-OH HEH	52.5	55.7	3.2
FeCoNiMo-OH	53.7	56.6	2.9
FeCoNiZn-OH	51.6	55.1	3.5
FeCoNi-OH	52.6	56.4	3.8

M_1 is the mass of bare NF and M_2 is the mass of the sample after electrodeposition. ΔM indicates the difference in mass before and after deposition.

Moreover, the long-term durability of catalysts is regarded as another crucial indicator of catalyst performance. Figure 3f shows the chronopotentiometry curve of the FeCoNiMoZn-OH HEH tested in an alkaline aqueous solution. The OER polarization potential was consistently maintained at about 1.51 V (vs. RHE) for 55 h at 10 mA cm^{-2}, demonstrating its satisfactory long-term durability. To reveal the change in morphology, structure, and composition of FeCoNiMoZn-OH HEH after stability testing for 55 h, a series of characterizations of the obtained samples were carried out. After stability testing for 55 h, FeCoNiMoZn-OH HEH remains the integrated 3D porous morphology consisting of plenty of nanosheets (Figure 4a–c), demonstrating its excellent structural stability. The elements of Fe, Co, Ni, Mo, Zn, and O are still observed and uniformly distributed in FeCoNiMoZn-OH HEH (Figure 4d), further implying excellent OER durability of FeCoNiMoZn-OH HEH.

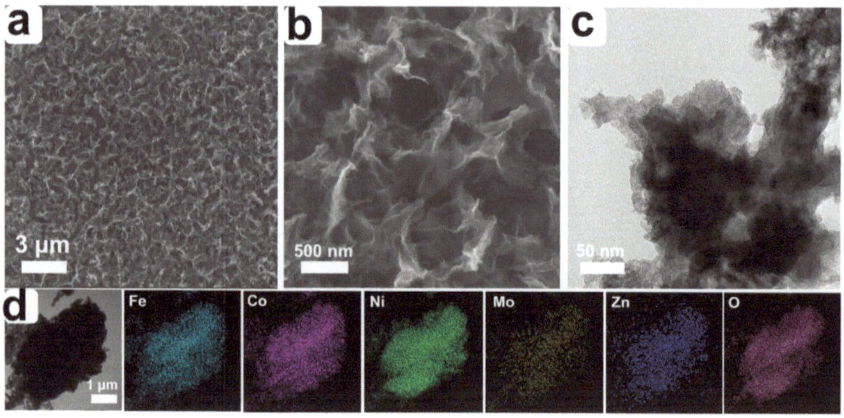

Figure 4. (**a**,**b**) FESEM images, (**c**) TEM image, and (**d**) corresponding elemental mappings of FeCoNiMoZn-OH HEH after stability testing for 55 h.

4. Comparison with the Literature

Table 2 shows the comparison of FeCoNiMoZn-OH HEH with the state-of-the-art catalysts reported in previous work. The obtained FeCoNiMoZn-OH HEH exhibits the extraordinary catalytic activity of OER (overpotential of 240 mV at 10 mA cm^{-2}, Tafel slope of 30 mV dec^{-1}) in 1 M KOH aqueous solution compared to the catalysts reported in previous work.

Table 2. Comparison of OER performance of FeCoNiMoZn HEH catalyst with the state-of-the-art electrocatalysts.

Catalysts	Electrolyte	η_{10} (mV)	Tafel Slope (mV dec^{-1})	Synthesis Method	Ref.
Co_3O_4-$MoSe_2$@C	1 M KOH	360	75.7	hydrothermal method	[46]
Co-Mo_2C@NC	1 M KOH	440	156	solvothermal method	[47]
$Co_2B/CoSe_2$	1 M KOH	320	56	wet chemical method	[48]
Ce-$MnCo_2O_4$	1 M KOH	390	125	coprecipitation and calcination	[49]
$La_{0.5}Sr_{0.5}Mn_{0.15}Fe_{0.15}Co_{0.4}Ni_{0.15}Cu_{0.15}O_3$	1 M KOH	309	93.4	electrospinning technology	[50]
MnFeCoNi HEA	1 M KOH	302	83.7	Mechanical alloying	[51]
$(Fe_{0.2}Co_{0.2}Ni_{0.2}Cr_{0.2}Mn_{0.2})_3O_4$	1 M KOH	275	50.3	solution combustion	[52]
CoFeNiMnMoPi	1 M KOH	270	74	high-temperature fly-through method	[53]
$(CoNiMnZnFe)_3O_{3.2}$	1 M KOH	336	47.5	mechanical alloying	[54]
FeCoNiCuIr NPs	1 M KOH	360	70.1	heat-up method	[55]
$(Cr_{0.2}Mn_{0.2}Fe_{0.2}Ni_{0.2}Zn_{0.2})_3O_4$	1 M KOH	295	53.7	solvothermal method	[56]
$Cu_{0.5}Fe_{0.5}NNi_2Co_{0.5}Fe_{0.5}$	1 M KOH	370	55	chemical solution deposition	[57]
AlNiCuCoFeY	1 M KOH	261	50	melt spinning	[58]
FeCoNiMnCu	1 M KOH	280	59	electrodeposition	[39]
FeCoNiMnW	1 M KOH	512	161	electrodeposition	[4]
$(Fe_{0.2}Co_{0.2}Ni_{0.2}Mn_{0.2}Cr_{0.2})_3O_4$@CC	1 M KOH	287	95.8	electrodeposition	[59]
CNFMPO	1 M KOH	252	44.3	electrodeposition	[60]
This work	1 M KOH	248	30	electrodeposition	–

5. Conclusions

In summary, a novel FeCoNiMoZn-OH-layered HEH was synthesized by electrodeposition under a mild atmosphere. The obtained FeCoNiMoZn-OH exhibits 3D porous nanosheet morphology and a polycrystalline structure. This structure possesses abundant lattice defects and a large number of surface unsaturated coordination active atoms, providing plenty of active sites. The FeCoNiMiZn-OH HEH demonstrates excellent OER catalytic performance under alkaline conditions, including low overpotential (248 mV at 10 mA cm^{-2}) and outstanding durability. The positive synergistic effect of Mo and Zn optimizes the electronic structure of HEH and thus modulates the adsorption process of intermediates, enhancing the intrinsic activity for OER. The obtained catalyst in this work exhibits excellent catalytic performance under laboratory conditions and is expected to be practically applied to the electrolysis of water. This study will afford an efficient strategy for the design and synthesis of novel high-entropy catalysts with high performance.

Supplementary Materials: The following supporting information can be downloaded at https://www.mdpi.com/article/10.3390/nano14100889/s1. Figure S1: SEM images of bare nickel foam; Figure S2: SEM image and line spectrum of FeCoNiMoZn-OH HEH; Figure S3: EDS spectrum of FeCoNiMoZn-OH HEH; Figure S4: XPS survey spectrum of FeCoNiMoZn-OH sample; Figure S5: Raman spectrum of FeCoNiMoZn-OH sample; Figure S6: FTIR spectrum of FeCoNiMoZn-OH sample; Figure S7: The CV curves of four samples at different scan rate (a) FeCoNiMoZn-OH, (b) FeCoNiMo-OH, (c) FeCoNiZn-OH, and (d) FeCoNi-OH; Figure S8: Histogram of electrochemically active area of four samples; Figure S9: The LSV curves normalized by ECSA; Figure S10: TOF curves of FeCoNiMoZn-OH HEH derived from LSV; Table S1: Weight and atomic ratios of the corresponding elements of FeCoNiMoZn-OH HEH; Table S2: Comparison of TOF at the corresponding overpotentials of FeCoNiMoZn-OH HEH catalyst with recently reported other electrocatalysts. References [61–67] are cited in the Supplementary Materials.

Author Contributions: Conceptualization, J.L. and Z.C.; methodology, Z.C., X.H., J.Z. and J.L.; validation, J.L.; investigation, Z.C. and X.H.; data curation, Z.C., X.H. and L.H.; writing—original draft preparation, Z.C. and J.L.; writing—review and editing, J.L.; supervision, C.Z.; project administration, J.L.; funding acquisition, J.L. and Z.W. All authors have read and agreed to the published version of the manuscript.

Funding: This research was funded by the National Natural Science Foundation of China (Nos. 52271224 and 51801134) and the Tianjin Natural Science Foundation (20JCQNJC01130).

Data Availability Statement: Data are contained within the article.

Conflicts of Interest: The authors declare no conflicts of interest.

References

1. Bian, H.; Wang, R.; Zhang, K.; Zheng, H.; Wen, M.; Li, Z.; Li, Z.; Wang, G.; Xie, G.; Liu, X.; et al. Facile electrodeposition synthesis and super performance of nano-porous Ni-Fe-Cu-Co-W high entropy alloy electrocatalyst. *Surf. Coat. Technol.* **2023**, *459*, 129407. [CrossRef]
2. Chen, Z.; Huang, K.; Zhang, T.; Xia, J.; Wu, J.; Zhang, Z.; Zhang, B. Surface modified CoCrFeNiMo high entropy alloys for oxygen evolution reaction in alkaline seawater. *Processes* **2023**, *11*, 245. [CrossRef]
3. Cao, W.; Yang, X.; Dai, W.; Wu, B.; Zhang, Y.; Zhao, C.; Sui, Y.; Huang, S. An amorphous FeCoNiMnCr high-entropy alloy supported by 2H-MoS_2 on carbon cloth as a highly efficient and robust electrocatalyst for water oxidation. *New J. Chem.* **2023**, *47*, 12670–12677. [CrossRef]
4. Chang, S.-Q.; Cheng, C.-C.; Cheng, P.Y.; Huang, C.-L.; Lu, S.-Y. Pulse electrodeposited FeCoNiMnW high entropy alloys as efficient and stable bifunctional electrocatalysts for acidic water splitting. *Chem. Eng. J.* **2022**, *446*, 137452. [CrossRef]
5. Chen, S.; Wang, Y.; Wang, Z.; Zhang, K. Advances in nonprecious metal catalysts for efficient water oxidation in alkaline media. *Ionics* **2023**, *29*, 9–32. [CrossRef]
6. Du, X.; Yin, L.; Zhang, W.; Zhang, M.; Su, K.; Li, Z. Synergistic coupling of $Ni_3ZnC_{0.7}$ decorated with homogeneous multimetal CoNiCuFe nitrogen-codoped carbon matrix as high-entropy catalysts for efficient overall water splitting. *J. Mater. Sci. Technol.* **2023**, *135*, 26–33. [CrossRef]
7. Battiato, S.; Bruno, L.; Pellegrino, A.L.; Terrasi, A.; Mirabella, S. Optimized electroless deposition of NiCoP electrocalysts for enhanced water splitting. *Catal. Today* **2023**, *423*, 113929. [CrossRef]
8. Cai, Z.-X.; Goou, H.; Ito, Y.; Tokunaga, T.; Miyauchi, M.; Abe, H.; Fujita, T. Nanoporous ultra-high-entropy alloys containing fourteen elements for water splitting electrocatalysis. *Chem. Sci.* **2021**, *12*, 11306–11315. [CrossRef]
9. Abdelhafiz, A.; Wang, B.; Harutyunyan, A.R.; Li, J. Carbothermal Shock Synthesis of High Entropy Oxide Catalysts: Dynamic Structural and Chemical Reconstruction Boosting the Catalytic Activity and Stability toward Oxygen Evolution Reaction. *Adv. Energy Mater.* **2022**, *12*, 2200742. [CrossRef]
10. Gao, G.J.; Xu, J.L.; Tang, J.; Zhang, L.W.; Ma, Y.C.; Luo, J.M. Plasma nitrided CoCrFeNiMn high entropy alloy coating as a self-supporting electrode for oxygen evolution reaction. *J. Mater. Res. Technol.* **2023**, *23*, 5357–5367. [CrossRef]
11. Liao, S.; Huang, T.; Wu, W.; Yang, T.; Hou, Q.; Sang, S.; Liu, K.; Yang, Y.; Liu, H. Engineering High-Entropy Duel-Functional nanocatalysts with regulative oxygen vacancies for efficient overall water splitting. *Chem. Eng. J.* **2023**, *471*, 144506. [CrossRef]
12. Park, H.K.; Ahn, H.; Lee, T.H.; Lee, J.Y.; Lee, M.G.; Lee, S.A.; Yang, J.W.; Kim, S.J.; Ahn, S.H.; Kim, S.Y.; et al. Grain Boundaries Boost Oxygen Evolution Reaction in NiFe Electrocatalysts. *Small Methods* **2021**, *5*, 2000755. [CrossRef]
13. Liu, S.; Wan, R.; Lin, Z.; Liu, Z.; Liu, Y.; Tian, Y.; Qin, D.-D.; Tang, Z. Probing the Co role in promoting the OER and Zn-air battery performance of NiFe-LDH: A combined experimental and theoretical study. *J. Mater. Chem. A* **2022**, *10*, 5244–5254. [CrossRef]
14. He, R.; Yang, L.; Zhang, Y.; Jiang, D.; Lee, S.; Horta, S.; Liang, Z.; Lu, X.; Moghaddam, A.O.; Li, J.; et al. A 3d-4d-5d high entropy alloy as a bifunctional oxygen catalyst for robust aqueous Zinc-air batteries. *Adv. Mater.* **2023**, *35*, e2303719. [CrossRef]
15. Duan, Y.; Yu, Z.-Y.; Hu, S.-J.; Zheng, X.-S.; Zhang, C.-T.; Ding, H.-H.; Hu, B.-C.; Fu, Q.-Q.; Yu, Z.-L.; Zheng, X.; et al. Scaled-up Synthesis of Amorphous NiFeMo Oxides and Their Rapid Surface Reconstruction for Superior Oxygen Evolution Catalysis. *Angew. Chem. Int. Ed.* **2019**, *58*, 15772–15777. [CrossRef]
16. Cui, M.; Yang, C.; Li, B.; Dong, Q.; Wu, M.; Hwang, S.; Xie, H.; Wang, X.; Wang, G.; Hu, L. High-entropy metal sulfide nanoparticles promise high-performance oxygen evolution reaction. *Adv. Energy Mater.* **2021**, *11*, 2002887. [CrossRef]
17. Yang, P.; Shi, Y.; Xia, T.; Jiang, Z.; Ren, X.; Liang, L.; Shao, Q.; Zhu, K. Novel self-supporting thin film electrodes of FeCoNiCrMn high entropy alloy for excellent oxygen evolution reaction. *J. Alloys Compd.* **2023**, *938*, 168582. [CrossRef]
18. Wang, Q.; Li, Y.; Yang, Y.; Zhang, X.; Guo, Y.; Jia, Z.; Shen, B. Self-supported AlFeNiCoMo high-entropy alloy with micropillar array structure for efficient oxygen evolution reaction. *APL Mater.* **2022**, *10*, 101106. [CrossRef]
19. Han, M.; Wang, C.; Zhong, J.; Han, J.; Wang, N.; Seifitokaldani, A.; Yu, Y.; Liu, Y.; Sun, X.; Vomiero, A.; et al. Promoted self-construction of beta-NiOOH in amorphous high entropy electrocatalysts for the oxygen evolution reaction. *Appl. Catal. B Environ.* **2022**, *301*, 120764. [CrossRef]
20. Lu, C.; Li, M.; Zhang, X.; Hou, H.; Li, X.; Yang, X.; Liu, X.; Ding, Y.; Hou, J.; Wang, Y. Low-temperature synthesized amorphous quasi-high-entropy carbonate electrocatalyst with superior surface self-optimization for efficient water oxidation. *Ceram. Int.* **2023**, *49*, 12156–12165. [CrossRef]
21. Huang, K.; Zhang, B.; Wu, J.; Zhang, T.; Peng, D.; Cao, X.; Zhang, Z.; Li, Z.; Huang, Y. Exploring the impact of atomic lattice deformation on oxygen evolution reactions based on a sub-5 nm pure face-centred cubic high-entropy alloy electrocatalyst. *J. Mater. Chem. A* **2020**, *8*, 11938–11947. [CrossRef]
22. Strotkoetter, V.; Krysiak, O.A.; Zhang, J.; Wang, X.; Suhr, E.; Schuhmann, W.; Ludwig, A. Discovery of High-Entropy Oxide Electrocatalysts: From Thin- FilmMaterial Libraries to Particles. *Chem. Mater.* **2022**, *34*, 10291–10303. [CrossRef] [PubMed]
23. Yan, X.; Zhang, X.; Wang, F.; Stockdale, T.; Dzenis, Y.; Nastasi, M.; Cui, B. Fabrication of ODS Austenitic Steels and CoCrFeNi High-Entropy Alloys by Spark Plasma Sintering for Nuclear Energy Applications. *JOM* **2019**, *71*, 2856–2867. [CrossRef]
24. Lindner, T.; Loebel, M.; Saborowski, E.; Rymer, L.-M.; Lampke, T. Wear and Corrosion Behaviour of Supersaturated Surface Layers in the High-Entropy Alloy Systems CrMnFeCoNi and CrFeCoNi. *Crystals* **2020**, *10*, 110. [CrossRef]

25. Yao, Y.; Huang, Z.; Xie, P.; Lacey, S.D.; Jacob, R.J.; Xie, H.; Chen, F.; Nie, A.; Pu, T.; Rehwoldt, M.; et al. Carbothermal shock synthesis of high-entropy-alloy nanoparticles. *Science* **2018**, *359*, 1489–1494. [CrossRef] [PubMed]
26. Redka, D.; Gadelmeier, C.; Winter, J.; Spellauge, M.; Eulenkamp, C.; Calta, P.; Glatzel, U.; Minar, J.; Huber, H.P. Sub-picosecond single-pulse laser ablation of the CrMnFeCoNi high entropy alloy and comparison to stainless steel AISI 304. *Appl. Surf. Sci.* **2021**, *544*, 148839. [CrossRef]
27. Park, J.; Kim, H.; Kim, S.Y.; Ahn, S.H. Empirical approach for configuring high-entropy catalysts in alkaline water electrolysis. *Int. J. Energy Res.* **2022**, *46*, 9938–9947. [CrossRef]
28. Cechanaviciute, I.A.; Antony, R.P.; Krysiak, O.A.; Quast, T.; Dieckhoefer, S.; Saddeler, S.; Telaar, P.; Chen, Y.-T.; Muhler, M.; Schuhmann, W. Scalable synthesis of multi-metal electrocatalyst powders and electrodes and their application for oxygen evolution and water splitting. *Angew. Chem. Int. Ed.* **2023**, *62*, e202218493. [CrossRef]
29. Abdelhafiz, A.; Tanvir, A.N.M.; Zeng, M.; Wang, B.; Ren, Z.; Harutyunyan, A.R.; Zhang, Y.; Li, J. Pulsed light synthesis of high entropy nanocatalysts with enhanced catalytic activity and prolonged stability for oxygen evolution reaction. *Adv. Sci.* **2023**, *10*, 2300426. [CrossRef]
30. He, L.; Wang, N.; Sun, B.; Zhong, L.; Yao, M.; Hu, W.; Komarneni, S. High-entropy FeCoNiMn (oxy)hydroxide as high-performance electrocatalyst for OER and boosting clean carrier production under quasi-industrial condition. *J. Clean Prod.* **2022**, *356*, 131680. [CrossRef]
31. Bian, H.; Wang, C.; Zhao, S.; Han, G.; Xie, G.; Qi, P.; Liu, X.; Zeng, Y.; Zhang, D.; Wang, P. Preparation of highly efficient high-entropy alloy catalysts with electrodeposition and corrosion engineering for OER electrocatalysis. *Int. J. Hydrogen Energy* **2024**, *57*, 651–659. [CrossRef]
32. Mei, Y.; Feng, Y.; Zhang, C.; Zhang, Y.; Qi, Q.; Hu, J. High-Entropy Alloy with Mo-Coordination as Efficient Electrocatalyst for Oxygen Evolution Reaction. *ACS Catal.* **2022**, *12*, 10808–10817. [CrossRef]
33. Huang, C.-L.; Lin, Y.-G.; Chiang, C.-L.; Peng, C.-K.; Raja, D.S.; Hsieh, C.-T.; Chen, Y.-A.; Chang, S.-Q.; Yeh, Y.-X.; Lu, S.-Y. Atomic scale synergistic interactions lead to breakthrough catalysts for electrocatalytic water splitting. *Appl. Catal. B Environ.* **2023**, *320*, 122016. [CrossRef]
34. Huang, J.; Wang, P.; Li, P.; Yin, H.; Wang, D. Regulating electrolytic $Fe_{0.5}CoNiCuZn_x$ high entropy alloy electrodes for oxygen evolution reactions in alkaline solution. *J. Mater. Sci. Technol.* **2021**, *93*, 110–118. [CrossRef]
35. Zhang, J.; Wang, K.; Fan, L.; Liu, H.; Zhu, H.; Yan, S. High-valence metal doped Co_2FeAl alloy as efficient noble-metal-free electrocatalyst for alkaline hydrogen evolution reaction. *J. Alloys Compd.* **2023**, *933*, 167613. [CrossRef]
36. Jin, Z.; Lyu, J.; Zhao, Y.-L.; Li, H.; Lin, X.; Xie, G.; Liu, X.; Kai, J.-J.; Qiu, H.-J. Rugged high-entropy alloy nanowires with in situ formed surface spinel oxide as highly stable electrocatalyst in Zn-air batteries. *ACS Mater. Lett.* **2020**, *2*, 1698–1706. [CrossRef]
37. Li, P.; Wan, X.; Su, J.; Liu, W.; Guo, Y.; Yin, H.; Wang, D. A Single-Phase FeCoNiMnMo High-Entropy Alloy Oxygen Evolution Anode Working in Alkaline Solution for over 1000 h. *ACS Catal.* **2022**, *12*, 11667–11674. [CrossRef]
38. Zhang, T.; Li, J.; Zhang, B.; Wang, G.; Jiang, K.; Zheng, Z.; Shen, J. High-entropy alloy CuCrFeNiCoP film of Cu-based as high-efficiency electrocatalyst for water splitting. *J. Alloys Compd.* **2023**, *969*, 172439. [CrossRef]
39. Huang, K.; Peng, D.; Yao, Z.; Xia, J.; Zhang, B.; Liu, H.; Chen, Z.; Wu, F.; Wu, J.; Huang, Y. Cathodic plasma driven self-assembly of HEAs dendrites by pure single FCC FeCoNiMnCu nanoparticles as high efficient electrocatalysts for OER. *Chem. Eng. J.* **2021**, *425*, 131533. [CrossRef]
40. Sharma, L.; Katiyar, N.K.; Parui, A.; Das, R.; Kumar, R.; Tiwary, C.S.; Singh, A.K.; Halder, A.; Biswas, K. Low-cost high entropy alloy (HEA) for high-efficiency oxygen evolution reaction (OER). *Nano Res.* **2022**, *15*, 4799–4806. [CrossRef]
41. Tang, J.; Xu, J.L.; Ye, Z.G.; Li, X.B.; Luo, J.M. Microwave sintered porous CoCrFeNiMo high entropy alloy as an efficient electrocatalyst for alkaline oxygen evolution reaction. *J. Mater. Sci. Technol.* **2021**, *79*, 171–177. [CrossRef]
42. Bai, Y.; Wu, Y.; Zhou, X.; Ye, Y.; Nie, K.; Wang, J.; Xie, M.; Zhang, Z.; Liu, Z.; Cheng, T.; et al. Promoting nickel oxidation state transitions in single-layer NiFeB hydroxide nanosheets for efficient oxygen evolution. *Nat. Commun.* **2022**, *13*, 6094. [CrossRef] [PubMed]
43. Triolo, C.; Moulaee, K.; Ponti, A.; Pagot, G.; Di Noto, V.; Pinna, N.; Neri, G.; Santangelo, S. Spinel-Structured High-Entropy Oxide Nanofibers as Electrocatalysts for Oxygen Evolution in Alkaline Solution: Effect of Metal Combination and Calcination Temperature. *Adv. Funct. Mater.* **2024**, *34*, 2306375. [CrossRef]
44. Ding, Y.; Wang, Z.; Liang, Z.; Sun, X.; Sun, Z.; Zhao, Y.; Liu, J.; Wang, C.; Zeng, Z.; Fu, L.; et al. A Monolayer High-Entropy Layered Hydroxide Frame for Efficient Oxygen Evolution Reaction. *Adv. Mater.* **2023**, 2302860. [CrossRef] [PubMed]
45. Li, J.; Gu, X.; Chang, J.; Wu, D.; Xu, F.; Jiang, K.; Gao, Z. Molybdenum oxide-iron, cobalt, copper alloy hybrid as efficient bifunctional catalyst for alkali water electrolysis. *J. Colloid Interface Sci.* **2022**, *606*, 1662–1672. [CrossRef] [PubMed]
46. Yu, W.; Zhang, S.; Gao, K.; Lin, X.; Han, Y.; Zhang, Z. Co_3O_4-$MoSe_2$@C nanocomposite as a multi-functional catalyst for electrochemical water splitting and lithium-oxygen battery. *Nanotechnology* **2022**, *33*, 505402. [CrossRef] [PubMed]
47. Wang, M.; Dipazir, S.; Lu, P.; Wang, Y.; Yuan, M.; Li, S.; Zhang, G. Synthesis of polyoxometalates derived bifunctional catalyst towards efficient overall water splitting in neutral and alkaline medium. *J. Colloid Interface Sci.* **2018**, *532*, 774–781. [CrossRef] [PubMed]
48. Guo, Y.; Yao, Z.; Shang, C.; Wang, E. Amorphous Co_2B Grown on $CoSe_2$ Nanosheets as a Hybrid Catalyst for Efficient Overall Water Splitting in Alkaline Medium. *ACS Appl. Mater. Interfaces* **2017**, *9*, 39312–39317. [CrossRef] [PubMed]

49. Huang, X.; Zheng, H.; Lu, G.; Wang, P.; Xing, L.; Wang, J.; Wang, G. Enhanced water splitting electrocatalysis over $MnCo_2O_4$ via introduction of suitable Ce content. *ACS Sustain. Chem. Eng.* **2019**, *7*, 1169–1177. [CrossRef]
50. Hao, J.; Ma, F.; Chen, Y.; Lu, S.; Duan, F.; Du, M.; Wang, C.; Zhang, W.; Zhu, H. Cocktail effect in high-entropy perovskite oxide for boosting alkaline oxygen evolution. *New J. Chem.* **2024**, *48*, 511–514. [CrossRef]
51. Dai, W.; Lu, T.; Pan, Y. Novel and promising electrocatalyst for oxygen evolution reaction based on MnFeCoNi high entropy alloy. *J. Power Sources* **2019**, *430*, 104–111. [CrossRef]
52. He, L.; Kang, H.; Hou, G.; Qiao, X.; Jia, X.; Qin, W.; Wu, X. Low-temperature synthesis of nano-porous high entropy spinel oxides with high grain boundary density for oxygen evolution reaction. *Chem. Eng. J.* **2023**, *460*, 141675. [CrossRef]
53. Qiao, H.; Wang, X.; Dong, Q.; Zheng, H.; Chen, G.; Hong, M.; Yang, C.-P.; Wu, M.; He, K.; Hu, L. A high-entropy phosphate catalyst for oxygen evolution reaction. *Nano Energy* **2021**, *86*, 106029. [CrossRef]
54. Zhang, Y.; Dai, W.; Zhang, P.; Lu, T.; Pan, Y. In-situ electrochemical tuning of $(CoNiMnZnFe)_3O_{3.2}$ high-entropy oxide for efficient oxygen evolution reactions. *J. Alloys Compd.* **2021**, *868*, 159064. [CrossRef]
55. Cai, C.; Xin, Z.; Zhang, X.; Cui, J.; Lv, H.; Ren, W.; Gao, C.; Cai, B. Facile Synthesis of FeCoNiCuIr High Entropy Alloy Nanoparticles for Efficient Oxygen Evolution Electrocatalysis. *Catalysts* **2022**, *12*, 1050. [CrossRef]
56. Xu, Y.; Sun, L.; Li, Q.; Huo, L.; Zhao, H. Co-prosperity of electrocatalytic activity and stability in high entropy spinel $(Cr_{0.2}Mn_{0.2}Fe_{0.2}Ni_{0.2}Zn_{0.2})_3O_4$ for the oxygen evolution reaction. *J. Mater. Chem. A* **2022**, *10*, 17633–17641. [CrossRef]
57. Zhu, L.; Li, C.; Zheng, R.; Cheng, W.; He, Y.; Gong, C.; Liu, M.; Huang, Y.; Zhu, X.; Sun, Y. Design of high-entropy antiperovskite metal nitrides as highly efficient electrocatalysts for oxygen evolution reaction. *Int. J. Hydrogen Energy* **2024**, *51*, 638–647. [CrossRef]
58. Yang, P.; An, Y.; Feng, C.; Liu, Y.; Liu, S.; Gao, L.; Zhou, Y.; Li, X.; Li, P.; Zeng, F. Heterogeneous high-entropy catalyst nanoparticles for oxygen evolution reaction: Impact of oxygen and fluorine introduction. *Int. J. Hydrogen Energy* **2024**, *51*, 1218–1228. [CrossRef]
59. Zhang, R.; Xu, Z.; Du, Z.; Wan, Y.; Yuan, S.; Zeng, F.; Xu, J.; Meng, Z.; Hu, X.; Tian, H. Electrodeposition of Self-Supported High-Entropy Spinel Oxides for Stable Oxygen Evolution. *Inorg. Chem.* **2023**, *62*, 19052–19059. [CrossRef]
60. Zhang, H.-M.; Zuo, L.; Gao, Y.; Guo, J.; Zhu, C.; Xu, J.; Sun, J. Amorphous high-entropy phosphoxides for efficient overall alkaline water/seawater splitting. *J. Mater. Sci. Technol.* **2024**, *173*, 1–10. [CrossRef]
61. Xie, Y.; Feng, Y.; Jin, S.; Li, C.; Li, C.; Sun, Y.; Luo, F.; Yang, Z. Nickel-doped iridium echinus-like nanosheets for stable acidic water splitting. *Chem. Commun.* **2023**, *59*, 8404–8407. [CrossRef] [PubMed]
62. Khiarak, B.N.; Zahraei, A.A.; Nazarzade, K.; Hasanjani, H.R.A.; Mohammadzadeh, H. Shape-controlled synthesis of thorn-like 1D phosphorized Co supported by Ni foam electrocatalysts for overall water splitting. *J. Mater. Sci. Mater. Electron.* **2021**, *32*, 18363–18370. [CrossRef]
63. Battiato, S.; Urso, M.; Cosentino, S.; Pellegrino, A.L.; Mirabella, S.; Terrasi, A. Optimization of Oxygen Evolution Reaction with Electroless Deposited Ni-P Catalytic Nanocoating. *Nanomaterials* **2021**, *11*, 3010. [CrossRef] [PubMed]
64. Shen, K.; Wang, T.; Li, C.; Chen, M.; Niu, L.; Gong, Y. Designing highly-efficient oxygen evolution reaction FeCoNiCrMnOx electrocatalyst via coexisted crystalline and amorphous Phases: Experiment and theory. *Appl. Surf. Sci.* **2024**, *650*, 159102. [CrossRef]
65. Liu, F.; Yu, M.; Chen, X.; Li, J.; Liu, H.; Cheng, F. Defective high-entropy rocksalt oxide with enhanced metal-oxygen covalency for electrocatalytic oxygen evolution. *Chin. J. Catal.* **2022**, *43*, 122–129. [CrossRef]
66. Zhu, W.; Wang, X.; Zhu, Y.; Fang, L.; Yao, C.; Song, X.; Chen, H.; Wang, X.; Zhu, G. Amorphous Fe/Co-based tannic acid salts as robust oxygen evolution pre-catalysts. *New J. Chem.* **2023**, *47*, 19087–19095. [CrossRef]
67. Liu, X.; Feng, L.; Li, Y.; Xia, T.; Sui, Z.; Chen, Q. Covalent Organic Frameworks Composites Containing Bipyridine Metal Complex for Oxygen Evolution and Methane Conversion. *Molecules* **2022**, *27*, 5193. [CrossRef]

nanomaterials

Article

Controllable Synthesis of Titanium Silicon Molecular Zeolite Nanosheet with Short b-Axis Thickness and Application in Oxidative Desulfurization

Tieqiang Ren [1,2], Yujia Wang [1,2], Lulu Wang [1,2], Lisheng Liang [3], Xianming Kong [2,*] and Haiyan Wang [1,2,*]

1 College of Chemistry and Chemical Engineering, China University of Petroleum (East China), Qingdao 266580, China; tqr_lnpu_121@126.com (T.R.); fsyzww@126.com (Y.W.); gracewangl@163.com (L.W.)
2 School of Petrochemical Engineering, Liaoning Petrochemical University, Fushun 113001, China
3 Petroleum Engineering Research Institute, Petrochina Dagang Oil Field Company, Tianjin 300280, China; 13312041240@163.com
* Correspondence: xmkong@lnpu.edu.cn (X.K.); fswhy@126.com (H.W.)

Abstract: Titanium silicon molecular zeolite (TS-1) plays an important role in catalytic reactions due to its unique nanostructure. The straight channel on TS-1 was parallel to the orientation of the short b-axis and directly exposed to the aperture of the 10-member ring with a diameter of 0.54 nm × 0.56 nm. This structure could effectively reduce the diffuse restriction of bulk organic compounds during the oxidative desulfurization process. As a kind of cationic polymer electrolyte, polydimethyldiallyl ammonium chloride (PDDA) contains continuous $[C_8H_{16}N^+Cl^-]$ chain segments, in which the Cl^- could function as an effective structure-directing agent in the synthesis of nanomaterials. The chain of PDDA could adequately interact with the [0 1 0] plane in the preparation process of zeolite, and then the TS-1 nanosheet with short b-axis thickness (6 nm) could be obtained. The pore structure of the TS-1 nanosheet is controlled by regulating the content of PDDA. Scanning electron microscopy (SEM), transmission electron microscopy (TEM), X-ray photoelectron spectroscopy (XPS), N_2 physical adsorption analysis, infrared absorption spectrum and ultraviolet–visible spectrum were used to determine the TS-1. The thinner nanosheets exhibit excellent catalytic performance in oxidative desulfurization of dibenzothiophene (DBT), in which the removal rate could remain at 100% after three recycles. Here, the TS-1 nanosheet with short b-axis thickness has a promising future in catalytic reactions.

Keywords: titanium silicon molecular zeolite; PDDA; nanosheets; tunable short b-axis thickness; desulfurization

Citation: Ren, T.; Wang, Y.; Wang, L.; Liang, L.; Kong, X.; Wang, H. Controllable Synthesis of Titanium Silicon Molecular Zeolite Nanosheet with Short b-Axis Thickness and Application in Oxidative Desulfurization. *Nanomaterials* **2024**, *14*, 953. https://doi.org/10.3390/nano14110953

Academic Editor: Diego Cazorla-Amorós

Received: 16 April 2024
Revised: 24 May 2024
Accepted: 27 May 2024
Published: 29 May 2024

1. Introduction

Fine desulfurization of fossil fuels has attracted considerable interest as an environmental pollution problem [1,2]. There are a lot of desulfurization methods, including oxidative desulfurization (ODS), which is an effective, high-efficiency technology used to obtain fuels with lower sulfur content under mild conditions [3]. TS-1 was first reported in 1983 [4]. Numerous research studies of TS-1 were developed in selective oxidation and epoxidation reactions of organic compounds, in which hydrogen peroxide (H_2O_2) and ter-butyl hydroperoxide (TBHP) were usually applied as oxidants. As an environmentally friendly feature, H_2O_2 is usually employed in catalytic reactions under mild reaction conditions [5].

TS-1 has been extensively applied for manufacturing oxygenates, in which H_2O_2 is usually employed as an oxidizing organic reactant, such as in the hydroxylation of benzene, epoxidation of olefins, and ammoxidation of cyclohexanone [6,7]. However, the small size (0.55 nm) of the micropore and framework structure in TS-1 zeolite hinders the diffusion of organic sulfur, especially for the molecules that are adsorbed on Ti-active centers. The microspore property of TS-1 further constrains the mass transfer of sulfides, which limits its application in oxidative desulfurization [8,9]. The catalytic oxidation performance of TS-1 is highly dependent on the content of Ti and the accessibility of the target in the active site of the catalyst.

The hierarchical TS-1 has attained considerable attention as an additional mesoporous pore formed between intercrystalline and interconnected intracrystalline, and is beneficial for mass transfer, which is a feature that could improve the efficiency of catalytic reactions [10,11]. The hierarchical structure in TS-1 zeolite could provide a mass transfer route and enhance the accessibility of the active sites in the TS-1 framework. Several methods [12–16] have been proposed for the preparation of mesoporous zeolites with hierarchical structures, such as the partial crystallization of amorphous mesoporous structure [17], zeolite seeds assisting in building primary structure units [18], the crystallization of mesocellular silica foam into a hierarchical structure [19], and a hydrothermal approach using a special surfactant [20]. As a kind of surfactant molecule, the cationic polymers can electrostatically adsorb onto nanocrystals and generate zeolite nanoparticles with hierarchical structure through a self-assembly process [21,22]. The spherical micelle, cylindrical micelle, and lamellar micelle could be formed through the self-assembly of surfactant molecules that could serve as a template for producing the mesopore structures of zeolite [23–25]. These nanoscopic micelle structures could incorporate meso-macropores into microporous zeolites with substantial structures [26,27].

Some researchers have confirmed that the single-crystalline zeolite beta with interconnected mesopores could be synthesized by using a commercial polymer, PDDA, as a dual-functional template [28]. The single-crystalline nature endowed Beta-MS with better hydrothermal stability and exhibited remarkably higher catalytic activity compared with surfactant-derived mesoporous zeolite Beta. PDDA has also been successfully applied in the regulation of hierarchical TS-1 zeolites with abundant intracrystalline mesopores and regular hexagonal morphology [29]. However, the manufacture of lamellar material with a hierarchical structure has not been reported, and the PDDA content should be a crucial factor.

In this work, a PDDA-assisted synthetic strategy was developed for constructing a hierarchical TS-1 stacked nanosheet with tunable short b-axis thickness, in which the PDDA mainly served as a mesopore template and structure-directing agent. TS-1 zeolite was successfully synthesized through a PDDA electrostatic adsorption route under conventional hydrothermal process, and the TS-1 nanosheets (PTS) showed superior catalytic performance compared with conventional TS-1 zeolite in the oxidative desulfurization of thiophenic sulfur [30].

2. Materials and Methods

2.1. Reagents and Chemicals

Tetraethyl orthosilicate (TEOS, 98%, Shanghai Macklin Biochemical Co., Ltd., Shanghai, China), tetrapropylammonium hydroxide (TPAOH, 25%, Beijing InnoChem Science & Technology Co., Ltd., Beijing, China), tetrabutyl titanate (TBOT, 98%, Guangfu Fine Chemical Research Institute, Tianjin, China), polydiallyldimethyl ammonium chloride (PDDA, Mw 100,000~200,000, 20%, Shanghai Yuanye Bio-Technology Co., Ltd., Shanghai, China), noctane (95%, Shanghai Macklin Biochemical Co., Ltd.), hydrogen peroxide solution (H_2O_2, 30%, Sigma-Aldrich chemicals Co., Ltd., St. Louis, MO, USA), dibenzothiophene (DBT, 98%, TCI Shanghai Development Co., Ltd., Shanghai, China), and ethanol (100%, Tianjin Wind Boat Chemical Reagent Technology Co., Ltd., Tianjin, China) were used in this experiment.

2.2. Synthesis of Conventional TS-1(CTS)

Conventional TS-1 zeolite was prepared using a hydrothermal method in which a mixture of $SiO_2/TiO_2/TPAOH/H_2O/EtOH$ with a molar ratio of 30/1/12/2100/120 was used as a starting gel. Typically, 5.1915 g TPAOH (25%) was mixed with deionized water under violent stirring for 30 min and then 3.32 g TEOS was added dropwise under stirring for another hour. Finally, the mixture of 0.181 g TBOT and 2.94 g ethanol was dispersed into the above solution under continuous stirring for 2 h. The resultant gel was transferred into a Teflon-lined stainless-steel autoclave. After crystallization at 150 °C for 48 h, the solid product was centrifuged twice, rinsed thoroughly with deionized water, and dried overnight at 80 °C. Then, the product was calcined at 550 °C for 8 h in air condition, and the conventional TS-1 (CTS) was obtained.

2.3. Synthesis of TS-1 Nanosheets (PTS)

The starting gel of SiO_2/TiO_2/TPAOH/H_2O/EtOH with a molar ratio of 30/1/12/2100/120 was prepared according to the above procedure, and then PDDA was dispersed in the starting gel under stirring for 1 h. The content of PDDA in the starting gel was set at x wt %, where x is 0.010, 0.015, 0.030, 0.090, and 0.180, respectively. Following the above procedure of crystallization and calcination, the TS-1 nanosheets with ultra-thin lamellar structure were obtained and named x PTS.

2.4. Characterizations

The X-ray diffraction (XRD) spectra were obtained with a Bruker D8 Advance diffractometer (Bruker, Billerica, MA, USA) equipped with Cu Kα (λ = 1.5418 Å, 40 kV, 40 mA). Transmission electron microscopy (TEM) images and selected area electron diffraction (SAED) patterns were observed on a JEM 2100F electron microscope (JEOL Ltd., Tokyo, Japan) with an accelerating voltage of 200 kV. Infrared (FT-IR) spectra were collected using a Nicolet iS50 infrared instrument (Nicolet Instrument Corporation, Madison, WI, USA) with a KBr compression method in the range of 400 cm^{-1} to 4000 cm^{-1}. The UV–vis diffuse reflectance spectra (DRS) were recorded in the range between 200 nm and 400 nm with an Agilent Cary5000 spectrophotometer (Agilent, Santa Clara, CA, USA). N_2 adsorption–desorption experiments were performed with an automated gas sorption analyzer of Quantachrome autosorb iQ_2. XPS spectroscopy (ESCALAB 250Xi+, Thermo Fisher Scientific, Waltham, MA, USA) was employed to obtain the core-level information of zeolite under the test condition of passing energy of 20 eV, residence time of 50 ms, step length of 0.05 eV, and 20 runs. The content of sulfur was measured with a TSN5000 sulfur analyzer (Jiangsu Electric Analysis Instrument, Jiangsu Skyray Instrument, Kunshan, China).

2.5. Catalytic Performance Test

The solution of noctane with DBT was used as the model fuel, in which the concentration of sulfur was 500 ppm. The oxidative desulfurization reaction was performed in a conical flask (50 mL) reactor at 60 °C and equipped with an air condenser tube. In a typical reaction, 20 mL of model fuel, 15 mL of H_2O, 0.05 mL of H_2O_2, and 0.15 g of a catalyst were added into the reactor. The oxidative desulfurization reaction was carried out at 60 °C for 105 min. The upper oil phase was extracted at a 15 min interval. After that, the concentration of organic sulfur in the oil phase was measured and the removal rate (%) was calculated based on the results.

3. Results and Discussion

3.1. Morphology and Characterization of CTS and PTS Zeolites

The TEM images of the TS-1 zeolites are shown in Figure 1. The bright gray areas in the images were due to the uniform micropores and hierarchical mesopores inside the zeolites. There is an obvious difference between the morphologies of zeolites prepared with different conditions. The bright gray regions with an irregular distribution of hierarchical pore structures of TS-1 zeolites were observed (Figure 1a–c). In contrast, the continuous mesopores were observed from the 0.030PTS, 0.090PTS, and 0.180PTS zeolites presented in the regular bright gray regions (Figure 1d–f). The high-magnification TEM image of 0.030PTS is shown in Figure 1g, in which the ultra-thin lamella structure (6 nm) was observed and continuous mesopores were marked with white lines. As shown in the SAED pattern (Figure 1h,i), the 0.030PTS zeolite showed single-crystalline properties and was presented in the distinguishable [0 1 0] plate. Three lattice planes of [1 0 1], [2 0 1], and [3 0 1] were observed from the 0.030PTS, and the straight channels of microporous structures were well exposed in 0.030PTS crystal along the short b-axis orientation. Based on the TEM images and SAED pattern, it could be concluded that PTS mesocrystals of nanosheets have an orderly stacked structure. The exposed [0 1 0] plates, ultra-thin lamellas (6 nm), and textural properties (Table 1) were beneficial for the oxidation desulfuration.

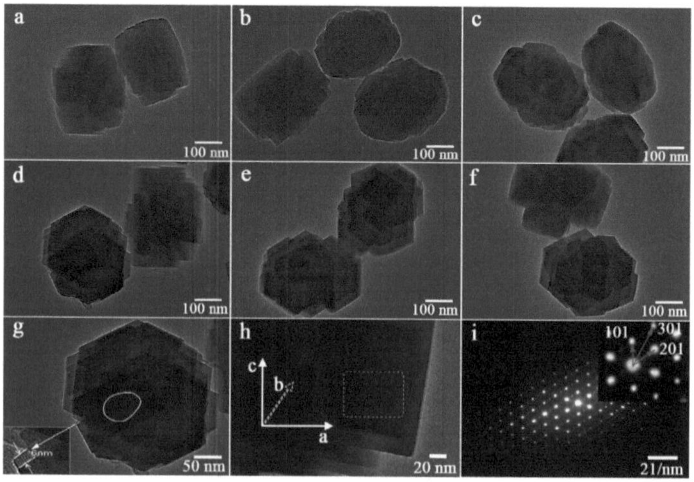

Figure 1. TEM images of CTS and PTS prepared with different amounts of PDDA: (**a**) CTS. (**b**) 0.010PTS. (**c**) 0.015PTS. (**d**) 0.030PTS. (**e**) 0.090PTS. (**f**) 0.180PTS. (**g**,**h**) High-magnification TEM for 0.030PTS. (**i**) The SAED pattern of selected region in (**h**) for 0.030PTS.

Table 1. The texture properties of CTS and PTS zeolites as prepared.

Sample	S^{a}_{total} (m^2/g)	S^{b}_{micro} (m^2/g)	$S^{c}_{external}$ (m^2/g)	V^{d}_{total} (cm^{-3}/g)	V^{e}_{micro} (cm^{-3}/g)	V^{f}_{meso} (cm^{-3}/g)	$d^{g}_{average}$ (nm)
CTS	509	423.9	85.11	0.4687	0.1914	0.2773	3.683
0.010PTS	508	368.6	139.5	0.4858	0.1797	0.3061	3.824
0.015PTS	509.3	396.1	113.2	0.3532	0.1825	0.1707	2.774
0.030PTS	504.6	386.8	117.8	0.3220	0.1722	0.1498	2.552
0.090PTS	503.2	397.8	105.4	0.3809	0.1837	0.1972	3.028
0.180PTS	516.3	406	110.3	0.3430	0.1880	0.1550	2.658

[a] Specific surface area calculated using BET equation. [b] Micropore area calculated using t-plot method. [c] External surface area. [d] Total pore volume at 0.995 of p/p_0. [e] Micropore volume calculated using t-plot method. [f] Mesopore volume resulting from total pore volume subtracting micropore volume. [g] Average pore diameter calculated by BET specific surface area and total pore volume of cylindrical pore model.

The N_2 adsorption–desorption isotherms of CTS and PTS zeolites are shown in Figure 2a. All samples presented a typical I adsorption isotherm based on the IUPAC classification, which indicated that the TS-1 zeolites have a typical microporous structure. The pore distribution of the cylindrical pore model for TS-1 zeolites was calculated using the density functional theory (DFT) method as shown in Figure 2b, in which the intrinsic micropores, hierarchical mesopores, and continuous mesopores were obtained. The micropore volumes of all zeolites ranged from 0.1722 to 0.1914 $cm^3 \cdot g^{-1}$ (Table 1, e), and the difference in micropore volumes was caused by the intercrystalline microchannel continuity. These results indicated that the intrinsic micropore of the MFI structure was almost unchanged with the different dosages of PDDA used in the starting gel.

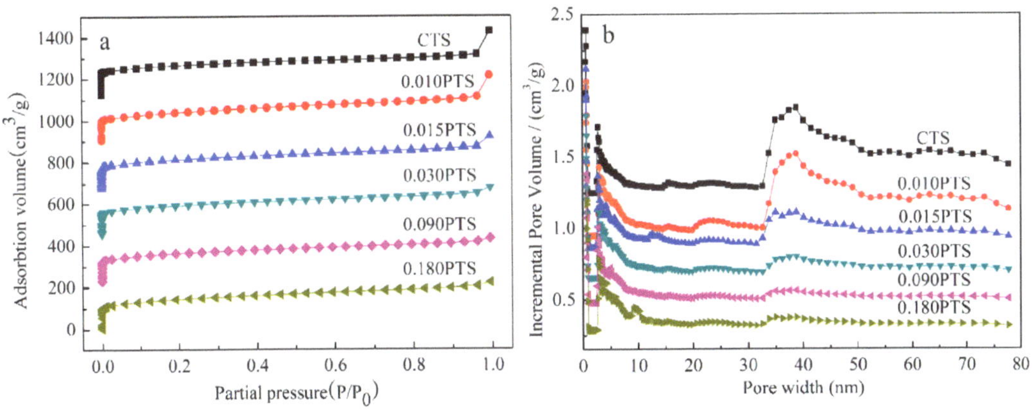

Figure 2. Nitrogen adsorption–desorption analysis of CTS and PTS zeolites: (**a**) Adsorption–desorption isotherm. (**b**) Pore size distribution of cylindrical pore model using DFT method.

The mesoporous and microporous surface areas of CTS zeolite were 85.11 $m^2 \cdot g^{-1}$ and 423.9 $m^2 \cdot g^{-1}$, respectively. In contrast, the surface areas of 0.010PTS zeolite were 139.5 $m^2 \cdot g^{-1}$ and 368.6 $m^2 \cdot g^{-1}$, as shown in (Table 1, b,c). The variation of the surface areas of 0.010PTS was due to the strong interaction between the polymer chain of PDDA and the growing crystal for zeolite, in which the PDDA functioned as a template agent for generating more mesoporous volume (Table 1, f). However, the external specific surface and mesoporous volume were decreased as the dosage of PDDA was further increased in the starting gel. As result, the 0.030PTS zeolite had a relatively high external specific surface and the lowest mesoporous volume. Obviously, The PDDA could generate more external surfaces, and the [0 1 0] plates of PTS zeolites were even exposed (Figure 1i). It was concluded that the PDDA was most likely adsorbed on the [0 1 0] plate and restricted the growth of crystals along the b-axis orientation.

The XRD patterns of CTS and PTS zeolites were measured and shown in Figure 3. Five diffraction peaks were observed at 7.9°, 8.8°, 23.1°, and 23.9°, which were assigned to the typical MFI structure of TS-1 zeolites (PDF#01-070-6302) [30,31]. There was no diffraction peak observed at 25.4°, which indicated the absence of the TiO_2 anatase phase. Furthermore, no additional peaks were observed, which confirmed the high phase purity of the zeolites. The diffraction peaks of CTS, 0.010PTS, and 0.180PTS zeolites showed higher intensity than the other samples because they had bigger crystal sizes of 40.8 nm, 37.0 nm, and 36.8 nm, as calculated by the Scherrer formula. Meanwhile, the other three zeolites had proximate and lower diffraction intensity, which indicated that the smaller crystal grains were formed. The 0.030PTS zeolite only showed a crystal size of 27.3 nm. It is noteworthy that the weak diffraction peaks of the [3 0 1] crystal plane could be observed at 14.926° from 0.015PTS, 0.030PTS, and 0.090PTS zeolites (Figure 3b). This had undoubtedly been observed in the TEM images (Figure 1i). The species adsorbed by PDDA between adjacent nanosheets could bring continuous mesoporous channels, accompanied by the changes in TS-1 zeolite aperture distributions (Figure 2b).

FT-IR spectra was used to determine the surface groups of TS-1 zeolites (Figure 4a), and six obvious peaks were observed at 450 cm^{-1}, 550 cm^{-1}, 800 cm^{-1}, 970 cm^{-1}, 1100 cm^{-1}, and 1230 cm^{-1}. The peaks at 550 cm^{-1}, 800 cm^{-1}, and 1230 cm^{-1} belonged to the stretching vibrations of the -O-Si-O- and -O-Ti-O- groups from the double five-membered rings [32]. The peak at 970 cm^{-1} was attributed to the stretching vibration of the Si-O-Ti bond and Si-O bond that was perturbed by the framework titanium species [33,34]. The intensity ratio of I970/I800 was commonly used to evaluate the relative content of Ti species in the framework structure, which was 1.18, 1.09, 1.14, 1.10, 1.13, and 1.12 corresponding to CTS, 0.010PTS, 0.015PTS, 0.030PTS, 0.090PTS, and 0.180PTS, respectively. The intensity ratios of I970/I800 indicated that the relative content of Ti in the framework structure was decreased by adding PDDA into the precursor gel. Meanwhile, it was also accompanied by the increment of surface Ti species content (Table 2, b).

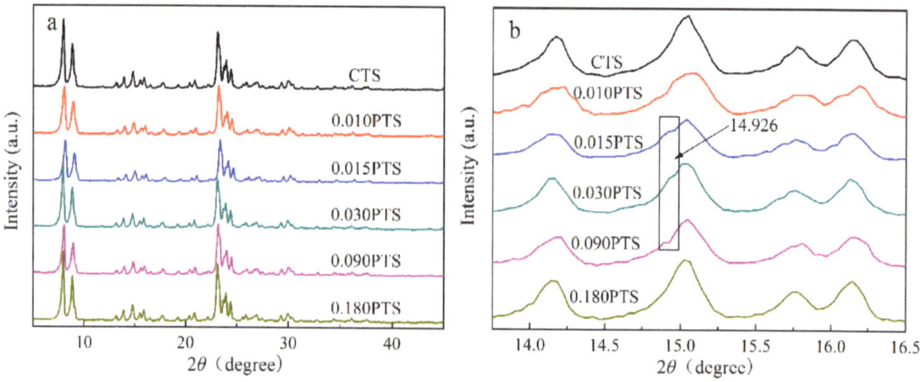

Figure 3. XPD powder patterns: (**a**) XRD characterization of CTS and PTS zeolites. (**b**) The [3 0 1] crystal plate of 0.030PTS zeolite.

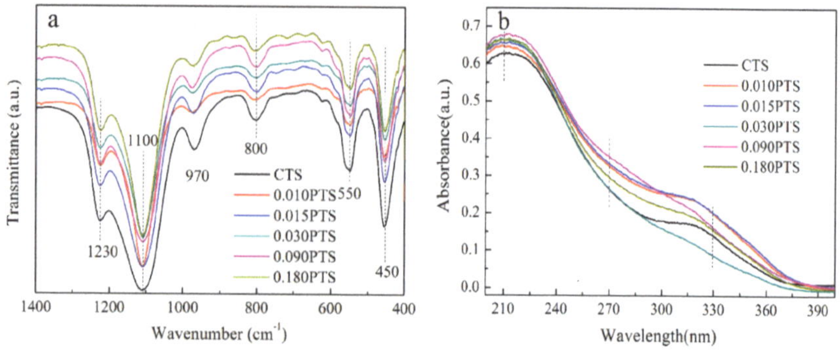

Figure 4. Structure composition analysis: (**a**) FT-IR spectra. (**b**) UV–vis spectra.

Table 2. Surface atomic composition by XPS and Ti relative content by FT-IR.

Sample	$Si^{a}_{Atomic\%}$	$Ti^{b}_{Atomic\%}$	$Si/Ti^{c}_{Atomicratio}$	$Non-Framework\ Ti^{d}_{Atomic\%}$	$I^{e}_{970/800}$
CTS	31.64	0.69	45.85	0.21	1.18
0.010PTS	30.73	0.79	38.89	0.29	1.09
0.015PTS	30.11	1.00	30.11	0.52	1.14
0.030PTS	31.03	0.95	32.66	0.48	1.10
0.090PTS	30.88	1.10	28.07	0.59	1.13
0.180PTS	28.74	1.08	26.61	0.58	1.12

[a], [b], [c], [d] Measured by XPS. [e] FT-IR intensity ratio of the bands at 970 cm^{-1} and 800 cm^{-1}.

The UV–vis DRS spectra were used to investigate the chemical environment of titanium species in TS-1 (Figure 4b). There are two major peaks presented at 210 nm and 330 nm. The strong peak at 210 nm was assigned to the charge transfer between titanium and oxygen atoms in tetrahedral coordination [8]. This result verified that the titanium atoms entered the framework of TS-1 zeolites. The weak peak around 270 nm corresponded to the partially polymerized titanium species with hexagonal coordination [35], and the absorption peak around 330 nm confirmed the existence of anatase TiO_2 in TS-1 zeolite.

However, there was no characteristic peak in the anatase phase of TiO_2 observed in the XRD powder patterns due to a smaller amount of incorporation or the small particle size of the anatase phase [36].

XPS spectroscopy was employed to investigate the content and the chemical state of the Ti species on the external surface of zeolites. The binding energy for carbon at 284.8 eV was used to make binding energy correction. The chemical state was observed after peak fit analysis of XPS spectra, as shown in Figure 5. The Ti $2p_{3/2}$ at 460.2 eV and Ti $2p_{1/2}$ at 465.0 eV were assigned to the framework Ti species in tetrahedral coordination. As the increment of PDDA, the Ti $2p_{3/2}$ at 460.2 eV split off a new band at 459.0 eV and became narrower, indicating the existence of new Ti species in hexahedral coordination (TiO_6) [37]. The oxidation mechanisms over TS-1 were investigated by means of DFT/$ONIOM_2$ calculations [38]. The calculated activation energy of oxidation with TS-1 containing non-framework Ti (TiO_6) was found to be remarkably lower than that without a non-framework Ti zeolite catalyst, and the intermediate of the double-toothed superoxide IM_1b (η^2) had more reaction activity and stability.

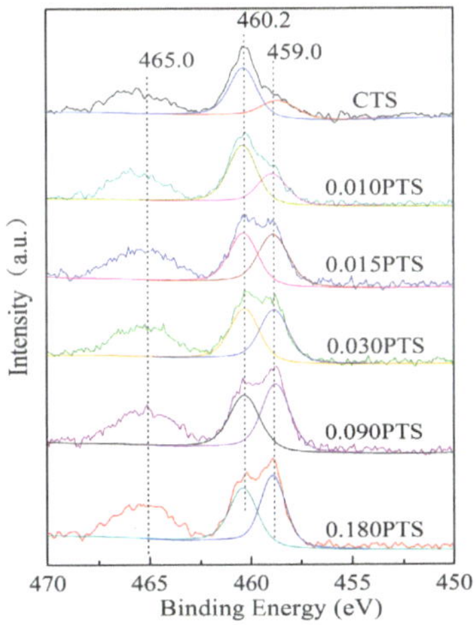

Figure 5. Ti 2p region for CTS and PTS zeolites.

Compared to CTS zeolite, the utilization of PDDA could enrich Ti species on the surface of PTS, which was also accompanied by the presentation of more than two-fold hexahedrally coordinated Ti species (Table 2, d), and it has been proved that the TiO_6 exhibited two~three-fold higher run activity than that of tetrahedral coordination. PDDA has numerous $C_8H_{16}N^+$ chains, which could effectively adsorb Si-OH and Ti-OH, and the Ti-OH exhibited a stronger interaction with PDDA compared with Si-OH [39]. The $C_8H_{16}N^+$ chain could not only enrich tetrahedrally coordinated Ti species on the surface but also improve the dispersion.

3.2. Formation Mechanism of Short b-Axis-Oriented TS-1 Nanosheets

The TS-1 nanosheets were successfully synthesized with a thickness of 6 nm. PDDA functioned as a mesoporous template and structure-directing agent during the preparation process of PTS. The texture property and chemical state of TS-1 zeolite could be controlled by adjusting the content of PDDA in the precursor gel. The ZSM-5 zeolite with a thinner b-axis was gradually formed in the growth process, and the fluoride anions functioned as a structure-directing agent in the TPA^+F^- ion pair, which conducted the process of microcrystal growth [40]. The abundant cations and anions in PDDA could match with TPA^+ and material species in the precursor gel.

The synthesis progress of nucleation, microcrystals, and nanosheets was investigated. In the initial stage, a network structure was formed between the TPA^+, silicon, and titanium species, which spontaneously developed into a large number of crystal nuclei. On one hand, the scale of the network structure began to mature and evolve into microcrystals. On the other hand, the [010] plate of microcrystals had the feature of energetically favorable orientation, which was inclined to adsorb additives because of the coordination between Si and O atoms on the [010] plate [41]. The precursor of 0.030PTS was also characterized by XPS, which showed the presence of Cl2p and N1s species on the surface of the precursor (Figure 6).

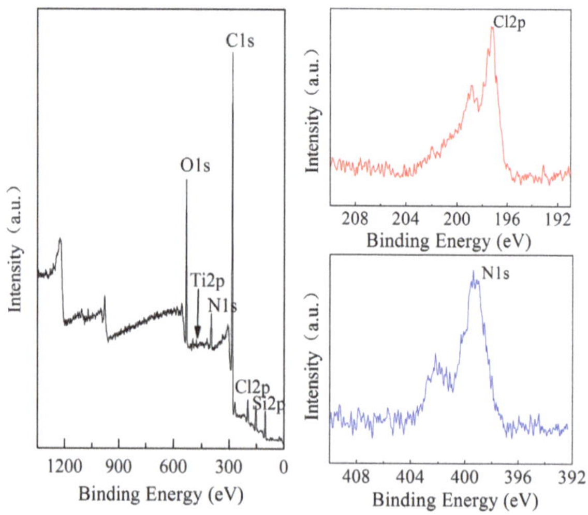

Figure 6. XPS for the precursor of 0.030PTS.

The $C_8H_{16}N^+$ and Cl^- could be adsorbed on the [010] plate of microcrystals and effectively prescribe the orientation to the TPA^+, silicon, and titanium species, and the PDDA could occupy the growing orientation along the b-axis. The TPA^+, silicon, and titanium species were fully dispersed along the a-axis and c-axis and then grew rapidly along the ac crystal plate (Figure 7). Moreover, polymer electrolyte was virtually located between the adjacent ac crystal plates, which could be easily removed through calcination and form a continuous mesoporous channel. Meanwhile, the abundant Ti species were incorporated into the crystal surface as revealed by XPS (Table 2, b,d). Thus, a bulky organic sulfur compound such as DBT with a diameter of 0.887 nm could easily reach the Ti site via continuous mesoporous channels.

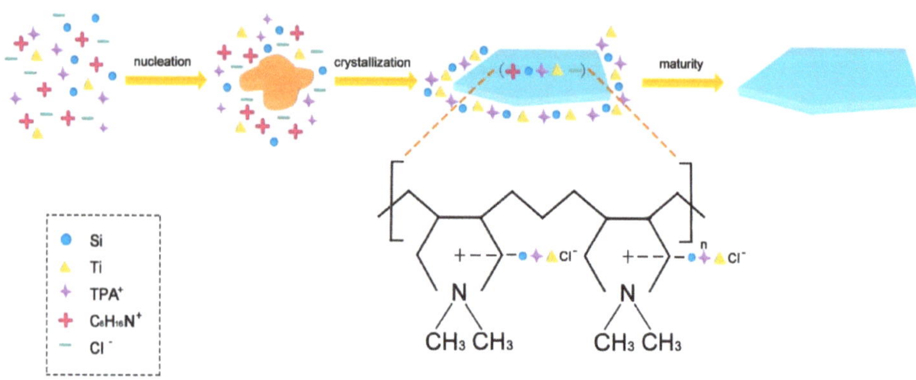

Figure 7. Mechanism of formation of PTS zeolite.

3.3. Catalytic Performance

The conventional CTS and PTS zeolites served as the heterogeneous catalysts for the oxidation of a sulfur compound in the model fuel. The PTS zeolites showed superior catalytic activity compared to the CTS zeolite in the oxidation of DBT [42]. The superior catalytic performance was due to the exposed external surface of PTS (105.4–139.5 m^2/g, Table 1c), which could greatly overcome the restricted access of bulky molecules. Thus, DBT was artificially mixed in noctane and used as a model fuel to evaluate the catalytic performance of the PTS zeolites. Hydrogen peroxide was employed as an oxidizing agent in the reaction. The removal rate of DBT was investigated with a sulfur analyzer (Figure 8a).

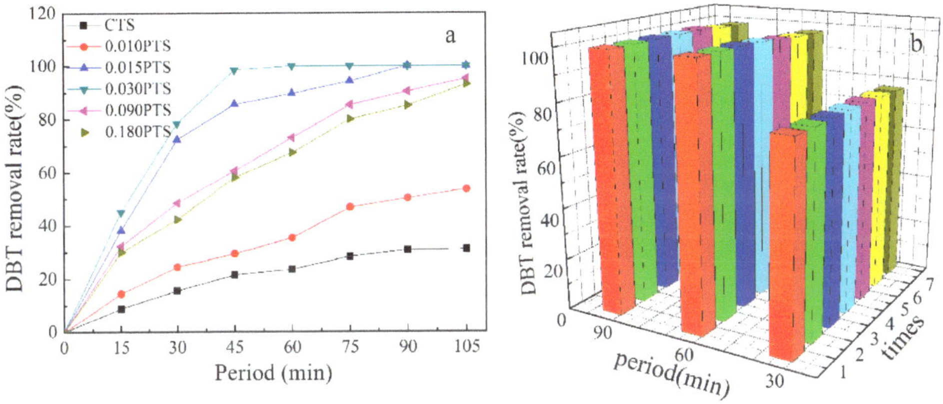

Figure 8. Catalytic test: (**a**) Oxidation of DBT over CTS and stacked nanoplate PTS zeolites. (**b**) Recycle tests for stacked nanosheet 0.030PTS zeolite.

The catalytic activity in oxidation was highly related to the diffusion of DBT in the catalyst. The removal rate of DBT could reach 98.5% in 45 min by using 0.030PTS, and the removal rate was 21.5%, 29.5%, 85.6%, 60.7%, and 58.2% corresponding to CTS, 0.010PTS, 0.015PTS, 0.090PTS, and 0.180PTS, respectively. The PDDA could function as a mesoporous template agent for the 0.010PTS zeolite, and the hierarchical intercrystal mesopores were formed after roasting. However, the lowest dosage of PDDA only incorporated a small amount of Ti species onto the surface of the 0.010PTS zeolite (0.79% and 0.29%, Table 2, b,d). As the increment of PDDA, continuous mesopores were formed in 0.015PTS, 0.030PTS, 0.090PTS, and 0.180PTS, and high titanium content was obtained, as shown in Table 2, b,d. It could be concluded that the texture characteristics, surface properties, and phase composition of PTS could be regulated by adjusting the dosage of PDDA. The catalytic activity of zeolite was highly reliant on the accessibility and enrichment degree of the Ti species on the exposed surface.

PDDA that was located between the ac crystal plates (Figure 7) was removed during the calcined process, and the Si and Ti species remained on the zeolite surface. The 0.030PTS zeolite presented the highest desulfurization performance among the prepared TS-1 zeolites because it possessed a higher external area, uniform pore diameter, and moderate surface Ti content. Moreover, the 0.030PTS zeolite could be easily recycled by centrifugation without any pretreatment and the removal rate of DBT was 98.6% in 90 min after seven recycles (Figure 8b). The results demonstrated that the PTS zeolite is a slightly effective and stable catalyst for oxidative desulfurization of DBT and could also be used as a catalyst for other oxidation reactions.

4. Conclusions

Herein, we developed a novel and efficient method for preparing TS-1 nanosheets along with short b-axis orientation by utilizing PDDA as a structure-directing agent, which could facilitate the incorporation of Ti species on the [0 1 0] plane of zeolites. The surface Ti content of the prepared TS-1 zeolite could be increased by 59%. The prepared surface of the Ti-rich zeolite catalysts possessed more hexahedrally coordinated Ti species and less anatase, which showed superior catalytic activity in the oxidation desulfurization of DBT compared to CTS zeolite. The PTS zeolites prepared in this study may bring more opportunities for oxidative desulfurization.

Author Contributions: Formal analysis, Y.W. and L.W.; Investigation, T.R.; Writing—original draft, T.R.; Writing—review & editing, X.K.; Supervision, H.W.; Funding acquisition, L.L. All authors have read and agreed to the published version of the manuscript.

Funding: The authors would like to acknowledge support from the Project of Liaoning Province Department of Education (No. L2020013/L2020014) and Chunhui Plan of Ministry of Education of China (No. 202200447).

Data Availability Statement: The original contributions presented in the study are included in the article, further inquiries can be directed to the corresponding authors.

Conflicts of Interest: Author Lisheng Liang was employed by the company Petrochina Dagang Oil Field Company. The remaining authors declare that the research was conducted in the absence of any commercial or financial relationships that could be construed as a potential conflict of interest.

References

1. Dou, J.; Zeng, H.C. Integrated Networks of Mesoporous Silica Nanowires and Their Bifunctional Catalysis–Sorption Application for Oxidative Desulfurization. *ACS Catal.* **2014**, *4*, 566–576. [CrossRef]
2. Xun, S.; Zhu, W.; Chang, Y.; Li, H.; Zhang, M.; Jiang, W.; Zheng, D.; Qin, Y.; Li, H. Synthesis of supported $SiW_{12}O_{40}$-based ionic liquid catalyst induced solvent-free oxidative deep-desulfurization of fuels. *Chem. Eng. J.* **2016**, *288*, 608–617. [CrossRef]
3. Fu, Z.; Shen, Z.; Fan, Q.; Hao, S.; Wang, Y.; Liu, X.; Tong, X.; Kong, X.; Yang, Z. Preparation of multi-functional magnetic–plasmonic nanocomposite for adsorption and detection of thiram using SERS. *J. Hazard. Mater.* **2020**, *392*, 122356. [CrossRef] [PubMed]
4. Taramasso, M.; Perego, G.; Notari, B. Preparation of Porous Crystalline Synthetic Material Comprised of Silicon and Titanium Oxides. U.S. Patent 4,410,501, 18 October 1983.
5. Wilde, N.; Pelz, M.; Gebhardt, S.G.; Gläser, R. Highly efficient nano-sized TS-1 with micro-/mesoporosity from desilication and recrystallization for the epoxidation of biodiesel with H_2O_2. *Green Chem.* **2015**, *17*, 3378–3389. [CrossRef]
6. Hu, Y.; Dong, C.; Wang, T.; Luo, G. Cyclohexanone ammoximation over TS-1 catalyst without organic solvent in a microreaction system. *Chem. Eng. Sci.* **2018**, *187*, 60–66. [CrossRef]
7. Schmidt, I.; Krogh, A.; Wienberg, K.; Carlsson, A.; Brorson, M.; Jacobsen, C.J.H. Catalytic epoxidation of alkenes with hydrogen peroxide over first mesoporous titanium-containing zeolite. *Chem. Commun.* **2000**, *21*, 2157–2158. [CrossRef]
8. Davis, T.M.; Drews, T.O.; Ramanan, H.; He, C.; Dong, J.; Schnablegger, H.; Katsoulakis, M.A.; Kokkoli, E.; McCormick, A.V.; Penn, R.L.; et al. Mechanistic principles of nanoparticle evolution to zeolite crystals. *Nat. Mater.* **2006**, *5*, 400–408. [CrossRef] [PubMed]
9. Kumar, M.; Berkson, Z.J.; Clark, R.J.; Shen, Y.; Prisco, N.A.; Zheng, Q.; Zeng, Z.; Zheng, H.; McCusker, L.B.; Palmer, J.C.; et al. Crystallization of Mordenite Platelets using Cooperative Organic Structure-Directing Agents. *J. Am. Chem. Soc.* **2019**, *141*, 20155–20165. [CrossRef] [PubMed]
10. Zhang, S.; Fan, Q.; Guo, J.; Jiao, X.; Kong, X.; Yu, Q. Surface-enhanced Raman spectroscopy tandem with derivatized thin-layer chromatography for ultra-sensitive on-site detection of histamine from fish. *Food Control.* **2022**, *138*, 108987. [CrossRef]
11. Wang, S.; Yu, Q.; Guo, J.; Yuan, C.; Kong, X. On-site detection of pyrene from mixture with ppb level sensitivity by plasmonic TLC-DSERS. *Spectrochim. Acta Part A Mol. Biomol. Spectrosc.* **2022**, *280*, 121547. [CrossRef]
12. Pérez-Ramírez, J.; Christensen, C.H.; Egeblad, K.; Christensen, C.H.; Groen, J.C. Hierarchical zeolites: Enhanced utilisation of microporous crystals in catalysis by advances in materials design. *Chem. Soc. Rev.* **2008**, *37*, 2530. [CrossRef] [PubMed]
13. Hartmann, M. Hierarchical Zeolites: A Proven Strategy to Combine Shape Selectivity with Efficient Mass Transport. *Angew. Chem. Int. Ed.* **2004**, *43*, 5880–5882. [CrossRef] [PubMed]
14. van Donk, S.; Janssen, A.H.; Bitter, J.H.; de Jong, K.P. Generation, Characterization, and Impact of Mesopores in Zeolite Catalysts. *Catal. Rev.* **2003**, *45*, 297–319. [CrossRef]
15. Čejka, J.; Mintova, S. Perspectives of Micro/Mesoporous Composites in Catalysis. *Catal. Rev.* **2007**, *49*, 457–509. [CrossRef]
16. Koekkoek, A.J.J.; Kim, W.; Degirmenci, V.; Xin, H.; Ryoo, R.; Hensen, E.J.M. Catalytic performance of sheet-like Fe/ZSM-5 zeolites for the selective oxidation of benzene with nitrous oxide. *J. Catal.* **2013**, *299*, 81–89. [CrossRef]
17. Trong On, D.; Kaliaguine, S. Large-Pore Mesoporous Materials with Semi-Crystalline Zeolitic Frameworks. *Angew. Chem. Int. Ed.* **2001**, *40*, 3248–3251. [CrossRef]
18. Xiao, F.-S.; Han, Y.; Yu, Y.; Meng, X.; Yang, M.; Wu, S. Hydrothermally Stable Ordered Mesoporous Titanosilicates with Highly Active Catalytic Sites. *J. Am. Chem. Soc.* **2002**, *124*, 888–889. [CrossRef] [PubMed]

19. Shakeri, M.; Shirzadeh Zarnagh, A. Crystallization of as-Mesocellular Silica Foam into Hierarchical TS-1 Zeolites by Conventional Hydrothermal and Dense-Gel Routes for Oxidation Reactions. *ACS Omega* **2020**, *5*, 9334–9338. [CrossRef] [PubMed]
20. Choi, M.; Cho, H.S.; Srivastava, R.; Venkatesan, C.; Choi, D.-H.; Ryoo, R. Amphiphilic organosilane-directed synthesis of crystalline zeolite with tunable mesoporosity. *Nat. Mater.* **2006**, *5*, 718–723. [CrossRef]
21. Du, Q.; Guo, Y.; Duan, H.; Li, H.; Chen, Y.; Liu, H. Synthesis of hierarchical TS-1 zeolite via a novel three-step crystallization method and its excellent catalytic performance in oxidative desulfurization. *Fuel* **2017**, *188*, 232–238. [CrossRef]
22. Gao, X.; An, J.; Gu, J.; Li, L.; Li, Y. A green template-assisted synthesis of hierarchical TS-1 with excellent catalytic activity and recyclability for the oxidation of 2,3,6-trimethylphenol. *Microporous Mesoporous Mater.* **2017**, *239*, 381–389. [CrossRef]
23. Zhu, Y.; Hua, Z.; Zhou, X.; Song, Y.; Gong, Y.; Zhou, J.; Zhao, J.; Shi, J. CTAB-templated mesoporous TS-1 zeolites as active catalysts in a desulfurization process: The decreased hydrophobicity is more favourable in thiophene oxidation. *RSC Adv.* **2013**, *3*, 4193. [CrossRef]
24. Du, S.; Li, F.; Sun, Q.; Wang, N.; Jia, M.; Yu, J. A green surfactant-assisted synthesis of hierarchical TS-1 zeolites with excellent catalytic properties for oxidative desulfurization. *Chem. Commun.* **2016**, *52*, 3368–3371. [CrossRef] [PubMed]
25. Zhang, W.; Jin, W.; Fukushima, T.; Saeki, A.; Seki, S.; Aida, T. Supramolecular Linear Heterojunction Composed of Graphite-Like Semiconducting Nanotubular Segments. *Science* **2011**, *334*, 340–343. [CrossRef] [PubMed]
26. Miszta, K.; de Graaf, J.; Bertoni, G.; Dorfs, D.; Brescia, R.; Marras, S.; Ceseracciu, L.; Cingolani, R.; van Roij, R.; Dijkstra, M.; et al. Hierarchical self-assembly of suspended branched colloidal nanocrystals into superlattice structures. *Nat. Mater.* **2011**, *10*, 872–876. [CrossRef] [PubMed]
27. Wang, T.; Zhuang, J.; Lynch, J.; Chen, O.; Wang, Z.; Wang, X.; LaMontagne, D.; Wu, H.; Wang, Z.; Cao, Y.C. Self-Assembled Colloidal Superparticles from Nanorods. *Science* **2012**, *338*, 358–363. [CrossRef] [PubMed]
28. Zhu, J.; Zhu, Y.; Zhu, L.; Rigutto, M.; van der Made, A.; Yang, C.; Pan, S.; Wang, L.; Zhu, L.; Jin, Y.; et al. Highly Mesoporous Single-Crystalline Zeolite Beta Synthesized Using a Nonsurfactant Cationic Polymer as a Dual-Function Template. *J. Am. Chem. Soc.* **2014**, *136*, 2503–2510. [CrossRef]
29. Du, S.; Sun, Q.; Wang, N.; Chen, X.; Jia, M.; Yu, J. Synthesis of hierarchical TS-1 zeolites with abundant and uniform intracrystalline mesopores and their highly efficient catalytic performance for oxidation desulfurization. *J. Mater. Chem. A* **2017**, *5*, 7992–7998. [CrossRef]
30. Khomane, R.B.; Kulkarni, B.D.; Paraskar, A.; Sainkar, S.R. Synthesis, characterization and catalytic performance of titanium silicalite-1 prepared in micellar media. *Mater. Chem. Phys.* **2002**, *76*, 99–103. [CrossRef]
31. Signorile, M.; Crocellà, V.; Damin, A.; Rossi, B.; Lamberti, C.; Bonino, F.; Bordiga, S. Effect of Ti Speciation on Catalytic Performance of TS-1 in the Hydrogen Peroxide to Propylene Oxide Reaction. *J. Phys. Chem.* **2018**, *122*, 9021–9034. [CrossRef]
32. Liu, C.; Huang, J.; Sun, D.; Zhou, Y.; Jing, X.; Du, M.; Wang, H.; Li, Q. Anatase type extra-framework titanium in TS-1: A vital factor influencing the catalytic activity toward styrene epoxidation. *Appl. Catal. A Gen.* **2013**, *459*, 1–7. [CrossRef]
33. Ricchiardi, G.; Damin, A.; Bordiga, S.; Lamberti, C.; Spanò, G.; Rivetti, F.; Zecchina, A. Vibrational Structure of Titanium Silicate Catalysts. A Spectroscopic and Theoretical Study. *J. Am. Chem. Soc.* **2001**, *123*, 11409–11419. [CrossRef] [PubMed]
34. Boccuti, M.R.; Rao, K.M.; Zecchina, A.; Leofanti, G.; Petrini, G. Spectroscopic Characterization of Silicalite and Titanium-Silicalite. *J. Catal.* **1989**, *48*, 133–144. [CrossRef]
35. Zhong, W.; Qiao, T.; Dai, J.; Mao, L.; Xu, Q.; Zou, G.; Liu, X.; Yin, D.; Zhao, F. Visible-light-responsive sulfated vanadium-doped TS-1 with hollow structure: Enhanced photocatalytic activity in selective oxidation of cyclohexane. *J. Catal.* **2015**, *330*, 208–221. [CrossRef]
36. Betiha, M.A.; Mahmoud, S.A.; Menoufy, M.F.; Al-Sabagh, A.M. One-pot template synthesis of Ti–Al-containing mesoporous silicas and their application as potential photocatalytic degradation of chlorophenols. *Appl. Catal. B Environ.* **2011**, *107*, 316–326. [CrossRef]
37. Zhang, M.; Wen, Y.; Zong, L.; Wei, H.; Wang, X. Improved Ti species distribution and hierarchical pores in TS-1: Towards regeneration of TS-1 deactivated due to alkali corrosion. *New J. Chem.* **2020**, *44*, 6394–6401. [CrossRef]
38. Wang, H.L.; Zhou, R.; Deng, Y.Q. Thiophene oxidation with H_2O_2 over defect and perfect titanium silicalite-1: A computational study. *React. Kinet. Mech. Catal.* **2018**, *124*, 45–60. [CrossRef]
39. Kwon, O.S.; Park, E.; Kweon, O.Y.; Park, S.J.; Jang, J. Novel flexible chemical gas sensor based on poly(3,4-ethylenedioxythiophene) nanotube membrane. *Talanta* **2010**, *82*, 1338–1343. [CrossRef]
40. Zhang, L.; Song, Y.; Li, G.; Zhang, Q.; Zhang, S.; Xu, J.; Deng, F.; Gong, Y. F-assisted synthesis of a hierarchical ZSM-5 zeolite for methanol to propylene reaction: A b-oriented thinner dimensional morphology. *RSC Adv.* **2015**, *5*, 61354–61363. [CrossRef]
41. Shan, Z.; Wang, H.; Meng, X.; Liu, S.; Wang, L.; Wang, C.; Li, F.; Lewis, J.P.; Xiao, F.-S. Designed synthesis of TS-1 crystals with controllable b-oriented length. *Chem. Commun.* **2011**, *47*, 1048–1050. [CrossRef]
42. Yamazaki, D.; Nishinaga, T.; Komatsu, K. Radical Cation of Dibenzothiophene Fully Annelated with Bicyclo[2.2.2]octene Units: X-ray Crystal Structure and Electronic Properties. *Org. Lett.* **2004**, *6*, 4179–4182. [CrossRef] [PubMed]

molecules

MDPI

Article

Theoretical Study of Copper Squarate as a Promising Adsorbent for Small Gases Pollutants

Celia Adjal [1,2], Nabila Guechtouli [1,3], Vicente Timón [2,*], Francisco Colmenero [4] and Dalila Hammoutène [1]

1 Laboratory of Thermodynamics and Molecular Modeling, Faculty of Chemistry, University of Science and Technology Houari Boumediene (USTHB), BP32, El Alia, Bab Ezzouar, Algiers 16111, Algeria; adjalcelia5@gmail.com (C.A.); n.guechtouli@univ-boumerdes.dz (N.G.); dhammoutene@yahoo.fr (D.H.)
2 Instituto de Estructura de la Materia, CSIC, Serrano 121, 28006 Madrid, Spain
3 Faculty of Sciences, University of M'hamed Bougara, (UMBB), Boumerdes 35000, Algeria
4 Centro de Investigaciones Energéticas, Medioambientales y Tecnológicas (CIEMAT), Avda/Complutense, 40, 28040 Madrid, Spain; francolm@ucm.es
* Correspondence: vicente.timon@csic.es

Abstract: Copper squarate is a metal–organic framework with an oxo-carbonic anion organic linker and a doubly charged metal mode. Its structure features large channels that facilitate the adsorption of relatively small molecules. This study focuses on exploring the potential of adsorbing small pollutants, primarily greenhouse gases, with additional investigations conducted on larger pollutants. The objective is to comprehend the efficacy of this new material in single and multiple molecular adsorption processes using theoretical methods based on density functional theory. Furthermore, we find that the molecular adsorption energies range from 3.4 $KJ \cdot mol^{-1}$ to 63.32 $KJ \cdot mol^{-1}$ depending on the size and number of adsorbed molecules. An exception is noted with an unfavorable adsorption energy value of 47.94 $KJ \cdot mol^{-1}$ for 4-nitrophenol. More importantly, we demonstrate that water exerts an inhibitory effect on the adsorption of these pollutants, distinguishing copper squarate as a rare MOF with hydrophilic properties. The Connolly surface was estimated to give a more accurate idea of the volume and surface accessibility of copper squarate. Finally, using Monte Carlo simulations, we present a study of adsorption isotherms for individual molecules and molecules mixed with water. Our results point out that copper squarate is an efficient adsorbent for small molecular pollutants and greenhouse gases.

Keywords: copper squarate; metal–organic framework; molecular adsorption; DFT; adsorption isotherms

Citation: Adjal, C.; Guechtouli, N.; Timón, V.; Colmenero, F.; Hammoutène, D. Theoretical Study of Copper Squarate as a Promising Adsorbent for Small Gases Pollutants. *Molecules* **2024**, *29*, 3140. https://doi.org/10.3390/molecules29133140

Academic Editors: Guangxu Lan and Yi-Nan Wu

Received: 29 May 2024
Revised: 24 June 2024
Accepted: 26 June 2024
Published: 2 July 2024

1. Introduction

Nowadays, scientists around the world are actively seeking new porous materials to tackle the pollution crisis caused by urbanization and emissions from motor vehicles and industry [1,2]. Extensive research has been conducted, resulting in the publication of in-depth studies on air emissions and the various types of pollutants present in our atmosphere, such as carbon dioxide (CO_2), the main contributor to the greenhouse effect, which is largely responsible for global warming. Additionally, methane (CH_4), tropospheric ozone (O_3), dinitrogen monoxide (N_2O), nitrogen trifluoride (NF_3), and sulfur hexafluoride (SF_6) are critical trace gases that contribute to climate change. These six compounds are among the main greenhouse gases identified in the Kyoto Protocol [3–6].

In addition, the combustion of coal, fossil fuels, or biomass also releases air pollutants, such as sulfur dioxide (SO_2) and volatile organic compounds (VOCs), which contribute significantly to industrial and vehicle exhaust pollution. At the same time, VOCs produce a variety of pollutants, including polycyclic aromatic hydrocarbons (PAHs) derivatives such as NPAH nitrates. This category includes, for example, 4-nitrophenol (4NP), which is considered one of the most hazardous pollutants in our daily lives [7–15]. Nitrophenols are classified as priority pollutants by the U.S. Environmental Protection Agency [16].

Due to their high toxicity, atmospheric pollutants have considerable negative effects on human health, wildlife, and global warming. For this reason, emphasis has been placed on experimental and theoretical studies concerning the application of separation, adsorption, gas capture and storage, and selective catalysis as potential solutions [17–22].

Recently, metal–organic frameworks (MOFs) have been in great demand for the capture and storage of gaseous pollutants due to their extraordinary porosity, surface area, crystallinity, controllable structure, and chemical separation capabilities [23–25]. Consequently, a plethora of work has been carried out on various materials, such as zeolitic imidazolate frameworks (ZIFs), clay minerals, lignite, anthracites, and more, to adsorb pollutants [26–30]. The presence of water has also been considered, providing valuable information on its impact on the adsorption process [31–33].

In this paper, we will focus on a type of MOF [34] that has not yet been explored for adsorption or capture processes: the copper squarate (C-S) metal–organic framework. The crystal structure of this material was first analyzed and described in detail by Dinnebier et al. [35,36]. C-S was subsequently investigated by Colmenero et al. [37] using theoretical solid-state methods based on density functional theory (DFT). This study firmly established the crystal structure of copper squarate, previously determined using a restricted rigid-body Rietveld refinement, which discovered that it exhibits the phenomenon of negative linear compressibility due to the presence of the empty channel structural motif. Colmenero et al. demonstrated that other squarate MOFs, such as zinc [36] and uranyl squarates [38], also exhibit negative linear compressibility. In fact, Colmenero [39] has shown that oxo-carbonic acids in the solid state exhibit extremely anomalous mechanical behavior, and recently, Qiu et al. [40] have revealed that lead and barium squarates exhibit zero linear compressibility and thermal expansion effects over a wide range of pressures. The presence of empty spaces inside these structures and their large pore size have led us to believe in the viability of these materials for the capture and storage of hazardous gases. As far as we know, squarate MOFs have only been recently studied for molecular adsorption applications. Namely, the use of cobalt squarate MOF has recently been investigated for CO_2/N_2 separation by Zhang et al. [41].

The aim of our work is to characterize the behavior of squarate materials in molecular adsorption applications, which have not yet been explored, using solid simulation techniques. Specifically, we will investigate the ability of copper squarate to absorb a series of different classes of gaseous pollutants in the presence and absence of water. We will closely examine its response to these pollutants, determining its preferred adsorption sites, energies, and mechanisms, as well as its adsorption isotherms. Unfortunately, to our knowledge, no experimental or theoretical studies have been published on adsorption in copper squarate. The present work is therefore the first to provide theoretical results that will serve as a starting point for further experiments. The exploration of new materials for the adsorption of contaminants is crucial to counter the wave of pollution caused by urbanization that we are currently experiencing.

2. Results and Discussion

The copper squarate crystal structure was re-optimized, and the results obtained are fairly consistent with those obtained by Colmenero et al. [36]. The powder X-ray diffraction (PXRD) patterns of copper squarate were derived from the computed and experimental structures using the program REFLEX included in the Materials Studio program suite [42]. The results are displayed in Figure S1 of the Supplementary Materials and, as can be seen, they are consistent. Because of the small size of the unit cell, it was doubled along the axis to facilitate the adsorption of the pollutant, as was conducted in previous works studying the adsorption of similar molecules in materials like zeolite [43]. The calculations were then continued with a series of optimizations of the double unit cell containing the pollutant gases. These gases were introduced inside using the Adsorption Locator tool of Materials Studio [42] in order to conduct the adsorption processes inside this material. The results are presented in Table 1, where the lattice parameters of the crystal unit and the double cell of copper squarate are provided, as well as the parameters of the cells distorted due to the capture of one or more pollutants and water molecules. The optimized geometries of different species are compared with the corresponding experimental data in Table S1 of the supporting information. The optimized structure of copper squarate is shown in Figure 1. The optimized copper squarate unit and double cell can be downloaded from the Supplementary Information (Table S2).

Table 1. Structural parameters of copper squarate (C-S) are in the pure form and contain adsorbed pollutant molecules and water separately. The experimental unit cell parameters are from Ref. [36].

Species	a(Å)	b(Å)	c(Å)	α(°)	β(°)	γ(°)	Vol. (Å^3)
C-S ($1 \times 1 \times 2$)	11.017	9.352	11.288	89.918	117.636	90.082	1030.39
Exp. ($1 \times 1 \times 1$)	10.686	8.939	5.674	90.0	117.600	90.0	480.455
1 CH_4	11.159	9.145	11.224	89.570	116.579	90.186	1024.40
4 CH_4	10.707	9.812	11.394	90.089	117.974	89.977	1057.25
1 CO_2	11.212	9.071	11.208	89.278	116.246	90.607	1022.29
2 CO_2	11.096	9.264	11.276	89.356	116.571	90.410	1036.68
5 CO_2	10.926	9.579	11.405	89.330	115.918	89.907	1073.61
1 O_3	11.062	9.275	11.253	90.081	117.058	89.949	1027.85
3 O_3	10.972	9.654	11.206	89.975	119.881	90.149	1029.24
5 O_3	10.937	9.563	11.409	90.103	116.515	90.254	1067.79
1 SF_6	10.904	9.754	11.226	89.641	118.322	89.846	1051.02
1 NF_3	11.106	9.225	11.261	90.333	116.512	89.765	1023.45
1 CF_4	11.015	9.610	11.192	89.801	119.085	90.008	1035.35
1 N_2O	10.925	9.683	11.217	90.014	120.345	90.007	1024.08
1 S_2O	11.139	9.167	11.215	90.160	116.500	90.140	1024.94
1 H_2O	11.214	9.015	11.187	90.279	116.117	89.714	1015.56
5 H_2O	11.070	9.431	11.155	90.185	117.025	89.544	1037.45
18 H_2O	10.907	9.524	12.334	93.334	106.045	89.533	1229.24
14NP	10.027	10.51	11.722	88.364	114.868	90.152	1120.26

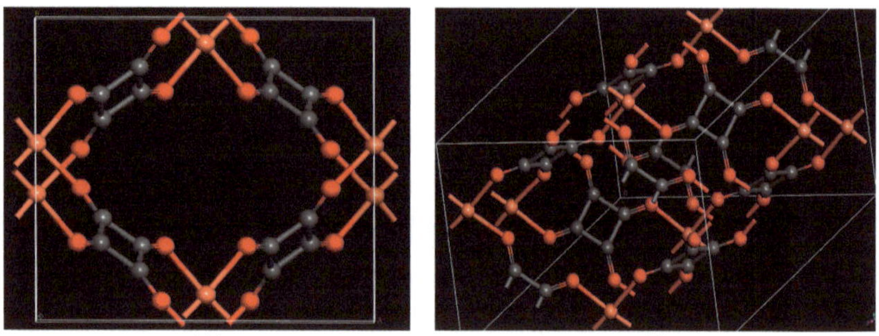

(A) (B)

Figure 1. Optimized crystal structure of copper squarate: (**A**) single unit cell; (**B**) double unit cell along edge. (Color code: Red: Oxygen/Orange: Copper/Gray: Carbon).

2.1. Crystal Bulk Structure and Analysis of Connolly Free Volume and Surface Area Properties

In this section, we employ another feature of Materials Studio 22.0 [42], the "Atoms Volume and Surfaces" to approximate the Connolly surface of copper squarate and the pollutants separately using the probe radius. This allows us to thoroughly explore the cavity and surface features of the internal complex structure after the adsorption process [44] and to obtain the occupied and free volumes, as well as surface area (Table 2). These parameters are then used to calculate the pore volume using the expression below [45]:

$$Pore\ volume\ (\mathrm{cm}^3 \cdot \mathrm{g}^{-1}) = \frac{Cell\ Free\ Volume\ (^3)}{Density\ (\mathrm{g} \cdot \mathrm{cm}^{-3}) \times Cell\ Volume\ (^3)} \quad (1)$$

Table 2. Connolly free volume, surface area of supercell sizes, and density of copper squarate.

Sample	Density (g·cm^{-3})	Accessible Surface Area (Å^2)	Free Volume (Å^3)	Pore Volume (g·cm^{-3})
C-S Exp. [36]	2.263 2.427	283.7	223.5	0.0958

From Table 2, we notice a good agreement between the experimental density (2.427 g·cm^{-3}) and the theoretical value (2.263 g·cm^{-3}), showing the accuracy of the DFT-D3 method in describing unit cell structures in crystalline materials. The total pore volume of copper squarate ($1 \times 1 \times 2$ cell) is expected to be 0.0958 g·cm^{-3}, i.e., 23% of the total cell volume reported in Table 1. This value is close to the experimental one (24.6%) approximated from available data (density and cell volume) in the work of Colmenero et al. [36] considering the ($1 \times 1 \times 2$) cell, as well as using the theoretical cell free volume. All these calculations use a probe radius of 1.0 Å and a grid interval of 0.6 Å [45].

Moreover, we employed the same grid interval of 0.6 Å for the Connolly surface parameters to estimate the volume and surface area of other adsorbed species. Meanwhile, we varied the probe radius from one molecule to another due to the electronic density of those molecules [32,45]. For H_2O, SO_2, N_2O, and CH_4, we use a probe radius of 1.37 Å due to their similar electronic cloud characteristic except for CH_4. The corresponding results were (18.99 Å^3/34.79 Å^2), (40.48 Å^3/58.08 Å^2), (28.87 Å^3/50.42 Å^2), and (27.47 Å^3/46.18 Å^2), respectively. In the case of 4NP, a slightly larger probe radius of 1.4 Å was employed to achieve a result of (115 Å^3/138.02 Å^2). Finally, for the more electronegative molecules CO_2, O_3, NF_3, SF_6, CF_4, and 2NP, a probe radius of 1.7 Å was chosen, considering their higher electron density due to hindrance, resulting in (32.28 Å^3/51.37 Å^2), (31.25 Å^3/50.37 Å^2), (40.71 Å^3/60.59 Å^2), (74.81 Å^3/89.45 Å^2), (50.04 Å^3/69.05 Å^2), and (110.49 Å^3/134.58 Å^2), respectively. These results indicate the ability of C-S to adsorb the pollutants listed, except 2NP and 4NP, which have almost the same Connolly surface and volume as the filter, potentially causing repulsive disturbances.

To delve deeper into the understanding of C-S capacity as a filter and elucidate the outcomes of the Connolly analysis, we employed the DFT-D3 veracity method. This method allows us to ascertain the maximum quantity of adsorbed pollutants within the bulk. Table 3 shows the structural parameters of the co-adsorption process.

Table 3. The structural parameters of copper squarate containing a co-adsorbed mixture of a pollutant with water.

Mixture	a(Å)	b(Å)	c(Å)	α(°)	β(°)	γ(°)	Vol. (Å^3)
1 CH_4 + 1 H_2O	11.088	9.246	11.231	89.983	116.856	89.765	1027.29
1 CO_2 + 1 H_2O	11.082	9.249	11.216	90.009	116.896	89.695	1025.20
1 O_3 +1 H_2O	11.134	9.192	11.229	89.511	116.154	90.203	1030.70
1 SF_6 +1 H_2O	10.902	9.796	11.247	90.335	118.839	90.090	1051.76
1 NF_3 +1 H_2O	11.021	9.376	11.298	89.924	116.777	89.956	1024.30
1 CF_4 +1 H_2O	11.014	9.573	11.187	89.728	118.839	90.084	1033.19
1 N_2O + 1 H_2O	11.053	9.533	11.185	90.004	119.398	90.076	1026.75
1 SO_2 +1 H_2O	11.095	9.232	11.221	90.499	116.896	89.693	1025.20
1 4NP + 1 H_2O	9.9750	10.55	11.774	88.224	116.634	90.125	1107.77

2.2. Adsorption Analysis

Achieving optimal DFT performance in this study involves a hierarchical approach. First, we locate the most stable configuration in conformational space, as was previously described using the Adsorption Locator module with COMPASS III force field [46], by dropping the molecule at various pressures employing the Metropolis algorithm. The DFT geometry is then optimized for the most favorable configuration. (Figure 2) shows the single final adsorption structure of pollutants and water in C-S and (Table 4) gives the energy data for the adsorption process. Negative adsorption energies indicate favorable capture of gases [31].

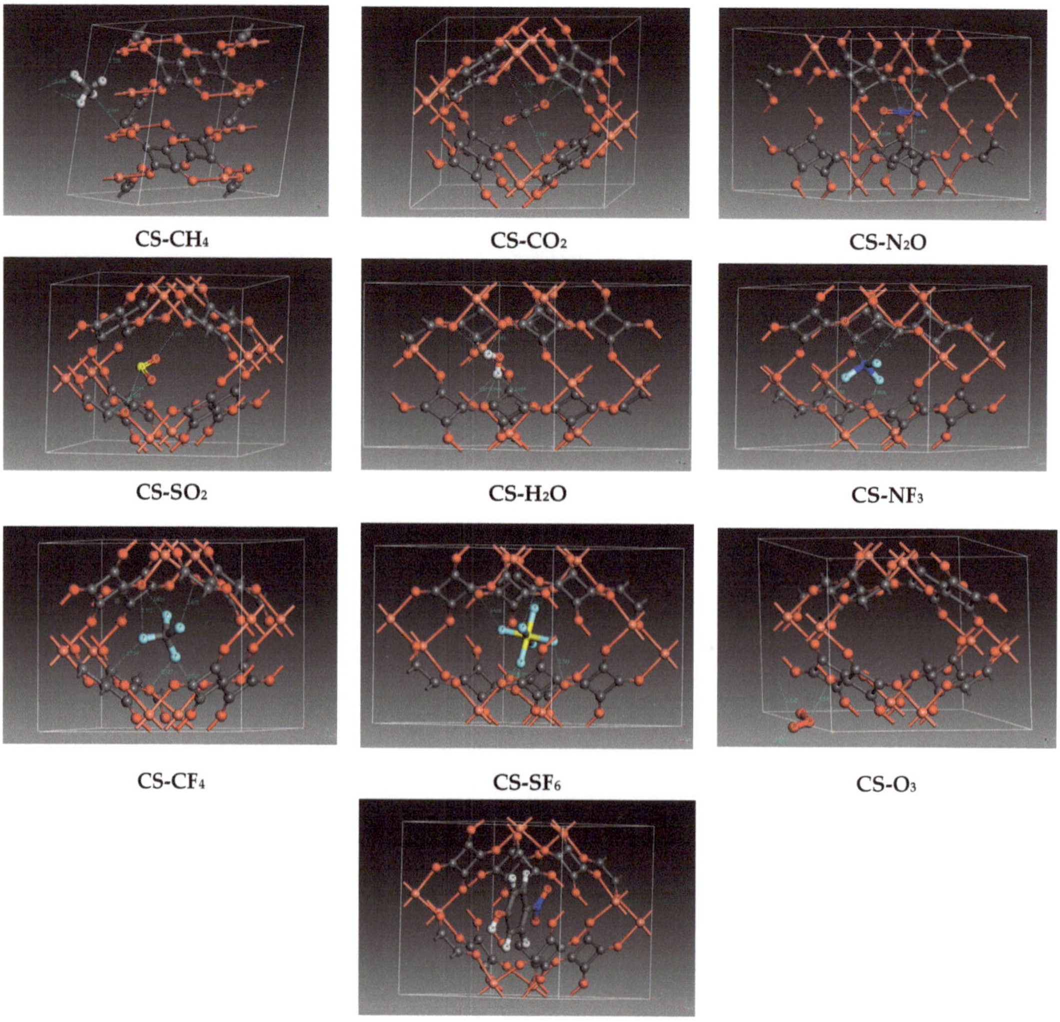

Figure 2. Zoomed interactions of optimized copper squarate containing single pollutants or water. (Color code: Red: Oxygen/Orange: Copper/Gray: Carbon/White: Hydrogen/Royal blue: nitrogen/Yellow: Sulfur/Light blue: Fluorine).

Table 4. Adsorption and co-adsorption energies in copper squarate calculated with CASTEP.

X	Pollutant	H_2O	Energy (eV)	ΔE_{ads}(KJ·mol^{-1})	ΔE_{ads}^{mult}(KJ·mol^{-1})	ΔE_{int}(KJ·mol^{-1})
1	CH_4	0	−28,377.13	−19.42		
4	CH_4	0	−29,036.92	−23.02		
1	CO_2	0	−29,174.10	−18.93		
2	CO_2	0	−30,190.92	−17.72		
5	CO_2	0	−33,241.66	−22.22		
1	O_3	0	−29,443.50	−30.51		

Table 4. *Cont.*

X	Pollutant	H_2O	Energy (eV)	ΔE_{ads}(KJ·mol^{-1})	ΔE_{ads}^{mult}(KJ·mol^{-1})	ΔE_{int}(KJ·mol^{-1})
3	O_3	0	−32,016.10	−50.05		
5	O_3	0	−34,588.30	−26.94		
1	SF_6	0	−32,355.32	−3.785		
1	NF_3	0	−33,375.60	−14.12		
1	CF_4	0	−30,908.98	−3.400		
1	N_2O	0	−29,137.72	−63.32		
1	SO_2	0	−29,322.81	−38.04		
1	H_2O	0	−28,621.07	−51.16		
5	H_2O	0	−30,476.14	−47.01		
18	H_2O	0	−36,505.20	−46.78		
1	4NP	0	−30,730.94	+47.94		
1	CH_4	1	−28,157.25		−33.11	+4.364
1	CO_2	1	−29,638.03		−40.39	−10.68
1	O_3	1	−29,907.21		−35.52	+10.63
1	SF_6	1	−32,818.70		−6.186	+42.67
1	NF_3	1	−30,839.30		−26.83	+11.01
1	CF_4	1	−31,373.09		−41.31	−27.96
1	N_2O	1	−29,601.17		−39.10	+36.30
1	SO_2	1	−29,786.61		−43.48	+1.952
1	4NP	1	−30,730.93		+47.93	−10.34

2.2.1. Pollutants and Water Adsorption

The results indicate favorable adsorption of CO_2, N_2O, SO_2, and H_2O molecules into C-S using force field loading, with up to five, one, one, and eighteen species per double unit cell, respectively. The accuracy of these findings was validated by CASTEP calculations, revealing adsorption energies of −18.93, −63.32, −38.04, and −51.16 KJ·mol^{-1}, respectively, for a single captured molecule. This suggests the existence of electrostatic interactions, as illustrated in the zoomed images in Figure 2. For all four species, we observe a short consistent interaction involving C-S oxygen with C (CO_2), N (N_2O), S (SO_2), and strong hydrogen [47] bound with H (H_2O), comprising a bound distance of 2.811, 2.954, 2.775, and 1.966 Å, respectively. In addition, a weaker interaction is evident with C (C-S), interacting with O (CO_2), N (N_2O), O (SO_2), and O (H_2O) at distances ranging from 2.847–3.080 Å.

In Table 4, the calculated energies are CH_4 (−219.678 eV), CO_2 (−1016.651 eV), O_3 (−1285.930 eV), SF_6 (−4198.029 eV), NF_3 (−2218.200 eV), CF_4 (−2751.691 eV), N_2O (−979.812 eV), SO_2 (−1165.160 eV), H_2O (−463.290 eV), 4NP (−2574.182 eV), and C-S (−28,157.254 eV), referring to the number of adsorbed pollutant and water molecules.

Generally, metal–organic frameworks (MOFs) are known to be unstable in humid environments. Zeolite has been highlighted in the literature for its hydrophobic nature [48–50]. However, recent experimental studies have identified exceptions, such as ZIF-8 and ZIF-90, which show stability in the presence of water [51,52]. In our study, we theoretically demonstrate that C-S exhibits stability in the presence of water, which is consistent with the hydrophilic behavior observed in some MOFs. In particular, we show that O-H bond lengths in water approach 0.966 Å, which closely matches the experimental value of 0.98 Å [52].

Methane is a non-polar molecule and has been the subject of previous adsorption studies, particularly on zeolites [45]. In our study, different quantities of CH_4 ranging from one to four were absorbed and optimized to identify the most favorable insertion sites. Our results are consistent with those of a previous study in which four methane molecules were observed inside the ZIF-4 cell [32]. By analyzing the modeled structures, we observed that all CH_4 molecules are in close proximity to the C-S, involving a hydrogen interaction with the oxygen atom of C-S, with a bonding distance of 2.737 Å, as illustrated in (Figure 2). This interaction represents a limiting case of hydrogen bonding, characterized by a weak donor to a strong acceptor due to their substantial contribution to dispersion [47]. This contributes to system stabilization, resulting in negative adsorption values ranging from -19.42 KJ·mol^{-1} to -23.02 KJ·mol^{-1} for one and four CH_4 uptake molecules, respectively.

From Table 4, we can observe high adsorption energies of -30.51, -50.05, and -26.94 KJ·mol^{-1} for the adsorption of one, three, and five O_3 molecules, respectively, including a strong uptake within the bulk. This is attributed to the reactive nature and high polarity of ozone due to the presence of oxygen atoms. Capture involves intermolecular Van der Waals forces, mainly O_{O3}-$O_{C\text{-}S}$ with an average bond length of 2.8 Å for all the optimized structures. In this case, the charged fluctuation of the oxygen and the electronic density leads to a sensitive instantaneous dipole moment, creating a type of London dispersion.

The vacuum inside the cavity allows the capture of one molecule with fluorine atoms due to the steric effect of the electronic density of the halogen, causing a repulsion with the oxo-carbonic anion of the copper squarate. We observed that the adsorption energy is -14.12 KJ·mol^{-1} for NF_3, while it decreases consistently for CF_4 and SF_6, with values of -3.40 and -3.78 KJ·mol^{-1}, respectively. We deduce that as the number of fluorine atoms increases, the adsorption energy decreases. Additionally, we notice that when the central atom is nitrogen, the adsorption energy is higher, as seen with N_2O. The optimized lengths are shown in Figure 2, where it can be seen that the carbon of C-S interacts with the fluorine of CF_4 and NF_3 at 2.7 and 2.8 Å, while a shorter value of 1.98 Å is noted with SF_6 because its octahedral form minimizes the distance between the pollutant and the bulk.

A positive adsorption energy of +47.94 KJ·mol^{-1} was noted for 4NP (Table 4), indicating unfavorable capture of the aromatic pollutant by the C-S. This is in contradiction with the adsorption locator calculation, which allows the uptake of one molecule. This implies the lack of accuracy of the force field and the foremost importance of using the DFT-D3 method to complement and achieve a reliable calculation.

2.2.2. Co-Adsorption of Pollutants and Water

In this section, we study the effect of a humid environment on the adsorption inside copper squarate bulk. The adsorption energy and interaction energy are calculated using Equations (2) and (3), respectively, and the results are presented in Table 4.

According to Table 4, the co-adsorption of the pollutant with water gives negative adsorption energy, revealing favorable adsorption energy, except for 4NP. The latter has a positive adsorption value in both dry and humid environments, which can be caused by the unavailability of sufficient space on the pores due to its large size and radius.

For CH_4, SO_2, and CO_2, the adsorption energies increased in the presence of water to reach -33.11, -43.48, and -40.39 KJ·mol^{-1}, respectively. This implies the same interaction as above between the pollutant and the bulk. In addition, weak hydrogen bonds are formed between the water and the pollutant, which further increases the adsorption energy. The same trend is noted for molecules with fluorine atoms, where the values of the mixed adsorption energy increased to reach -26.83, -41.31, and -6.186 KJ·mol^{-1} for NF_3, CF_4, and SF_6, respectively. The important impact on the adsorption energetic value is due to the multiple interactions caused by the reactive sites of these types of pollutants. In contrast, for ozone and dinitrogen monoxide, the adsorption energy decreases slightly but remains favorable to the adsorption process. Most of the interaction energies are positive, indicating repulsion between the species and far adsorption between the small pollutants and the water inside the pore. This is due to the electron density, as depicted in Figure 3.

Finally, the adsorption of 4NP in the presence of water does not counterbalance capture towards a favorable process. In previous work on nitrophenol on activated carbon bulk under humid conditions, the authors observed a possibility of adsorption by water as it moved to place itself between 4NP and the surface bulk, creating hydrogen bonds and electrostatic interactions. This suggests a lack of space inside the C-S pore, which can create repulsion due to the high charge density of water, 4NP, and C-S cell, as can be seen in Figure 3 where water is placed in the cell extremity, not inside. We also noticed this in the previous section on the Connolly free volume, where we noted that 4NP and 2NP have almost the same volume. The interaction energy here is negative due to the hydrogen bonding of the aromatic compound to the copper squarate.

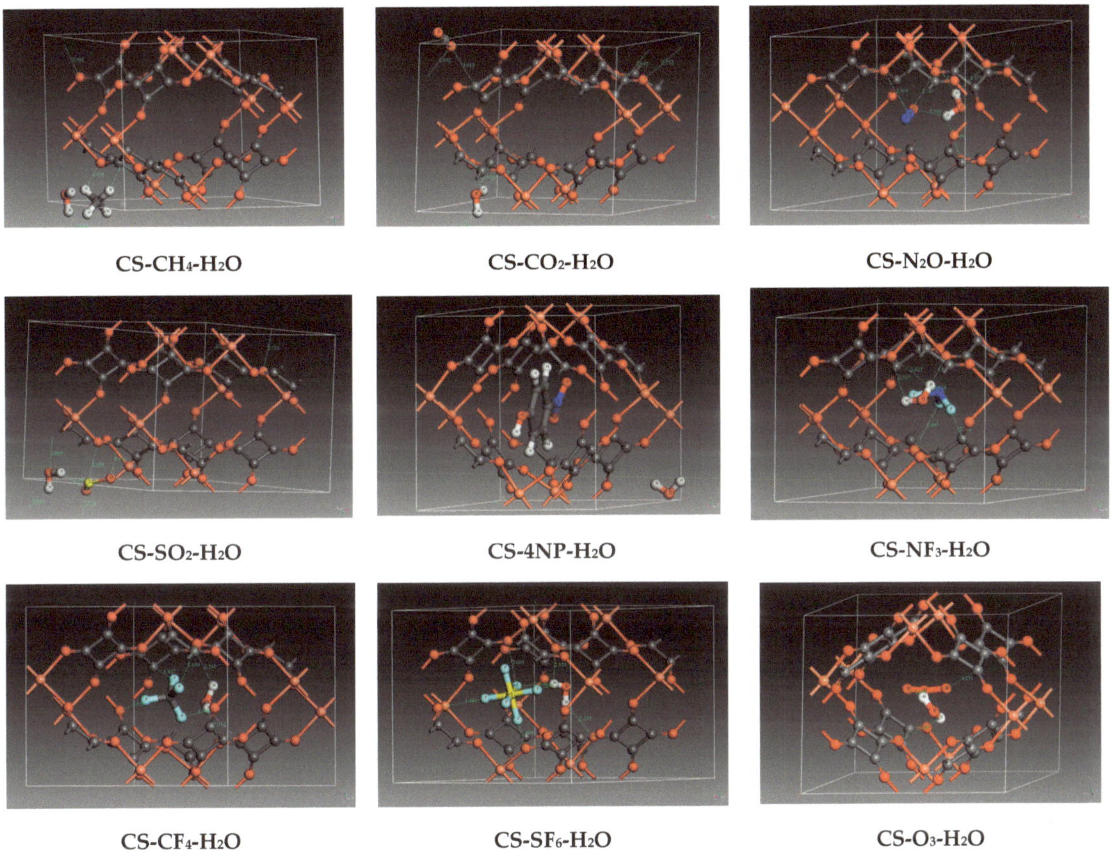

Figure 3. Zoomed interactions of optimized copper squarate containing co-adsorbed pollutants with water. (Color code: Red: Oxygen/Orange: Copper/Gray: Carbon/White: Hydrogen/Royal blue: nitrogen/Yellow: Sulfur/Light blue: Fluorine).

These fluctuations in no way interfere with copper squarate's favorable ability to capture these small pollutants. On the contrary, they positively highlight the fact that the adsorbent can capture them in both dry and humid states, without the presence of water playing a counterbalancing role. It is important to remember that the small pollutants, as well as water, are energetically stable separately in the C-S.

The main geometric parameters of the co-adsorption (water/pollutant) were analyzed and are reported in Table 5 above. The results indicate that there is no significant change in the bond distance and valence angle between them compared with the isolated adsorbed form. This is due to the presence of more free space, allowing more water molecules to be captured, confirming the observation that the interaction energy is positive. The presence of additional adsorbed molecules can lead to an increase in distortion due to space limitations. This phenomenon has already been observed in solid simulations, notably in studies of methane, where the angle of increased from 109° to 114° [45].

Table 5. Main geometrical parameters for the optimized co-adsorption (1 X −1 H_2O) structures (distances in Å and angles in degrees) on copper squarate X = (CH_4, CO_2, O_3, SF_6, NF_3, CF_4, N_2O, and SO_2).

X	H_2O	$d_{H\text{-}O}$		One Distance of (X)	Angles (X)
CH_4	1	0.961	105.9	C-H = 1.094	109.255
CO_2	1	0.961	105.7	C-O = 1.165	179.344
O_3	1	0.961	103.4	O-O = 1.267	117.303
SF_6	1	0.960	104.8	S-F = 1.584	89.787
NF_3	1	0.969	104.0	N-F = 1.410	102.390
CF_4	1	0.961	106.6	C-F = 1.345	109.673
N_2O	1	0.961	105.9	N-N = 1.140 N-O = 1.149	179.782
SO_2	1	0.962	103.8	S-O = 1.440	118.205

2.3. Sorption Isotherms

The isothermal sorption of pollutants and water in singular and mixed forms was studied at 298 K using the "sorption isotherm" module implemented in Materials Studio 22.0 [42]. This study is based on random moves to perform energy changes for the new conformation. This is by measuring the chemical potential, expressed as the gas fugacity with pressure inside the cell bulk. The results are shown in Figure 4, indicating the average adsorption capacity of gas molecules by the copper squarate.

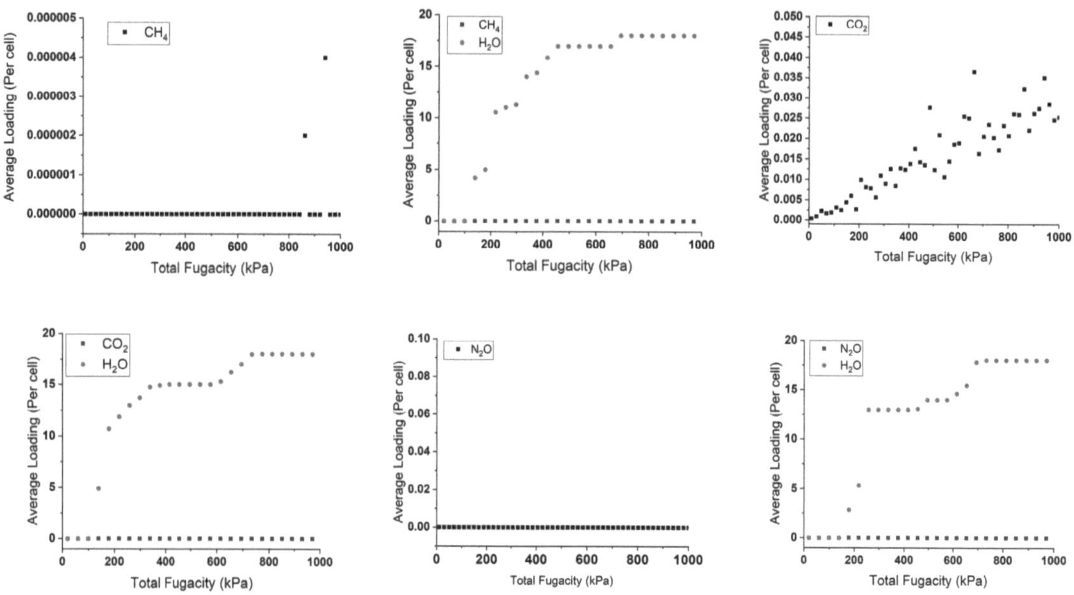

Figure 4. *Cont.*

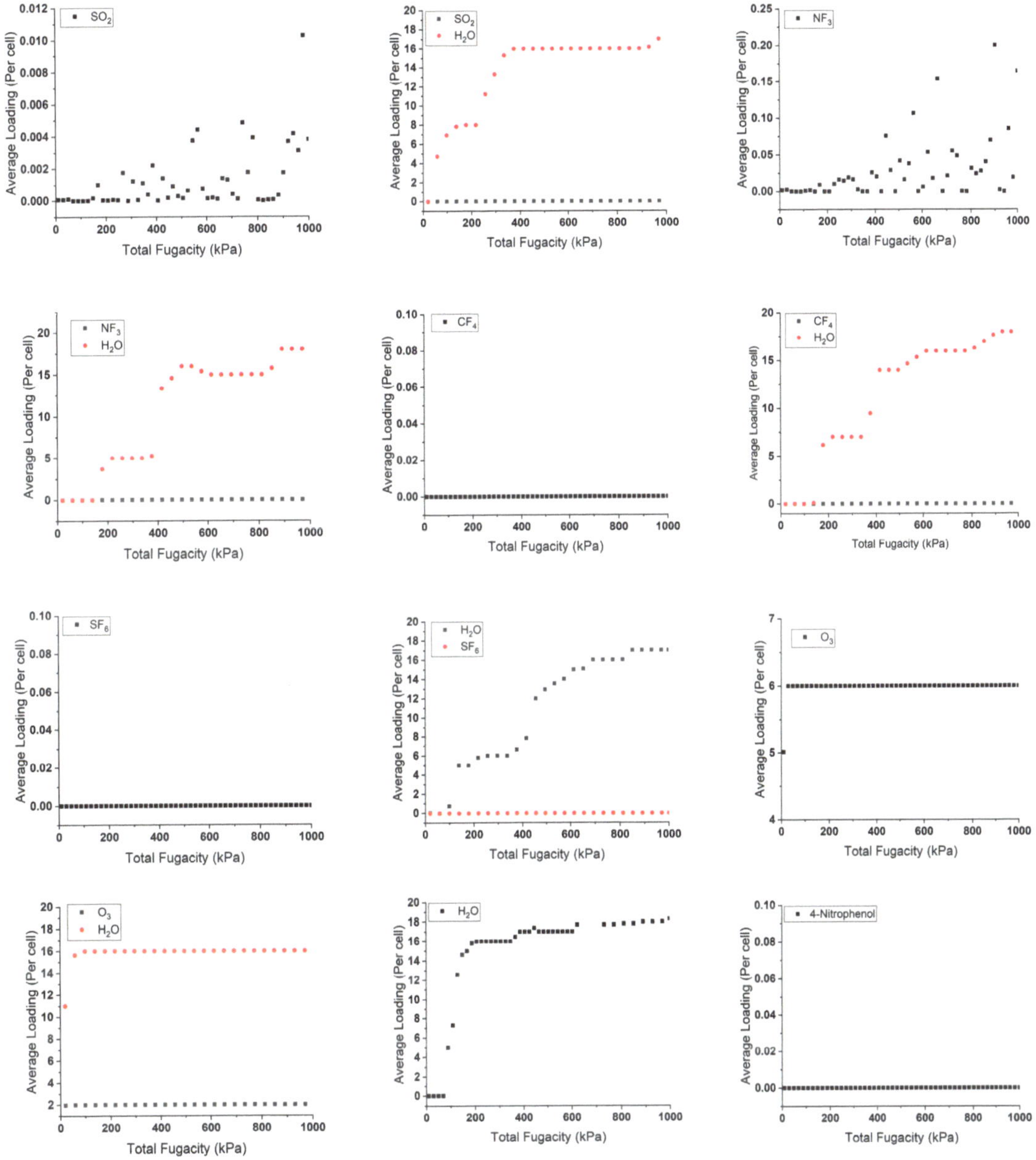

Figure 4. Sorption isotherms of single and mixture adsorbed pollutants with and without water at 298 K.

By analyzing the adsorbed singular graphs in Figure 4, we observe that most of the pollutants present a consistent loading profile with molecules inserted around zero or zero per the considered unit cell in low- or high-pressure variations. For example, N_2O shows zero loaded molecules, while the adsorption study shows high negative adsorption energy of −63.32 kJ/mol, indicating a favorable capture. This aligns also with the Connolly studies, which show available space inside copper squarate. These particular results contradict the precise results obtained by the DFT-D3 method. These strange results can be attributed to the limitations of the force field used in this C-S model, which includes pressure and is unable to accurately describe the unusual properties of the material, in particular its negative linear compressibility [36], highlighting the necessity of complementing it with ab initio calculations. In contrast, water is well described, showing an uptake loading of 15 molecules per cell, which is in good agreement with CASTEP calculations. This agreement may indicate the established hydrophilic nature of the solid. Similarly, ozone molecules show comparable behavior due to their common nature with water.

In the presence of water, the line graph representing the pollutant does not change. There are two reasons for this. The first reason is the same as when they are loaded individually. The second reason may be due to the hydrophilicity of water, which dominates the pollutants and competes with them, reducing their loading capacity in the cell. This phenomenon is evident with ozone, which loads around three molecules instead of six in a dry environment. This is due to the stronger hydrogen bonds between the water and the copper squarate to the intermolecular interactions between the pollutants and the bulk.

3. Quantum Computational Methods

All calculations for the solid-state compounds considered in this work were performed with CASTEP code [53] using theoretical methods based on DFT using norm-conserving pseudopotentials and plane wave basis sets [54]. Due to the large pore size of MOFs, which contain hundreds of atoms or more, quantum mechanical methods such as post-HF are impractical for such large systems [55]. The geometry optimizations were performed employing the Broyden–Fletcher–Goldfarb–Shanno scheme (BFGS) [55,56]. We treated the exchange-correlation energy term using the generalized gradient approximation (GGA) with Perdew–Burke–Ernzerhof (PBE) density functionals [57]. Since the pollutant molecules are packed together inside the unit cell of copper squarate, their description requires the inclusion of Van der Waals forces. For this purpose, Grimme's empirical dispersion correction method (DFT-D3) for dispersive vdW interactions was used in calculations [58,59]. Customized atom pseudopotentials [60] generated with the CASTEP code and a plane wave kinetic energy cut-off of 1000 eV were utilized. Convergence criteria were set at 2×10^{-5} eV/atom for energy, 0.05 eV/Å for interatomic forces, 0.1 GPa for maximum stress, and 0.002 Å for displacements. Atomic forces and charges were evaluated at the minimum in potential energy surface to predict the harmonic vibrational infrared spectrum using density functional perturbation theory [61,62].

After a full geometry optimization of all CH_4, CO_2, O_3, SF_6, NF_3, CF_4, N_2O, SO_2, 4NP, water, and copper squarate, the most stable adsorption sites of the pollutants inside the solid material were determined in order to minimize the first adsorption energy. The adsorption sites and mechanisms within the copper squarate were determined by Monte Carlo simulations [63], implemented in the Adsorption Locator module using COMPASS III [46] force field in Materials Studio 22.0 [42] code. Simulations involve a fraction of 10,000 steps following a number of cycles of 3 with 15,000 steps per cycle. Subsequent to the initial results, Density Functional Theory (DFT) calculations were conducted to optimize performance by adjusting the generic force fields within the Adsorption Locator module. More specifically, it was already proved that force fields such as COMPASS III [46] and Dreiding [64] tend to overestimate the number of molecules adsorbed in simulations [65], as was demonstrated in previous studies involving ZIF-4 or ZIF-6 [32,45].

The adsorption process occurs through chemical or physical binding, facilitated by the attractive forces between the solid cavity of the copper squarate and the pollutants, in both dry and humid conditions. The adsorption energies were calculated according to equations [66]:

$$\Delta E_{ads}(X) = [E_T - (E_{C-S} + nE_X)]/n(X = Polluants\ or\ H_2O) \quad (2)$$

$$\Delta E_{ads}^{mult}(X, Y, \cdots) = \frac{[E_T - (E_{C-S} + nE_X + mE_Y + \cdots + tE_Z)]}{(n + m + \cdots + t)} / (X, Y, Z \cdots = Polluants\ or\ H_2O) \quad (3)$$

In these equations, ΔE_{ads} refers to the isolated adsorption energy for each individual species (X) and ΔE_{ads}^{mult} denotes the mixed adsorption energy for species ($X, Y, \cdots\ldots$). E_T stands for the total energy of copper squarate (C-S) with gas molecules included, E_{C-S} corresponds to the energies of pure C-S, and E_X is the energy of the respective pollutants. In addition, n, m, t represent the quantities of loaded molecules. The assessment of pollutants and water is conducted using a 10 Å supercell. A negative adsorption energy implies favorable adsorption.

The interaction energies between the co-adsorbed pollutants on the C-S material were calculated with the help of the energetic database obtained from DFT calculations using equation [45]:

$$E_{int} = E_{(n)-pollutants+Bulk} + E_{Bulk} - \sum E_{n-pollutants+Bulk} \quad (4)$$

where $E_{(n)-pollutants+Bulk}$ represents the total energy of the solid material after adsorption of the n pollutants by copper squarate bulk, and $\sum E_{n-pollutants-Bulk}$ is the sum of the total energy of the previously adsorbed species calculated separately.

The Grand Canonical Monte Carlo (GCMC) method [67] implemented in the Sorption module of Materials Studio 22.0 [42] was used to determine adsorption isotherms with the Metropolis algorithm and COMPASS III. The gas molecules were randomly loaded at a fixed temperature of 298 K into the fixed-volume copper squarate structure based on the gas fugacity. The simulations involve 100,000 equilibration steps followed by 1,000,000 Monte Carlo steps, assuming periodic boundary conditions. Long-range interactions were treated using the Ewald and group summation method with an Ewald accuracy of 0.0001 kcal/mol and a cut-off distance of 15.5 Å. The results are presented as a line graph (N = f (μ)), where f is the density of the number of loaded molecules N, expressed as a function of chemical potential.

4. Conclusions

In this work, we focus our interest on a new kind of MOF, namely copper squarate, whose adsorption behavior has not been explored in previous studies. For this reason, the adsorption of a series of small pollutants and some aromatic compounds was investigated using DFT calculations in a dry and humid environment. As a first observation, we note that small pollutants such as CH_4, CO_2, O_3, SF_6, NF_3, CF_4, N_2O, and SO_2 favor the adsorption by different amounts depending on their chemical composition.

Squarate demonstrates excellent stability in the presence of the considered pollutants, exhibiting a favorable capture of the latter that results in a negative adsorption energy value. This phenomenon can be attributed to the existence of electrostatic interaction forces between pollutants and squarate. These interactions are particularly pronounced at short distances, occurring between the oxygen of squarate and the carbon atoms, nitrogen, and sulfur in the molecules CO_2, N_2O, SO_2, and H_2O, respectively. The presence of other weak parallel interactions further reinforces the binding of these molecules of the C-S. Additionally, the non-polarity of CH_4 gas makes it a favorable insert within squarate, accentuated by a hydrogen bond. Finally, the dipole nature of O_3 imparts characteristic properties, enabling it to react very selectively as an electrophile. These characteristic properties enable them to react very selectively as an electrophile. This characteristic facilitates its strong adsorption with squarate, driven by the intermolecular forces of Van der Waals and London dispersion.

Pollutants with higher volume and surface density are adsorbed in smaller quantities; for example, molecules with fluorine o nitrogen atoms. Indeed, adsorption in NF_3, CF_4, and SF_6 molecules is influenced by the number of fluorine atoms used and the adapted geometry. The smaller this number, the more the adsorption process is promoted. Furthermore, the more it adopts a complex structure, the more the interactions between pollutants and squarate are minimized. In contrast to all previous gases, nitrophenols are not able to be adsorbed in this bulk because of the small pore size.

The adsorption energy of the molecule containing nitrogen as N_2O in a dry environment is revelaed to be the highest one (−63.02), even if the maximal insertion is one molecule per cell compared to four insert molecules of CH_4 with a maximum value of −23.02. This indicates the strong interaction involved in the first adsorption. Globally, the addition of water does not interfere with the adsorption, keeping the adsorption energy negative.

The co-adoption of CH_4, SO_2, and CO_2 with H_2O implies an increase in adsorption energy. The same trend is observed for molecules containing fluorine atoms, such as NF_3, CF_4, and SF_6. This impact on adsorption energy values is attributed to multiple interactions caused by the reactive sites of these pollutants, as well as the formation of weak bonds between water molecules.

In a humid environment, the same order is found for O_3 and N_2O molecules as in the dry state, with a very slight decrease in adsorption energies. Meanwhile, 4NP is an exception to the other pollutants studied. On the other hand, it is characterized by an unfavorable capture on C-S in both humid and dry environments, resulting in positive adsorption energy. This is attributed to its large size and the unavailability of sufficient space in the pore. However, C-S exhibits a high capacity to insert water inside its cavity, revealing its hydrophilic nature. This behavior stands in contrast to enormous MOFs well-documented in the literature for their highly hydrophobic characteristics. Such hydrophilicity is particularly advantageous, especially considering that gas molecules are typically present in humid conditions, allowing us to regard C-S as a revolutionary MOF for adsorption, making it a novel target in this field.

The number of molecules fixed in the C-S decreases from the dry to the wet state, indicating that water molecules compete with pollutants and form strong hydrogen bonds with the squarate, compared to the interactions between pollutants and the bulk.

In future work, it is recommended to simulate a molecular work with a reduced part of copper squarate to understand deeper what is happening inside the cell cavity and comprehend with better precision to adsorption nature and the type of electrostatic interaction, including Van der Waals or hydrogen bonds in both dry and humid environments, as well as to determine the factors that govern this adsorption phenomenon. In addition, to complete this, a molecular dynamics (MD) trajectory will be helpful to understand dynamic adsorption.

Supplementary Materials: The following supporting information can be downloaded at: https://www.mdpi.com/article/10.3390/molecules29133140/s1, Figure S1: Comparison between the Powder X-ray diffraction patterns derived from the computed and experimental structures [35]; Table S1: Comparison between the theoretical data on the geometry of molecules and the experimental data obtained from NIST database [68]; Table S2: The reported calculated structure in *.cif format, used in the DFT-PBE calculations.

Author Contributions: Conceptualization, C.A., N.G., V.T., F.C. and D.H.; Methodology, C.A.; Validation, V.T., F.C. and D.H.; Formal analysis, C.A. and D.H.; Investigation, C.A. and N.G.; Resources, V.T.; Writing—original draft, C.A.; Writing—review and editing, N.G., V.T., F.C. and D.H.; Supervision, N.G., V.T., F.C. and D.H.; Project administration, V.T.; Funding acquisition, V.T. and D.H. All authors have read and agreed to the published version of the manuscript.

Funding: The authors acknowledge the Ministerio de Ciencia, Innovación y Universidades of Spain through the grants PID2020-112887GB-I00 and PID2020-113084GB-I00/AEI/10.13039/501100011033.

Institutional Review Board Statement: Not applicable.

Informed Consent Statement: Not applicable.

Data Availability Statement: The original contributions presented in the study are included in the article and Supplementary Materials; further inquiries can be directed to the corresponding authors.

Acknowledgments: This project has received funding from the European Union's Horizon 2020 research and innovation programme under the Marie Skłodowska-Curie grant agreement No. 872081. The authors acknowledge the Ministerio de Ciencia, Innovación y Universidades of Spain through the grants PID2020-112887GB-I00 and PID2020-113084GB-I00. The authors acknowledge the CTI (CSIC) and CESGA and the "Red Española de Computación" for the grant RES-AECT-2022-3-0006 and computing facilities.

Conflicts of Interest: The authors declare no conflicts of interest.

References

1. De Lotto, R. Assessment of Development and Regeneration Urban Projects: Cultural and Operational Implications in Metropolization Context. *Int. J. Energy Environ.* **2008**, *2*, 25–34.
2. Molina, M.J.; Molina, L.T. Megacities and Atmospheric Pollution. *J. Air Waste Manag. Assoc.* **2004**, *54*, 644–680. [CrossRef] [PubMed]
3. Chen, G.; Jin, Y.; Lu, J. Experimental Study on Adsorption of SO_2 and DCM from Air Pollutants by Modified Biochar. *Biomass Convers. Biorefin* **2023**, 1–15. [CrossRef]
4. Zhang, J.; Zhou, J.Z.; Xu, Z.P.; Li, Y.; Cao, T.; Zhao, J.; Ruan, X.; Liu, Q.; Qian, G. Decomposition of Potent Greenhouse Gas Sulfur Hexafluoride (SF6) by Kirschsteinite-Dominant Stainless Steel Slag. *Environ. Sci. Technol.* **2014**, *48*, 599–606. [CrossRef] [PubMed]
5. Yao, Z.; Zheng, X.; Dong, H.; Wang, R.; Mei, B.; Zhu, J. A 3-Year Record of N_2O and CH_4 Emissions from a Sandy Loam Paddy during Rice Seasons as Affected by Different Nitrogen Application Rates. *Agric. Ecosyst. Environ.* **2012**, *152*, 1–9. [CrossRef]
6. Ehhalt, D.; Prather, M.; Dentener, F.; Derwent, R.; Dlugokencky, E.; Holland, E.; Isaksen, I.; Katima, J.; Kirchhoff, V.; Matson, P.; et al. Atmospheric Chemistry and Greenhouse Gases. In *Climate change 2001: The Scientific Basis*; Intergovernmental Panel on Climate Change: Geneva, Switzerland, 2001.
7. Arani, M.H.; Jaafarzadeh, N.; Moslemzadeh, M.; Ghalhari, M.R.; Arani, S.B.; Mohammadzadeh, M. Dispersion of NO_2 and SO_2 Pollutants in the Rolling Industry with AERMOD Model: A Case Study to Assess Human Health Risk. *J. Environ. Health Sci. Eng.* **2021**, *19*, 1287–1298. [CrossRef]
8. García-Berríos, Z.I.; Arce, R. Photodegradation Mechanisms of 1-Nitropyrene, an Environmental Pollutant: The Effect of Organic Solvents, Water, Oxygen, Phenols, and Polycyclic Aromatics on the Destruction and Product Yields. *J. Phys. Chem. A* **2012**, *116*, 3662–3664. [CrossRef]
9. Wang, H.L.; Nie, L.; Li, J.; Wang, Y.F.; Wang, G.; Wang, J.H.; Hao, Z.P. Characterization and Assessment of Volatile Organic Compounds (VOCs) Emissions from Typical Industries. *Chin. Sci. Bull.* **2013**, *58*, 724–730. [CrossRef]
10. Inomata, S.; Yamada, H.; Tanimoto, H. Investigation on VOC Emissions from Automobile Sources by Means of Online Mass Spectrometry. *Curr. Pollut. Rep.* **2016**, *2*, 188–199. [CrossRef]
11. Lin, Y.; Qiu, X.; Ma, Y.; Ma, J.; Zheng, M.; Shao, M. Concentrations and Spatial Distribution of Polycyclic Aromatic Hydrocarbons (PAHs) and Nitrated PAHs (NPAHs) in the Atmosphere of North China, and the Transformation from PAHs to NPAHs. *Environ. Pollut.* **2015**, *196*, 164–170. [CrossRef]
12. Keyte, I.J.; Albinet, A.; Harrison, R.M. On-Road Traffic Emissions of Polycyclic Aromatic Hydrocarbons and Their Oxy- and Nitro- Derivative Compounds Measured in Road Tunnel Environments. *Sci. Total Environ.* **2016**, *566–567*, 1131–1142. [CrossRef] [PubMed]
13. Supong, A.; Bhomick, P.C.; Sinha, U.B.; Sinha, D. A Combined Experimental and Theoretical Investigation of the Adsorption of 4-Nitrophenol on Activated Biocarbon Using DFT Method. *Korean J. Chem. Eng.* **2019**, *36*, 2023–2034. [CrossRef]
14. Ipek, I.; Kabay, N.; Yüksel, M. Separation of Bisphenol A and Phenol from Water by Polymer Adsorbents: Equilibrium and Kinetics Studies. *J. Water Process Eng.* **2017**, *16*, 206–211. [CrossRef]
15. Plaza-Medina, E.F.; Rodríguez-Córdoba, W.; Morales-Cueto, R.; Peon, J. Primary Photochemistry of Nitrated Aromatic Compounds: Excited-State Dynamics and NO• Dissociation from 9-Nitroanthracene. *J. Phys. Chem. A* **2011**, *115*, 577–585. [CrossRef] [PubMed]
16. Tian, M.; Bakovic, L.; Chen, A. Kinetics of the Electrochemical Oxidation of 2-Nitrophenol and 4-Nitrophenol Studied by in Situ UV Spectroscopy and Chemometrics. *Electrochim. Acta* **2007**, *52*, 6517–6524. [CrossRef]
17. Bertinelli, L.; Camacho, C.; Zou, B. Carbon Capture and Storage and Transboundary Pollution: A Differential Game Approach. *Eur. J. Oper. Res.* **2014**, *237*, 721–728. [CrossRef]
18. Shan, Y.; Du, J.; Zhang, Y.; Shan, W.; Shi, X.; Yu, Y.; Zhang, R.; Meng, X.; Xiao, F.S.; He, H. Selective Catalytic Reduction of NOx with NH3: Opportunities and Challenges of Cu-Based Small-Pore Zeolites. *Natl. Sci. Rev.* **2021**, *8*, nwab010. [CrossRef] [PubMed]
19. Inomata, S.; Fushimi, A.; Sato, K.; Fujitani, Y.; Yamada, H. 4-Nitrophenol, 1-Nitropyrene, and 9-Nitroanthracene Emissions in Exhaust Particles from Diesel Vehicles with Different Exhaust Gas Treatments. *Atmos. Environ.* **2015**, *110*, 93–102. [CrossRef]
20. Yang, L.; Zhang, L.; Chen, L.; Han, C.; Akutagawa, T.; Endo, O.; Yamauchi, M.; Neroda, A.; Toriba, A.; Tang, N. Polycyclic Aromatic Hydrocarbons and Nitro-Polycyclic Aromatic Hydrocarbons in Five East Asian Cities: Seasonal Characteristics, Health Risks, and Yearly Variations. *Environ. Pollut.* **2021**, 287. [CrossRef]
21. Zeng, Y.; Zou, R.; Zhao, Y. Covalent Organic Frameworks for CO_2 Capture. *Adv. Mater.* **2016**, *28*, 2855–2873. [CrossRef]
22. Wang, S.M.; Mu, X.T.; Liu, H.R.; Zheng, S.T.; Yang, Q.Y. Pore-Structure Control in Metal–Organic Frameworks (MOFs) for Capture of the Greenhouse Gas SF6 with Record Separation. *Angew. Chem. -Int. Ed.* **2022**, *61*, e202207066. [CrossRef]

23. Jang, Y.; Sekaran, B.; Singh, P.P.; Misra, R.; D'Souza, F. Accelerated Intramolecular Charge Transfer in Tetracyanobutadiene- and Expanded Tetracyanobutadiene-Incorporated Asymmetric Triphenylamine-Quinoxaline Push-Pull Conjugates. *J. Phys. Chem. A* **2023**, *127*, 4455–4462. [CrossRef] [PubMed]
24. Han, Y.; Chen, Y.; Ma, Y.; Bailey, J.; Wang, Z.; Lee, D.; Sheveleva, A.M.; Tuna, F.; McInnes, E.J.L.; Frogley, M.D.; et al. Control of the Pore Chemistry in Metal-Organic Frameworks for Efficient Adsorption of Benzene and Separation of Benzene/Cyclohexane. *Chem* **2023**, *9*, 739–754. [CrossRef]
25. Zhang, S.; Yang, Q.; Liu, X.; Qu, X.; Wei, Q.; Xie, G.; Chen, S.; Gao, S. High-Energy Metal-Organic Frameworks (HE-MOFs): Synthesis, Structure and Energetic Performance. *Coord. Chem. Rev.* **2016**, *307*, 292–312. [CrossRef]
26. Thi Thuy Huong, T.; Ngoc Thanh, P.; Thi Xuan Huynh, N.; Ngoc Son, D. Metal-Organic Frameworks: State-of-the-Art Material for Gas Capture and Storage. *VNU J. Sci. Math. Phys.* **2016**, *32*, 67–85.
27. Paudel, H.P.; Shi, W.; Hopkinson, D.; Steckel, J.A.; Duan, Y. Computational Modelling of Adsorption and Diffusion Properties of CO_2 and CH_4 in ZIF-8 for Gas Separation Applications: A Density Functional Theory Approach. *React. Chem. Eng.* **2021**, *6*, 990–1001. [CrossRef]
28. Miyazaki, R.; Nakatani, N.; Levchenko, S.V.; Yokoya, T.; Nakajima, K.; Hara, K.; Fukuoka, A.; Hasegawa, J.Y. DFT Mechanistic Study on the Complete Oxidation of Ethylene by the Silica-Supported Pt Catalyst: C=C Activation via the Ethylene Dioxide Intermediate. *J. Phys. Chem. C* **2019**, *123*, 12706–12715. [CrossRef]
29. Zhang, J.; Wang, J.; Zhang, C.; Li, Z.; Lu, B.; Zhu, J. Molecular Simulation of $C_2H_4/CO_2/N_2/O_2$ Adsorption Characteristics in Lignite and Anthracite. *AIP Adv.* **2021**, *11*, 085205. [CrossRef]
30. Zhang, S.; Wang, J.; Zhang, Y.; Ma, J.; Huang, L.; Yu, S.; Chen, L.; Song, G.; Qiu, M.; Wang, X. Applications of Water-Stable Metal-Organic Frameworks in the Removal of Water Pollutants: A Review. *Environ. Pollut.* **2021**, *291*, 118076. [CrossRef]
31. Adjal, C.; Timón, V.; Guechtouli, N.; Boussassi, R.; Hammoutène, D.; Senent, M.L. The Role of Water in the Adsorption of Nitro-Organic Pollutants on Activated Carbon. *J. Phys. Chem. A* **2023**, *127*, 8146–8158. [CrossRef]
32. Timón, V.; Senent, M.L.; Hochlaf, M. Structural Single and Multiple Molecular Adsorption of CO2 and H2O in Zeolitic Imidazolate Framework (ZIF) Crystals. *Microporous Mesoporous Mater.* **2015**, *218*, 33–41. [CrossRef]
33. Furukawa, H.; Gándara, F.; Zhang, Y.B.; Jiang, J.; Queen, W.L.; Hudson, M.R.; Yaghi, O.M. Water Adsorption in Porous Metal-Organic Frameworks and Related Materials. *J. Am. Chem. Soc.* **2014**, *136*, 4369–4381. [CrossRef]
34. Guo, S.; Huang, X.; Situ, Y.; Huang, Q.; Guan, K.; Huang, J.; Wang, W.; Bai, X.; Liu, Z.; Wu, Y.; et al. Interpretable Machine-Learning and Big Data Mining to Predict Gas Diffusivity in Metal-Organic Frameworks. *Adv. Sci.* **2023**, *10*, 2301461. [CrossRef] [PubMed]
35. Dinnebier, R.E.; Nuss, H.; Jansen, M. Anhydrous CuC4O4, a Channel Structure Solved from X-Ray Powder Diffraction Data. *Z. Anorg. Allg. Chem.* **2005**, *631*, 2328–2332. [CrossRef]
36. Colmenero, F.; Lobato, Á.; Timón, V. Compressing the Channels in the Crystal Structure of Copper Squarate Metal-Organic Framework. *Solids* **2022**, *3*, 374–384. [CrossRef]
37. Colmenero, F. Negative Linear Compressibility in Nanoporous Metal-Organic Frameworks Rationalized by the Empty Channel Structural Mechanism. *Phys. Chem. Chem. Phys.* **2021**, *23*, 8508–8524. [CrossRef]
38. Colmenero, F.; Cobos, J.; Timón, V. Negative Linear Compressibility in Uranyl Squarate Monohydrate. *J. Phys. Condens. Matter* **2019**, *31*, 175701. [CrossRef]
39. Colmenero, F. Addendum: Anomalous Mechanical Behavior of the Deltic, Squaric and Croconic Cyclic Oxocarbon Acids. *Mater. Res. Express* **2019**, *6*, 045610. [CrossRef]
40. Qiu, W.; Zeng, Q.; Li, C.; Hao, J.; Li, Y. Theoretical Investigation of Zero Linear Compressibility on Metal Squarates MC4O4 (M = Pb and Ba). *J. Phys. Chem. C* **2023**, *127*, 9957–9963. [CrossRef]
41. Zhang, L.; He, Z.; Liu, Y.; You, J.; Lin, L.; Jia, J.; Chen, S.; Hua, N.; Ma, L.A.; Ye, X.; et al. A Robust Squarate-Cobalt Metal-Organic Framework for CO_2/N_2 Separation. *ACS Appl. Mater. Interfaces* **2023**, *15*, 30394–30401. [CrossRef]
42. BIOVIA. Dassault Systemes, Materials Studio 2022. Dassault Systemes: San Diego, CA, USA, 2022.
43. Delgado, M.R. Structure and Stability of Gas Adsorption Complexes in Periodic Porous Solids as Studied by VTIR Spectroscopy: An Overview. *Appl. Sci.* **2020**, *10*, 8589. [CrossRef]
44. Manak, M.; Jirkovsky, L.; Kolingerova, I. Interactive Analysis of Connolly Surfaces for Various Probes. *Comput. Graph. Forum* **2017**, *36*, 160–172. [CrossRef]
45. El, N.; Timón, V.; Boussessi, R.; Dalbouha, S.; Senent, M.L. DFT Studies of Single and Multiple Molecular Adsorption of CH4, SF6 and H2O in Zeolitic-Imidazolate Framework (ZIF-4 and ZIF-6). *Inorganica Chim Acta* **2019**, *490*, 272–281. [CrossRef]
46. Akkermans, R.L.C.; Spenley, N.A.; Robertson, S.H. COMPASS III: Automated Fitting Workflows and Extension to Ionic Liquids. *Mol. Simul.* **2021**, *47*, 540–551. [CrossRef]
47. Desiraju, G.R.; Gautam, R.; Steiner, T. The Weak Hydrogen Bond. In *Structural Chemistry and Biology*; Oxford University Press: Oxford, UK, 1999; ISBN 9780198509707.
48. Park, K.S.; Ni, Z.; Cô, A.P.; Choi, J.Y.; Huang, R.; Uribe-Romo, F.J.; Chae, H.K.; O'keeffe, M.; Yaghi, O.M. Exceptional Chemical and Thermal Stability of Zeolitic Imidazolate Frameworks. *Proc. Natl. Acad. Sci. USA* **2006**, *103*, 10186–10191. [CrossRef] [PubMed]
49. Tanaka, S.; Tanaka, Y. A Simple Step toward Enhancing Hydrothermal Stability of ZIF-8. *ACS Omega* **2019**, *4*, 19905–19912. [CrossRef] [PubMed]
50. Shieh, F.K.; Wang, S.C.; Leo, S.Y.; Wu, K.C.W. Water-Based Synthesis of Zeolitic Imidazolate Framework-90 (ZIF-90) with a Controllable Particle Size. *Chem.-A Eur. J.* **2013**, *19*, 11139–11142. [CrossRef]

51. Nalaparaju, A.; Zhao, X.S.; Jiang, J.W. Molecular Understanding for the Adsorption of Water and Alcohols in Hydrophilic and Hydrophobic Zeolitic Metal-Organic Frameworks. *J. Phys. Chem. C* **2010**, *114*, 11542–11550. [CrossRef]
52. Benedict, W.S.; Gailar, N.; Plyler, E.K. Rotation-Vibration Spectra of Deuterated Water Vapor. *J. Chem. Phys.* **1956**, *24*, 1139–1165. [CrossRef]
53. Clark, S.J.; Segall, M.D.; Pickard, C.J.; Hasnip, P.J.; Probert, M.J.; Refson, K. MC Payne. *First Princ. Methods Using Castep. Z. Für Krist. Cryst. Mater.* **2005**, *220*, 567–570.
54. Hai, G.; Wang, H. Theoretical Studies of Metal-Organic Frameworks: Calculation Methods and Applications in Catalysis, Gas Separation, and Energy Storage. *Coord. Chem. Rev.* **2022**, *469*, 214670. [CrossRef]
55. Payne, M.C.; Teter, M.P.; Ailan, D.C.; Arias, T.A.; Joannopouios, J.D. Iterative Minimization Techniques for Ab Initio Total-Energy Calculations: Molecular Dynamics and Conjugate Gradients. *Rev. Mod. Phys.* **1992**, *64*, 1045. [CrossRef]
56. Pfrommer, B.G.; Cô, M.; Louie, S.G.; Cohen, M.L. Relaxation of Crystals with the Quasi-Newton Method. *J. Comput. Phys.* **1997**, *131*, 233–240. [CrossRef]
57. Perdew, J.P.; Burke, K.; Ernzerhof, M. Generalized Gradient Approximation Made Simple Phys. *Rev. Lett.* **1996**, *77*, 3865–3868. [CrossRef] [PubMed]
58. Grimme, S. Semiempirical GGA-Type Density Functional Constructed with a Long-Range Dispersion Correction. *J. Comput. Chem.* **2006**, *27*, 1787–1799. [CrossRef] [PubMed]
59. Ehrlich, S.; Moellmann, J.; Reckien, W.; Bredow, T.; Grimme, S. System-Dependent Dispersion Coefficients for the DFT-D3 Treatment of Adsorption Processes on Ionic Surfaces. *ChemPhysChem* **2011**, *12*, 3414–3420. [CrossRef] [PubMed]
60. Troullier, N.; Martins, J.L. EfFicient Pseudopotentials for Plane-Wave Calculations-II. Phys. *Rev. B* **1991**, *8*, 1993. [CrossRef] [PubMed]
61. Baroni, S.; De Gironcoli, S.; Corso, A.D.; Giannozzi, P. Phonons and Related Crystal Properties from Density-Functional Perturbation Theor. *Rev. Mod. Phys.* **2001**, *73*, 515. [CrossRef]
62. Refson, K.; Tulip, P.R.; Clark, S.J. Variational Density-Functional Perturbation Theory for Dielectrics and Lattice Dynamics. *Phys. Rev. B Condens. Matter Mater. Phys.* **2006**, *73*, 155114. [CrossRef]
63. Khaled, K.F. Monte Carlo Simulations of Corrosion Inhibition of Mild Steel in 0.5 M Sulphuric Acid by Some Green Corrosion Inhibitors. *J. Solid State Electrochem.* **2009**, *13*, 1743–1756. [CrossRef]
64. Mayo, S.L.; Olafson, B.D.; Goddard, W.A., III. DREIDING: A Generic Force Field for Molecular Simulations. *J. Phys. Chem.* **1990**, *94*, 8897–8909. [CrossRef]
65. Babarao, R.; Dai, S.; Jiang, D.E. Effect of Pore Topology and Accessibility on Gas Adsorption Capacity in Zeolitic-Imidazolate Frameworks: Bringing Molecular Simulation Close to Experiment. *J. Phys. Chem. C* **2011**, *115*, 8126–8135. [CrossRef]
66. Gao, H.; Pishney, S.; Janik, M.J. First Principles Study on the Adsorption of CO_2 and H_2O on the K_2CO_3 (001) Surface. *Surf. Sci.* **2013**, *609*, 140–146. [CrossRef]
67. Smit, B.; Frenkel, D. *Understanding Molecular Simulation*, 2nd ed.; Elsevier: Amsterdam, The Netherlands, 2023; Volume 1.
68. NIST Computational Chemistry Comparison and Benchmark Database. NIST Standard Reference Database Number 101. Release 22, May 2022, Editor: Russell D. Johnson III. Available online: https://cccbdb.nist.gov/ (accessed on 23 May 2024).

materials

MDPI

Article

Adsorption Properties and Mechanisms of Methylene Blue by Modified Sphagnum Moss Bio-Based Adsorbents

Junpeng Ren [1], Shijiang Zhang [1], Yu Wang [1,*] and Hengxiu Yang [2]

[1] School of Chemistry and Materials Science, Guizhou Education University, Guiyang 550018, China; rjunpeng@126.com (J.R.); 18212717435@163.com (S.Z.)
[2] Guizhou Provincial Key Laboratory of Computational Nano-Material Science, Guizhou Education University, Guiyang 550018, China; hxyang@gznc.edu.cn
* Correspondence: wangyugeu@126.com; Tel.: +86-0851-85816467; Fax: +86-0851-85816467

Abstract: The abundant pore structure and carbon composition of sphagnum peat moss render it a bio-based adsorbent for efficient methylene blue removal from wastewater. By utilizing sphagnum moss sourced from Guizhou, China, as raw material, a cost-effective and highly efficient bio-based adsorbent material was prepared through chemical modification. The structure and performance of the modified sphagnum moss were characterized using SEM, EDS, FTIR, and TGA techniques. Batch adsorption experiments explored the effects of contact time, adsorbent dosage, pH, initial dye concentration, and temperature on adsorption performance. Kinetics, isotherm models, and thermodynamics elucidated the adsorption behavior and mechanism. The modified sphagnum moss exhibited increased surface roughness and uniform surface modification, enhancing active site availability for improved adsorption. Experimental data aligned well with the Freundlich isotherm model and pseudo-second-order kinetic model, indicating efficient adsorption. The study elucidated the adsorption mechanism, laying a foundation for effective methylene blue removal. The utilization of modified sphagnum moss demonstrates significant potential in effectively removing MB from contaminated solutions due to its robust adsorption capability and efficient reusability.

Keywords: sphagnum moss; modification; adsorption; methylene blue; adsorption mechanisms

Citation: Ren, J.; Zhang, S.; Wang, Y.; Yang, H. Adsorption Properties and Mechanisms of Methylene Blue by Modified Sphagnum Moss Bio-Based Adsorbents. *Materials* **2024**, *17*, 4329. https://doi.org/10.3390/ma17174329

Academic Editor: Alessandro Pegoretti

Received: 21 June 2024
Revised: 5 August 2024
Accepted: 28 August 2024
Published: 31 August 2024

1. Introduction

Water pollution is a grave environmental issue, posing significant threats to both human health and aquatic organisms as well as ecosystems. The presence of organic dye pollutants in wastewater has become a global concern, particularly originating from industries such as printing, textiles, cosmetics, and pharmaceuticals [1,2]. These pollutants are characterized by their complex composition, high toxicity levels, and resistance to biodegradation [3]. If left untreated before discharge into the environment, they can pose severe risks to both the ecosystem and human well-being. In fact, large quantities of organic pollutants are discharged into water bodies annually [4].

Organic synthetic dyes are extensively used due to their vibrant colors and light sensitivity [5]. However, the uncontrolled release of organic dye wastewater (e.g., methyl orange, malachite green, methylene blue, methyl red) from textile manufacturing processes or other related industries significantly contributes to pollution levels. Such untreated waste liquids not only disrupt the activity of aquatic organisms but also lead to cell carcinogenesis in individuals who consume contaminated water, ultimately inflicting damage upon essential organs like the liver, kidneys, and eyes, thus posing substantial risks to human well-being [6].

Notably, methylene blue (MB), a cationic dye favored for its chemical stability and color properties, finds application across multiple sectors, including textile, medical, and scientific research [7,8]. Despite its benefits, MB's high solubility and toxicity classify it as a detrimental pollutant capable of ascending the food chain via direct contact or consumption, thereby endangering both human and animal health. Moreover, the chemical structure of methylene blue contains a benzene ring and methylene group, which confer high stability, making it resistant to natural degradation [9]. Therefore, effective treatment measures for dye-containing wastewater are crucial in preventing major health risks associated with environmental discharge. Consequently, studying efficient removal methods for dyes from wastewater holds great significance for environmental protection and resource recycling. Currently, predominant approaches for wastewater decolorization comprise adsorption [10–12], membrane filtration [13], photocatalytic degradation [14,15], advanced oxidation [16], etc., However, the drawbacks of these methods, such as non-biodegradability, high cost, and process complexity, have significantly limited their application in treating large-scale production wastewater. Conversely, adsorption emerges as a straightforward, cost-effective, and eco-friendly technique for water reclamation. This process entails the adhesion of ions or molecules from the aqueous phase onto the surface of an adsorbent material [17]. This selective attraction and retention make it an ideal choice for many industrial wastewater flows based on dye chemical properties. Bio-based adsorption materials have gained wide popularity in water pollution treatment due to their low cost, high removal efficiency, easy recovery/reusability of the adsorbent material without causing secondary pollution issues, and versatile applications. Several low-cost bio-based materials like coconut wastes [18], cactus [19], rice husk [20], luffa fiber [21], cotton husk bracts [7], sugar cane residue [22], betel nut shell [6], and corn straw [12] have been utilized to reduce dye concentrations and explore possibilities for water treatment purposes. Nevertheless, traditional adsorbents suffer from limitations, including low efficiency, high costs, poor reusability, and complex operations; hence, there is a pressing need to explore and develop new types of adsorption materials.

Sphagnum moss, a plant thriving in damp environments, primarily comprises cellulose, lignin, and other constituents, exhibiting a porous structure and a significant specific surface area [23]. Found in countries like Brazil, Ireland, China, and the United Kingdom, it is biocompatible, biodegradable, non-toxic, and cost-efficient. Sphagnum moss contains abundant oxygen-containing functional groups and aromatic structural substances, which provide numerous active sites for organic matter adsorption. Additionally, it exhibits a high cation exchange capacity, enabling the absorption of diverse water pollutants through ion exchange and complexation reactions [24]. While extensive research has been conducted on sphagnum moss-based adsorption materials for the effective removal of heavy metals like Cd, Pb, Cu, Ni, and Cr [25–28], there are limited studies on dye wastewater adsorption, particularly regarding methylene blue dye's adsorption process and mechanism [29]. This paper utilizes sphagnum moss sourced from Guizhou Province in China as raw material to enhance its adsorption efficiency through appropriate chemical modification of organic compounds. The study systematically investigates the adsorption performance of methylene blue dye by conducting batch adsorption experiments while exploring various experimental parameters to determine their impact on the adsorption capacity. Furthermore, the study delves into the thermodynamics, kinetics, and isotherms of the adsorptive process to elucidate its mechanism, thereby providing novel ideas and methods for the efficient treatment of dye wastewater. This research represents the first detailed study in the literature on organic pollutant removal using Sphagnum moss.

2. Experimental Section

2.1. Materials

In this research, sphagnum moss (SM) collected from Guizhou Province, China, served as the adsorbent material. Before application, the moss underwent a pretreatment process involving sodium hydroxide (NaOH) and hydrogen peroxide (H_2O_2), sourced from Aladdin Reagent (Shanghai, China) Co., Ltd. Additional modifying agents such as silane, silicon dioxide, and sulfuric acid (H_2SO_4) and nitric acid (HNO_3), were obtained from Sinopharm Beijing Co., Ltd. (Beijing, China). Ethanol, a chemically pure substance, was acquired from Tianjin Hengxing Chemical Co., Ltd. (Tianjin, China), while methylene blue and potassium bromide were supplied by Tianjin Kemeiou Chemical Reagent Co., Ltd. (Tianjin, China). All chemicals utilized in the study were of analytical grade and were used as received without further purification.

2.2. Preparation of the Materials

The sphagnum moss was crushed and dried in preparation of the materials. Subsequently, 10 g of this moss was added to a 1 L three-necked flask, combined with 800 mL of a sodium hydroxide solution at 0.0375 mol/L concentration, and the mixture was agitated. This blend was magnetically stirred at 95 °C within a water bath. Subsequently, hydrogen peroxide (2.0 mL, 30%) was incrementally introduced, with stirring continuing for an hour. After the reaction ended, the mixture was allowed to cool to ambient temperature, and its pH was balanced to neutral using a dilute sulfuric acid solution (0.1 mol/L). Vacuum filtration facilitated the separation of the treated sphagnum moss, which was then vacuum-dried at 50 °C for six hours.

The pretreated sphagnum moss (6 g) was placed in a beaker containing nitric acid (60 mL, 14.4 mol/L) and sulfuric acid (120 mL, 18.4 mol/L), maintaining a 1:2 volume ratio of nitric to sulfuric acid. The mixture was stirred for five minutes, then heated in a water bath at 31 °C for 20 min. Once the reaction was complete, the material was filtered and rinsed until a neutral pH was achieved in the filtrate. Subsequently, it was dried at 50 °C for 24 h to obtain the adsorption material of modified sphagnum moss.

2.3. Analytical Methods

Various analytical techniques were applied to assess the morphology and characteristics of sphagnum moss samples before and after modification. Surface morphology and structure were examined using scanning electron microscopy (SEM; Gemini 300, Zeiss Oberkochen, Germany), while electron diffraction scattering (EDS) analysis was conducted using Smart EDX, Zeiss, Germany. The infrared spectrum of sphagnum moss before and after modification was analyzed using an FT-IR spectrometer (Tensor 27, Bruker, Karlsruhe, Germany). The sample was mixed with potassium bromide in a mass ratio of 1:200 and pressed into a pellet. FT-IR spectra were collected in the 4000–400 cm^{-1} range for each sample. The thermal stability of the sphagnum moss was characterized using thermogravimetric analysis (TGA) (209 F1, Netzsch, Sable, Germany). The analysis took place in a nitrogen atmosphere with a heating rate of 10 $°C{\cdot}min^{-1}$, covering a temperature range from 25 °C to 800 °C.

2.4. Batch Adsorption Experiments

A stock solution (300 mg/L) was prepared by dissolving the appropriate amount of methylene blue (MB) in deionized water. Batch experiments were conducted to investigate the impact of contact time (0–210 min), pH of the dye solution (4–14), adsorbent dosage (0.3–2 g/L), initial dye concentration (20–200 mg/L), and temperature (298–328 K). In each experiment, except for one that aimed to test the effect of adsorbent dosage, 0.05 g of modified sphagnum moss was added to 30 mL of MB solution after adjusting its pH. The mixture was then agitated at a rate of 150 r/min for 180 min until reaching equilibrium conditions at room temperature. Separation of adsorbate from adsorbent occurred through sand core funnel filtration. The supernatant solution was collected by removing it from the filter bottle, and its residual concentrations of MB were measured using a UV–visible spectrophotometer to determine absorbance levels (UV-1800, Shimadzu, Kyoto, Japan; Specord210 plus, Analytik Jena AG, Jena, Germany). The residual concentration could be determined from the calibration curve using Beer–Lambert law principles. Each sample underwent three tests, and an average value was taken to ensure accuracy and repeatability. Removal efficiency R and adsorption capacity q_e (mg/g) were determined using Equations (1) and (2), respectively,

$$R = \frac{C_0 - C_e}{C_0} \times 100\% \tag{1}$$

$$q_e = \frac{V}{m}(c_0 - c_e) \tag{2}$$

Initial MB concentration (C_0) in mg/L, equilibrium concentrations (C_e) in mg/L, batch solution volume (V) in L, and the adsorbent mass (m) in g were utilized.

2.5. Adsorption Isotherms

An amount of 0.05 g of the modified sphagnum moss was added to centrifuge tubes containing 30 mL of MB solutions at the optimal pH. The initial concentrations of the MB solutions were adjusted to 100, 120, 140, 160, 180, 200, and 220 mg/L. The centrifuge tubes were placed in a temperature-controlled water bath shaking box and agitated for a duration of 12 h at a constant temperature of 298 K and a speed of rotation at 150 r/min. The adsorption processes followed the method described in Section 2.4 above. Subsequently, the temperature was adjusted to both 308 K and 318 K for further repetitions of the aforementioned procedure. The experimental data were fitted using both Langmuir and Freundlich models to determine the relevant parameters associated with these adsorption isotherm models [7,12]. The formulas were presented by Equations (3) and (4):

Langmuir:

$$\frac{c_e}{q_e} = \frac{c_e}{q_m} + \frac{1}{K_L q_m} \tag{3}$$

Freundlich:

$$\ln q_e = \ln K_F + \frac{1}{n}\ln c_e \tag{4}$$

The equilibrium adsorption capacity (q_e) is represented in mg/g, while q_m represents the saturation adsorption capacity in mg/g, C_e represents the concentration of MB in the solution at adsorption equilibrium (mg/L). K_L denotes the Langmuir constant, which relates to the adsorption-free energy and affinity of adsorption sites. The Freundlich constants, K_F and n, represent the system's adsorption capacity and strength, respectively. By linearly fitting ($\frac{c_e}{q_e}$) against (c_e), K_L and q_m can be determined from the slope and intercept of the fitted line, respectively. For favorable adsorption, the Freundlich constant should fall within a range of 1–10. These parameters can be obtained from plotting $\ln q_e$ vs. $\ln C_e$.

2.6. Adsorption Kinetics

In this experiment, the modified sphagnum moss, weighing 0.05 g, was individually placed into 100 mL conical flasks containing 30 mL of methylene blue (MB) solution with varying initial concentrations of 150 mg/L and 200 mg/L at the optimal pH and temperature of 298 K. Subsequently, the flasks underwent adsorption experiments at room temperature in a shaking water bath. The time-dependent adsorption capacity of the modified sphagnum moss was monitored using a UV–visible spectrophotometer, and kinetic mechanisms were analyzed employing various models (Equations (5)–(7)) [13,30].

Pseudo-first-order kinetic model:

$$\ln(q_e - q_t) = \ln q_e - k_1 t \tag{5}$$

Pseudo-second-order kinetic model:

$$\frac{t}{q_t} = \frac{1}{k_2 q_e^2} + \frac{t}{q_e} \tag{6}$$

Intra-particle diffusion model:

$$q_t = k_{id} t^{\frac{1}{2}} + C \tag{7}$$

The adsorption time (t) is expressed in minutes, while the equilibrium adsorption amount (q_e) and the amount adsorbed at time t (q_t) are measured in mg/g. The rate constants for the pseudo-first-order model (k_1) and pseudo-second-order model (k_2) are also denoted. Additionally, C represents a constant parameter associated with the thickness of the boundary layer, while k_{id} corresponds to a constant parameter related to the intra-particle diffusion rate constant.

2.7. Thermodynamics

The thermodynamic adsorption isotherm experiments elucidate the variation in Gibbs free energy of the adsorbent throughout the wastewater adsorption process. The alteration in free energy can be determined utilizing the subsequent two equations: [6,31].

$$K_D = \frac{q_e}{c_e} \tag{8}$$

$$\Delta G^0 = -RT\ln K_D \tag{9}$$

The change in enthalpy ΔH^0 and the change in entropy ΔS^0 can be calculated based on the aforementioned equations. The values of ΔH^0 and ΔS^0 can be determined by analyzing the intercept and slope of the $\ln K_D$ versus $\frac{1}{T}$ plot.

$$\ln K_D = \frac{1}{R}\left(\Delta S^0 - \frac{\Delta H^0}{T}\right) \tag{10}$$

2.8. Adsorption Regeneration Experiment

The regenerative performance was demonstrated by conducting adsorption–desorption cycles using 0.05 g of modified sphagnum moss in a 30 mL solution containing 200 mg/L of MB at the optimal pH and temperature of 298 K for the adsorption experiments. Stirring occurred at 150 r/min for 3 h at room temperature. Subsequently, the modified sphagnum moss was filtered, dried, and subjected to desorption with absolute ethanol for 2 h to extract MB from the sample. Following desorption, the obtained modified sphagnum moss was washed and dried for reuse in subsequent rounds of adsorption experiments. This cycle was repeated five times, with the removal rate of MB by the sphagnum moss adsorbent in each experiment serving as an indicator to evaluate its regeneration performance.

3. Result and Discussion

3.1. Analyses of Morphologies

The surface morphologies of both raw and modified sphagnum moss are depicted in Figure 1. As observed from the original morphology of sphagnum moss (Figure 1a,b), the material's surface is abundantly porous, displaying a smooth and pristine appearance that is highly suitable for adsorption material preparation [29]. Upon comparing the raw material with its modified counterpart, it becomes evident that the surface of the modified sphagnum moss exhibits increased coarseness and blurred edges, forming a porous network structure that offers numerous active sites for subsequent grafting (Figure 1d,e). Moreover, the analysis utilizing the EDS spectrum has revealed the co-existence of N elements alongside C and O elements on the surface of the modified sphagnum moss, indicating a modification in its surface composition (Figure 1c,f). This modification likely improves the interaction between the modified sphagnum moss and the dye, potentially boosting the adsorption capacity. Analyzing both the SEM image of the modified sphagnum moss and its corresponding elemental mapping images of C, O, and N presented in Figure 2 allows us to conclude that uniform surface modification reactions using concentrated sulfuric acid and concentrated nitric acid have occurred on the surface of sphagnum moss.

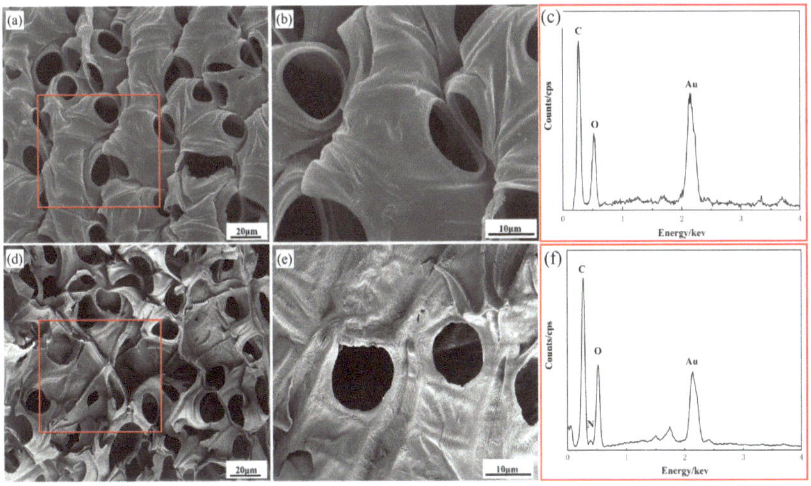

Figure 1. SEM images of sphagnum moss before and after modification: (**a**,**b**) original sphagnum moss; (**c**) EDS analysis of original sphagnum moss corresponding to the red frame labeled in (**a**); (**d**,**e**) sphagnum moss after modification; (**f**) EDS analysis of modified sphagnum moss corresponding to the red frame labeled in (**d**).

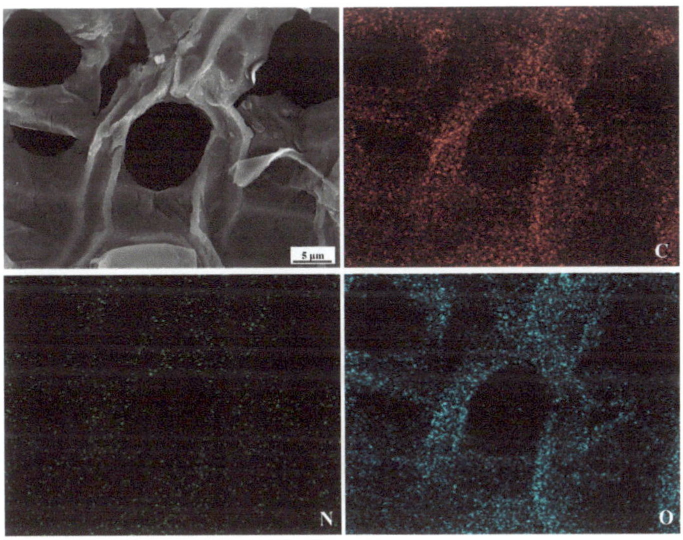

Figure 2. The SEM image of sphagnum moss through modification and its corresponding elemental mapping images of C, O, and N.

3.2. Infrared Spectroscopic Analysis

In order to assess the changes in surface information of samples before and after modification, infrared spectrometry was employed to analyze the functional groups present in the materials. Figure 3 presents the FT-IR spectra of both raw and modified sphagnum moss within the 400–4000 cm^{-1} range.

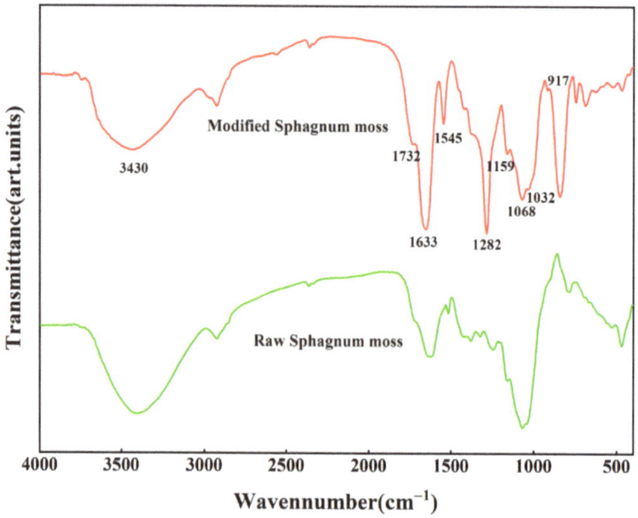

Figure 3. Infrared spectra of sphagnum moss before and after modification.

Within the high-frequency region, a prominent peak at 3430 cm^{-1} was observed in raw and modified sphagnum moss samples, indicating the presence of hydroxyl (-OH) groups. However, the adsorption peak corresponding to -OH on modified sphagnum moss exhibited significant attenuation, suggesting a reaction with HNO_3 and substitution by nitro groups (-NO_2). The introduction of nitrogen-containing functional groups onto the surface of sphagnum moss through nitric acid functionalization is confirmed by observing a characteristic nitro (-NO_2) peak at 1545 cm^{-1} [32]. The characteristic peak for aromatic -C=C- bonds is observed at 1032 cm^{-1}, while bands within the range of 1650–1600 cm^{-1} correspond to carbon-based carboxyl bonds [24]; additionally, a distinctive peak formed by polysaccharide carbonyl can be seen at 1633 cm^{-1}. Due to the disruption of the internal structure during the modification process, chemical bonding is altered, leading to changes in the types of surface functional groups present. At 1732 cm^{-1}, there exists a carbon-based bond associated with carboxylic acid compound (-COOH) and ester functionality, an absorption peak related to stretching vibration for -C=C- occurs at 1418 cm^{-1} [12]; furthermore, a strong absorption peak attributed to contraction vibration for -C-O- appears at 1282 cm^{-1}. Simultaneously, a new characteristic peak emerges between 1300 and 1050 cm^{-1}, which corresponds to stretching vibrations for ester's -C-O- group, and ether's C-O-C group stretching vibrations are also evident within this range [8,33]. Lastly, the adsorption peak detected at a wavenumber value of 917 cm^{-1} arises from C-H bending vibration.

3.3. Thermogravimetric Analysis

The thermal stability of the sphagnum moss before and after modification was confirmed through thermal weight analyses, as depicted in Figure 4. TGA curves exhibited slight mass loss below 200 °C, which was attributed to continuous volatilization of water, hydrocarbons, and other volatile organic compounds [19,34]. The sphagnum moss material primarily comprises plant-based cellulose and lignin tissue, with a pyrolysis temperature range of 300–373 °C [35].

Evaluation of thermal stability involved comparing temperatures T_{10} (10% mass loss) and Tdm (maximum decomposition rate) before and after modification. The surface modification increased T_{10} from 212.2 °C to 272.0 °C and T_{dm} from 331.1 °C to 342.2 °C, indicating enhanced thermal stability. At a temperature of 600 °C, there was minimal change in mass with some residual sample remaining. The modified adsorbents exhibited a maximum weight loss temperature range between 270 and 340 °C, demonstrating their stability within operational temperatures.

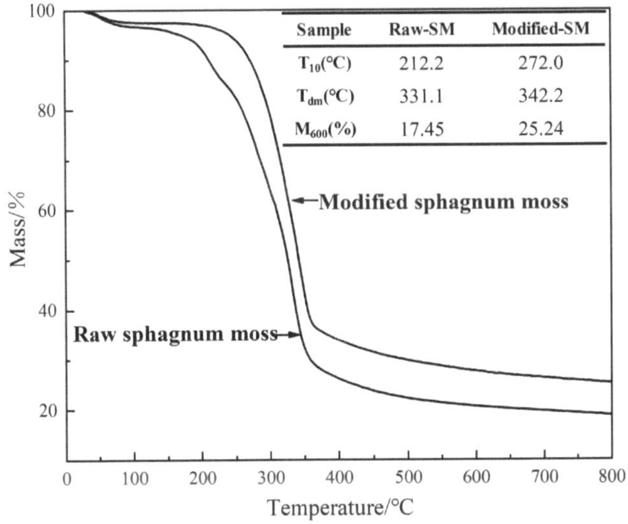

Figure 4. Thermogravimetric analysis of raw sphagnum moss and modified sphagnum moss.

3.4. Adsorption Experiments

3.4.1. The Influence of Contact Time and Initial MB Concentration on Adsorption Effect

To examine the effects of contact duration and starting concentration on the adsorption capabilities of sphagnum moss, experiments were conducted using methylene blue (MB) solutions, with results displayed in Figure 5. Figure 5a shows a comparison of adsorption capacities between untreated and chemically modified sphagnum moss at initial MB concentrations of 150 mg/L and 200 mg/L. The modified sphagnum moss exhibited significantly enhanced adsorption efficiency compared to the original sphagnum moss, with an increase in adsorption capacity from 68.528 mg/g to 88.767 mg/g and from 98.775 mg/g to 116.898 mg/g for the respective concentrations. Moreover, the adsorption efficiency improved from 76.1% and 82.31% to 98.63% and 97.41%, respectively. Furthermore, it can be observed that there is a rapid increase in MB uptake by the modified sphagnum moss material within the first 35 min of contact time, followed by a gradual approach towards equilibrium without any substantial rise thereafter. This swift initial adsorption is attributed to the large surface area of the moss, the strong interaction between MB molecules and sphagnum moss surface, and an abundance of accessible active sites on the modified moss [36]. The overall process of MB adsorption onto modified sphagnum moss can be divided into two distinct stages. In the first stage, there is evident electrostatic adsorption between MB solution and active sites on the adsorbent's surface, with the reaction primarily occurring at the surface. The second stage involves diffusion of the MB solution absorbed on the surface into the pores of the adsorbent. As a result of significant internal diffusion resistance, as adsorption progresses, saturation gradually occurs at active sites while reducing concentration differences between MB inside and outside the adsorbent until reaching equilibrium.

Figure 5b illustrates a proportional rise in the adsorption capacity of modified sphagnum moss across a range of MB concentrations from 20 mg/L to 200 mg/L, suggesting a strong linear correlation. The observed findings can be attributed to the interaction between the MB solution and the active sites of modified sphagnum moss. A higher initial concentration of MB enhances the driving force for migration towards these active adsorption sites [37], thereby increasing collision probability between MB molecules and modified sphagnum moss adsorption sites. Consequently, this promotes greater entry of methylene blue molecules into unoccupied adsorbent pores. Therefore, increasing the initial concentration of MB potentially boosts the adsorption efficiency of MB wastewater treatment.

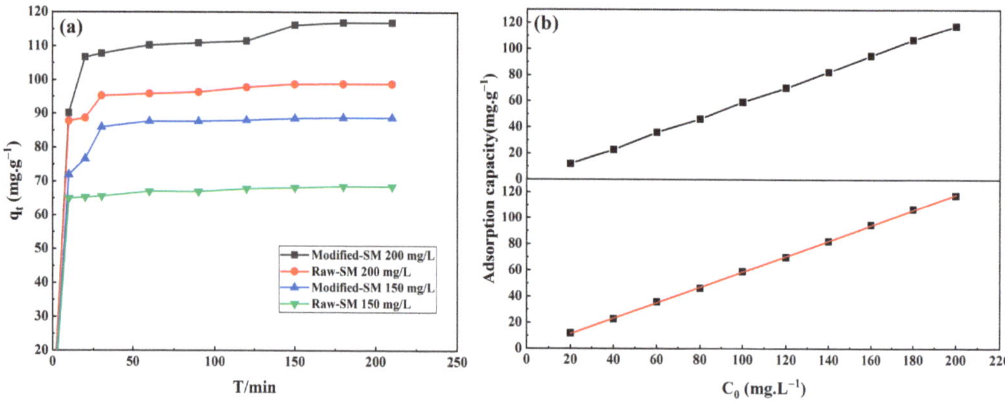

Figure 5. The effect of contact time and initial MB concentration on the adsorption capacity of sphagnum moss before and after modification: (**a**) the curve of adsorption capacity over time; (**b**) the effect of MB concentration on adsorption capacity corresponding to the grey line and the influence of MB concentration on adsorption capacity were linearly fitted corresponding to the red line.

3.4.2. The Influence of the MB Solution pH on Adsorption Effect

The impact of pH on the adsorption performance of modified sphagnum moss at varying concentrations of MB is depicted in Figure 6a,b. The adsorbent's efficacy in removing MB is contingent upon the pH values. The removal efficiency of MB exhibits an ascending trend up to pH 8, followed by a decline beyond this point.

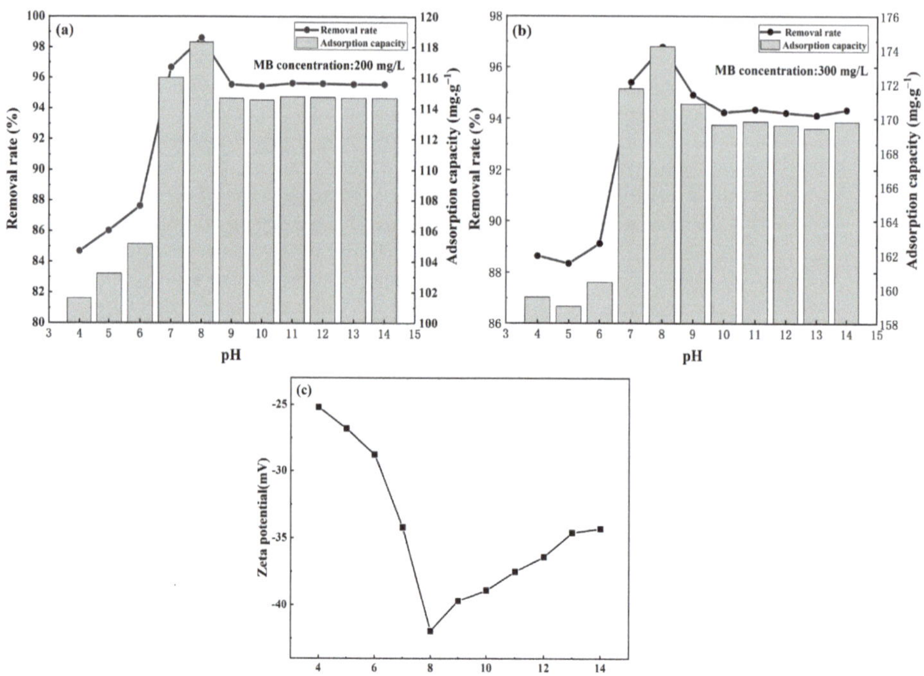

Figure 6. Effect of sphagnum moss after modification on MB adsorption under different pH: (**a**) initial MB concentration of 200 mg/L; (**b**) initial MB concentration of 300 mg/L; (**c**) zeta potential change in modified sphagnum moss with pH.

The reasons for this phenomenon are as follows: As can be seen from Figure 6c, different acid–base environments will have a different impact on the surface charge of the adsorbent. Zeta potential measurements clearly indicate that the surface-modified sphagnum moss adsorbent maintains a negative zeta potential across a broad pH range of 4–14, facilitating strong electrostatic attraction with methylene blue molecules and resulting in relatively superior adsorption performance under various pH conditions [30]. However, at low pH, the negative surface charge of the adsorbent decreases, leading to reduced adsorption efficiency. This is due to competition between H^+ ions and methylene blue molecules in acidic solutions, causing partial protonation of the adsorbent and a reduction in its surface negative charge [6,8]. Conversely, as the pH increases, the adsorbent undergoes deprotonation, which increases the negative surface charge. This enhancement in negative charge increases the number of adsorption sites on the adsorbent, thereby improving adsorption efficiency. However, when pH exceeds 8, deprotonation occurs, causing methylene blue to gradually transition from a positive to a negative charge state, which leads to a decline in the electrostatic adsorption mechanism and a subsequent decrease in the overall performance of dye wastewater removal by this system [38,39]. Therefore, optimizing the pH of the solution is crucial for maximizing MB adsorption onto modified sphagnum moss, as it directly influences the availability and accessibility of surface binding sites on the adsorbent.

3.4.3. The Influence of Adsorbent Dosage on Adsorption Effect

The results depicted in Figure 7 demonstrate the MB removal efficiency and adsorption capacity of modified sphagnum moss at different dosages. In this investigation, the removal efficiency exhibited an increase from 84.05% to 98.63% as the dosage of adsorbent was elevated from 0.3 to 2 g/L. This can be attributed to the augmentation of surface binding sites per unit volume with higher doses of adsorption. Nevertheless, there was a notable decline in adsorption capacity, which decreased from 378.227 to 73.53 mg/g.

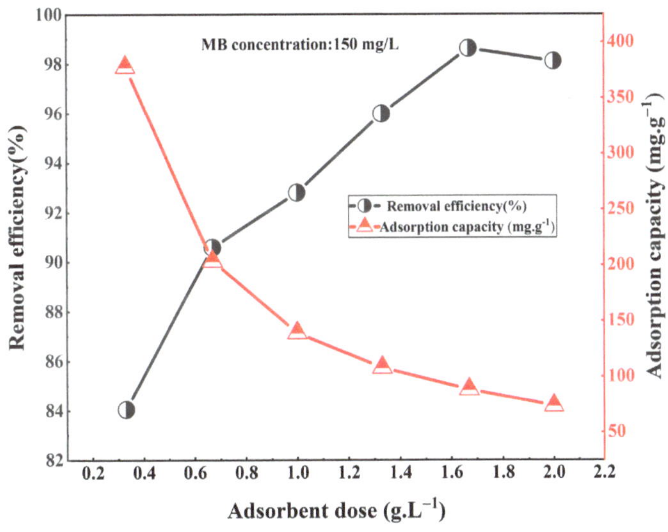

Figure 7. Effect of adsorbent dose on MB adsorption.

The trend observed suggests that at lower doses, the surface adsorption groups on modified sphagnum moss effectively interacted with MB in solution, efficiently occupying active sites and enhancing the adsorbent's capacity per unit mass. However, with an increase in the adsorbent mass, there was a decrease in contact between each unit mass of modified sphagnum moss and MB dye in solution, and the adsorption active point of the modified sphagnum moss did not reach saturation, which brought down the utilization of adsorption sites, lowering the adsorbed amount [12]. Additionally, excessive suspended adsorbents may have formed aggregates [31], resulting in a decreased specific surface area that hindered complete contact with MB dye and consequently impacted overall performance. In line with our objective to enhance both adsorption efficiency and cost-effectiveness while reducing expenses, an optimal dosage value for modified sphagnum moss was determined as 1.67 g/L through experimental analysis.

3.4.4. Adsorption Isotherms

Adsorption isotherms are vital for the analysis and understanding of the thermodynamic parameters associated with adsorption, offering insights into the chemical interactions and mechanisms between pollutant molecules and the adsorbent surfaces [6,40]. A series of experiments utilizing both the Langmuir and Freundlich isotherm models was conducted to explore the adsorption behavior of modified sphagnum moss on MB (refer to Figure 8 and Table 1). The Langmuir model is applied to explain monolayer adsorption on homogeneous surfaces, whereas the Freundlich model addresses multilayer adsorption on heterogeneous surfaces. These experiments facilitate a deeper comprehension of the adsorption phenomena.

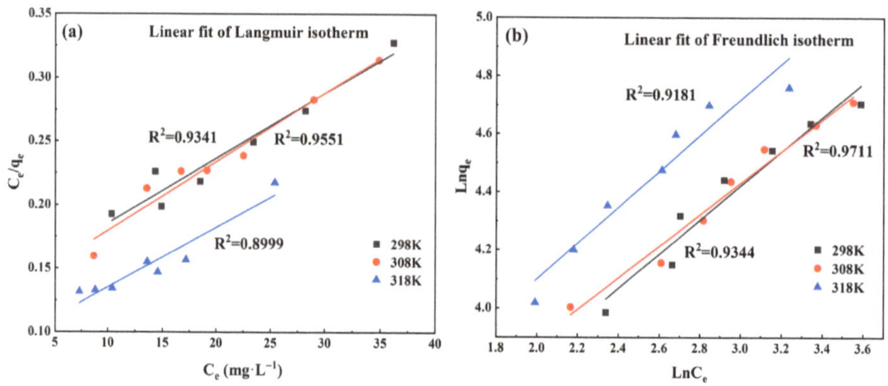

Figure 8. Langmuir model and Freundlich model of isothermal adsorption under different concentrations: (**a**) Linear fit of Langmuir isotherm; (**b**) Linear fit of Freundlich isotherm.

Table 1. Langmuir and Freundlich isothermal adsorption model parameters under different solution concentrations.

Adsorption Isotherm	Parameter	298 K	308 K	318 K
Langmuir	$q_m/(\text{mg}\cdot\text{g}^{-1})$	195.3125	184.5018	221.3142
	$K_L/(\text{L}\cdot\text{mg}^{-1})$	0.0381	0.0432	0.0535
	R^2	0.9341	0.9551	0.8999
Freundlich	$K_F(\text{mg}\cdot\text{g}^{-1})(\frac{\text{L}}{\text{mg}})^{\frac{1}{n}}$	14.2116	16.3634	17.5152
	$\frac{1}{n}$	0.5891	0.5453	0.6182
	R^2	0.9344	0.9711	0.9181

From the data presented in Table 1, it is evident that the R^2 value for the Freundlich model surpasses that of the Langmuir model, indicating a superior ability of the Freundlich model to describe the adsorption process of modified sphagnum moss adsorbent on MB solution. Moreover, this suggests that multilayer adsorption predominantly occurs [11]. By plotting $\ln C_e$ against $\ln Q_e$, a linear Freundlich adsorption isotherm (Figure 8b) was observed. The slope and intercept of this equation provide insights into the parameters

$\frac{1}{n}$ and K_F, respectively. A $\frac{1}{n}$ value less than 1 indicates surface non-uniformity of the adsorbent, which enhances adsorption efficacy and promotes the adsorption process.

3.4.5. Adsorption Kinetics

The investigation of adsorption kinetics can offer a more comprehensive depiction of the rate at which adsorption occurs and establish a reasonable understanding of the adsorption mechanism. To investigate the kinetic mechanism of MB adsorption by surface-modified sphagnum moss adsorbent, various kinetic models, including pseudo-first-order, pseudo-second-order, and intra-particle diffusion models, were employed. Figure 9a–c illustrates the fitting results from these models, while Table 2 lists the corresponding parameters.

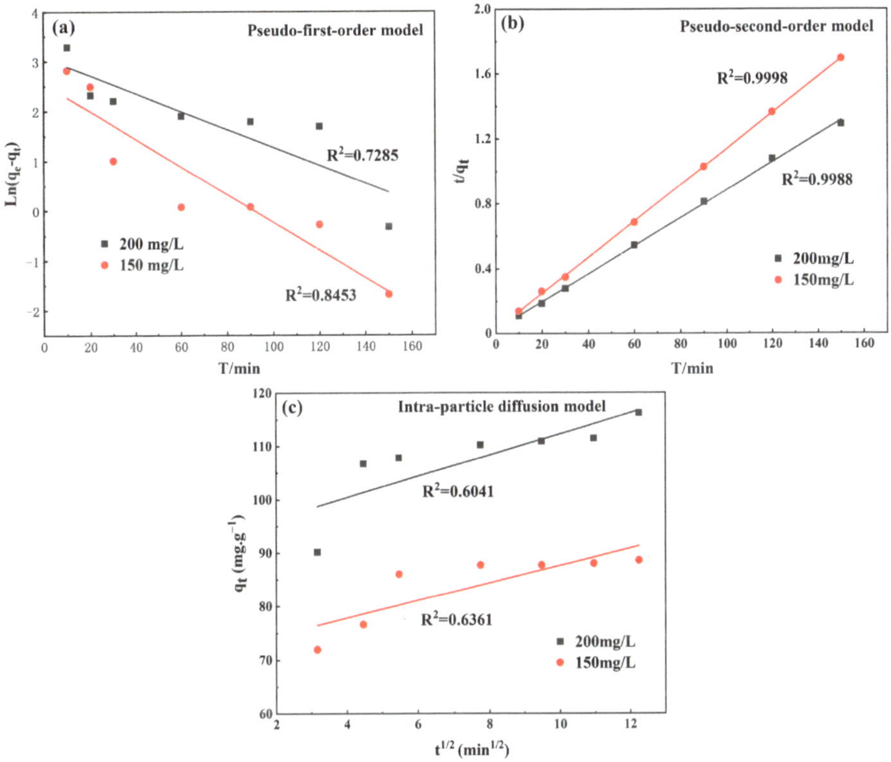

Figure 9. Adsorption kinetics of MB adsorbed by modified sphagnum moss: (**a**) pseudo-first-order kinetics model; (**b**) pseudo-second-order kinetics model; (**c**) intra-particle diffusion model.

The superior performance of the pseudo-second-order kinetics, as evidenced by higher correlation coefficient (R^2) values and a close match between calculated and experimentally observed equilibrium adsorption capacities (Figure 9a,b; Table 2), indicates that the MB adsorption process conforms to this model. This suggests a chemisorption process involving electron transfer between adsorbents and adsorbates [41]. Notably, at low concentrations, k_2 surpasses its counterpart at high concentrations, which explains the rapid uptake observed at lower concentrations. The parameter k_2: $0 < k_2 < 1$ signifies favorable absorption. Furthermore, according to the intra-particle diffusion model, intraparticle diffusion significantly influences the reaction rate. Based on the fitting results shown in Figure 9c, it is evident that there is a relatively low R^2 value and poor linearity in the fitting for internal diffusion model analysis, indicating the minimal impact of intraparticle diffusion factors on MB adsorption by the adsorbent. Thus, intra-particle diffusion does not significantly impact the overall adsorption process.

Table 2. Adsorption kinetic parameters for adsorption of MB onto the modified sphagnum moss.

q_e (exp.) (mg·g^{-1})		Concentration (mg·L^{-1})	q_e (cal.) (mg·g^{-1})	k_1 (min^{-1})	R^2
88.76748	Pseudo-first-order	150	12.8689	0.0278	0.8453
116.89756		200	21.6427	0.0180	0.7285
		Concentration (mg·L^{-1})	**q_e(cal.) (mg·g^{-1})**	**k_2 (g·mg^{-1}·min^{-1})**	**R^2**
	Pseudo-second-order	150	90.9090	0.0047	0.9988
		200	116.2791	0.0029	0.9998
		Concentration (mg·L^{-1})	**k_{id} (mg·g^{-1}·min$^{-\frac{1}{2}}$)**	**C (mg·g^{-1})**	**R^2**
	Intra-particle diffusion model	150	1.6260	71.3677	0.6361
		200	1.9789	92.4940	0.6041

3.4.6. Adsorption Thermodynamics

The thermodynamic parameters of adsorption—change in enthalpy (ΔH^0), Gibbs free energy (ΔG^0), and entropy (ΔS^0)—are crucial in analyzing the adsorption process. These parameters determine the driving force and direction of the adsorption process, as well as elucidate its microscopic mechanism. By plotting the relationship between $\ln K_D$ and $\frac{1}{T}$, ΔH^0 and ΔS^0 were derived from the line's slope and intercept, respectively (refer to Figure 10). Additionally, ΔG^0 was computed using Equation (9), and the resulting values are detailed in Table 3.

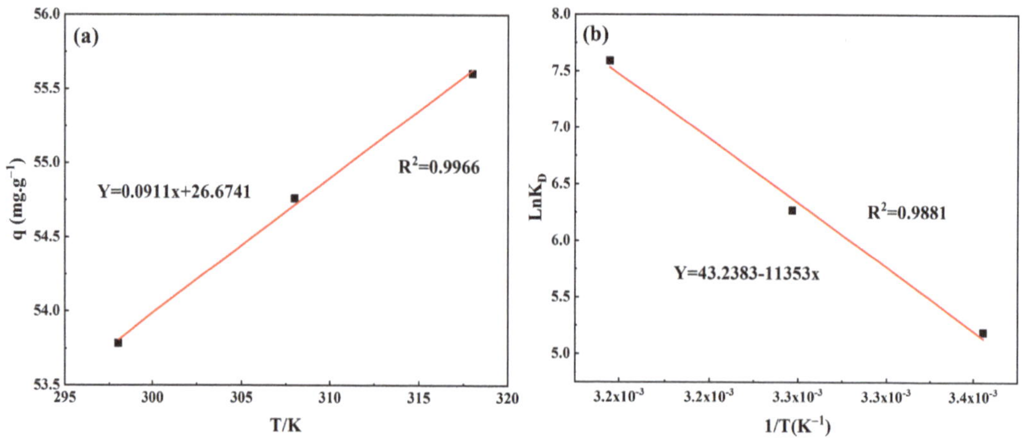

Figure 10. Thermodynamics fitting of modified sphagnum moss: (**a**) adsorption capacity temperture diagram; (**b**) $\ln K_D$ and temperature reciprocal linear fitting curve.

Table 3. Thermodynamic parameters of modified sphagnum moss.

ΔH^0 (kJ·mol^{-1})	ΔS^0 (J·mol^{-1}K^{-1})	ΔG^0 (J·mol^{-1})		
		298 K	308 K	318 K
94.3884	359.4830	−4080.5043	−4700.6398	−5358.9559

According to Figure 10 and Table 3, an increase in temperature correlates with an enhanced adsorption capacity of the modified sphagnum moss adsorbent for MB. The positive ΔH^0 suggests that the adsorption is endothermic, while a positive ΔS^0 indicates irreversible adsorption and increased disorder at the interface between the adsorbent and MB. These findings imply that the adsorption is chemically driven, necessitating heat for effective dye uptake and promoting rapid diffusion to the adsorbent's surface [8,42]. A negative value for ΔG^0 indicates that external conditions are not necessary to promote the spontaneous feasibility of this absorption reaction. Additionally, it is observed that higher temperatures lead to decreased values of ΔG^0, implying greater spontaneity within this absorption process. Furthermore, an increase in temperature can improve the adsorption performance of dyes.

3.4.7. Adsorption–Desorption Cycle Experiments

Reusability is a critical parameter for assessing the feasibility and stability of adsorbent materials. The results presented in Figure 11 demonstrate the performance of modified sphagnum moss adsorbents during five cycles of adsorption–desorption experiments. Following five cycles, there was only a marginal decrease in the adsorption efficiency, with the removal rate decreasing from 98.1% to 97%. These results indicate that the efficiency of adsorption remained within a reasonable range. After undergoing surface modification, the hydrophobicity of sphagnum moss was enhanced. During the removal of MB, the cationic organic molecule selectively enters the hydrophobic cavity of sphagnum moss to form the inclusion complex and facilitate rapid adsorption of MB molecules [43]. However, a slight reduction in effectiveness may be ascribed to hindrance from intermediate adsorbed molecules obstructing active sites and impeding the adsorption process of methylene blue dye. Additionally, the addition of ethanol as a desorption agent enhanced the hydrophilicity of the material and disrupted its hydrophobic cavities, consequently diminishing the adsorbent's affinity for methylene blue wastewater and thereby influencing its adsorption performance [6,12,44]. Consequently, this leads to decreased affinity between the adsorbent material and MB wastewater, thereby reducing overall performance. Nevertheless, based on these cyclic experiments, it can be concluded that the modified sphagnum moss adsorbent demonstrates robust stability and noteworthy regenerative capability in MB adsorption.

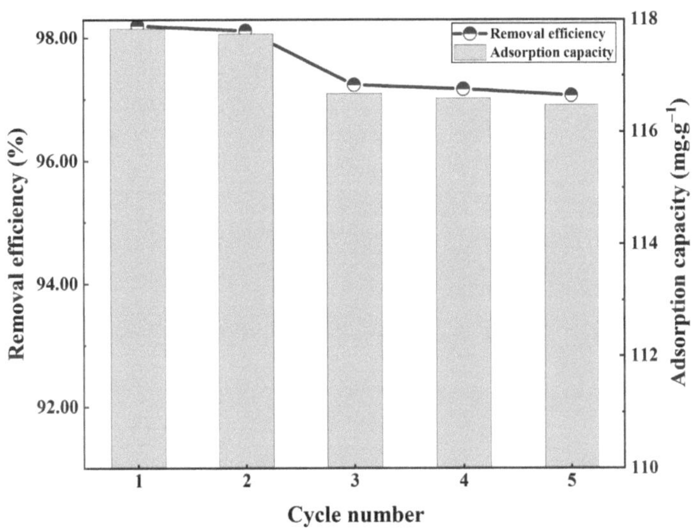

Figure 11. Adsorption–desorption regeneration experiment results of modified sphagnum moss for MB solution.

3.5. *Adsorption Mechanism*

The adsorption mechanism involves interactions between the adsorbate molecules and the active sites on the adsorbent's surface. Sphagnum peat moss, with its abundant cellulose, hemicellulose, lignin, and other components, holds great promise as a bio-based material. Biomass materials generally exhibit effective adsorption onto adsorbents through electrostatic interactions, hydrogen bonding, n-π and π-π interactions, as well as van der

Waals forces [45]. The exceptional adsorption performance of modified sphagnum moss adsorbents can be attributed to several factors: Firstly, the surfaces of sphagnum moss possess numerous pores that facilitate MB wastewater absorption by filling these pores. However, due to limited surface adsorption sites, there are restrictions on its capacity for MB wastewater. By roughening the surface of modified sphagnum moss and increasing micropore numbers simultaneously, the penetration of methylene blue molecules into the absorbent material is enhanced via pore-filling effects, thereby promoting contact with modified surface adsorption sites and improving overall adsorption efficiency [31,46]. Secondly, hydrogen bonding plays a significant role in influencing the adsorption performance. Following surface modification, the surface structure of the modified sphagnum moss undergoes alterations. Apart from the successfully attached nitro group, there are also functional groups containing oxygen, including carboxyl and hydroxyl groups, present on the surface. During the process of MB adsorption, facile formation of hydrogen bonds occurs with nitrogen-containing functional groups present in methylene blue molecules, thereby enhancing the adsorption performance of the modified sphagnum moss [19,30,47]. Simultaneously, the introduction of nitro groups (-NO_2) enhances the hydrophobicity of the material while effectively blocking water molecule adsorption, thereby improving the adsorption efficiency of methylene blue dye. On the other hand, the hydrophilic component in the material, like unreacted -OH on the surface of modified sphagnum moss, provided a possibility for the formation of hydrogen bonds among adsorbent particles as well as with water molecules (Figure 12b). Thus, agglomeration between adsorbent particles can be achieved after adsorption, which enables material recycling. Thirdly, robust electrostatic interactions occur between the modified sphagnum moss adsorbent and MB molecules. The modified adsorbent surface carries a substantial negative charge, facilitating robust electrostatic interactions with methyl blue (MB) molecules that are positively charged. This enhances their affinity for binding to the modified sphagnum moss. Furthermore, the surface of the modified sphagnum moss adsorbents possesses various functional groups, such as -NO_2 and -COOH, which act as active sites for binding functional groups like S^+ or N^+ in the MB dye [6,7,12]. Fourthly, the presence of strong π-π conjugation due to the characteristic aromatic ring structure of the modified sphagnum moss contributes significantly to improved adsorption performance [48,49]. The carbon-carbon double bond on the modified sphagnum moss adsorbent interacts with the benzene ring of MB through π-π interactions, thereby promoting efficient adsorption. Lastly, it is worth noting the potential role of n-π interactions, where carbon-based oxygen in ester derivatives acts as an electron donor and the dye's aromatic ring acts as an electron acceptor, may also play a significant role in enhancing MB adsorption.

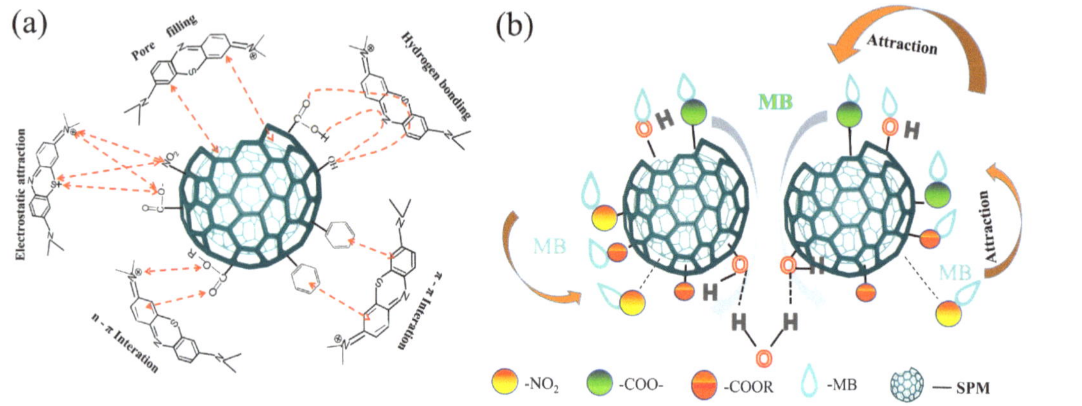

Figure 12. Mechanism of chemically modified sphagnum moss. (**a**) Mechanism of modified sphagnum moss for MB; (**b**) Diagram of the rapid adsorption of MB by modified sphagnum moss.

4. Conclusions

The bio-adsorbent was prepared by surface modification of sphagnum moss sourced from Guizhou Province, China, using concentrated sulfuric acid and concentrated nitric acid. Its adsorption capacity for methylene blue was investigated. The modified sphagnum moss exhibited a rough surface structure with porous and irregular textures due to the replacement of hydroxyl groups by nitro groups, resulting in an increased number of adsorption sites and significantly improved adsorption performance and recyclability. Optimal adsorption conditions were established by varying factors such as adsorbent dosage, pH, contact time, and initial dye concentration. Adsorption isotherms and kinetics conformed well to the Freundlich model and pseudo-second-order kinetic model, respectively. Thermodynamic analysis indicated spontaneous and endothermic adsorption. Mechanistically, the interaction between the modified sphagnum moss adsorbent and methylene blue involved electrostatic attraction, hydrogen bonding, and π-π interactions. In summary, modified sphagnum moss offers an efficient, cost-effective, environmentally friendly, and reusable solution for methylene blue removal from wastewater.

Author Contributions: Conceptualization, Y.W. and J.R.; Methodology, J.R., Y.W. and S.Z.; Validation, H.Y.; Data curation, S.Z.; Writing—original draft, J.R.; Writing—review & editing, Y.W. and J.R.; Supervision, Y.W. and J.R. All authors have read and agreed to the published version of the manuscript.

Funding: This research was funded by the Science and Technology Foundation of Guizhou Province, grant numbers: QKHJC-ZK [2021]254, and the Youth Science and Technology Talent Development Program of Guizhou Provincial Department of Education, grant number: QJH-KY [2022]299.

Institutional Review Board Statement: Not applicable.

Informed Consent Statement: Not applicable.

Data Availability Statement: The original contributions presented in the study are included in the article, further inquiries can be directed to the corresponding author.

Conflicts of Interest: The authors declare no conflict of interest.

References

1. El-Kammah, M.; Elkhatib, E.; Gouveia, S.; Cameselle, C.; Aboukila, E. Enhanced removal of indigo carmine dye from textile effluent using green cost-efficient nanomaterial: Adsorption, kinetics, thermodynamics and mechanisms. *Sustain. Chem. Pharm.* **2022**, *29*, 100753. [CrossRef]
2. Zhang, L.; Zaoui, A.; Sekkal, W. Adsorption efficiency of highly methylene blue dye concentrations with multilayer chitosan-modified clays for a precise nanofiltration performance of polluted water. *J. Water Process. Eng.* **2024**, *57*, 104651. [CrossRef]
3. Dilshad, S.; Muslim, M.; Ahmed, A.; Ali, A.; Firdaus, S.; Alam, M.J.; Ahmad, S.; Ahmad, M.; Ahmad, A. Design and synthesis of a new coordination polymer of zinc (ii): Crystal structure and evaluation for its adsorption studies in removing perilous organic dye pollutants in aqueous medium. *J. Mol. Struct.* **2024**, *1301*, 137350. [CrossRef]
4. Guezzen, B.; Medjahed, B.; Benhelima, A.; Guendouzi, A.; Didi, M.A.; Zidelmal, S.; Abdelkrim Boudia, R.; Adjdir, M. Improved pollutant management by kinetic and box-behnken design analysis of HDTMA-modified bentonite's adsorption of indigo carmine dye. *J. Ind. Eng. Chem.* **2023**, *125*, 242–258. [CrossRef]
5. Khan, I.; Saeed, K.; Zekker, I.; Zhang, B.; Hendi, A.H.; Ahmad, A.; Ahmad, S.; Zada, N.; Ahmad, H.; Shah, L.A.; et al. Review on methylene blue: Its properties, uses, toxicity and photodegradation. *Water* **2022**, *14*, 242. [CrossRef]
6. Gupta, S.V.; Kulkarni, V.V.; Ahmaruzzaman, M. Fabrication of a bio-adsorbent material by grafting CeO_2 quantum dots (QDTs) over areca nut shell biochar using saccharum officinarum extract as a solvent/capping agent for adsorption of methylene blue dye: Synthesis, material analyses, adsorption kinetics and isotherms studies. *Colloids Surf. A Physicochem. Eng. Asp.* **2024**, *680*, 132611.
7. Zavala-Flores, E.; Flores-López, L.Z.; Alonso-Nuñez, G.; Espinoza-Gómez, H. Removal and adsorption kinetics of methylene blue dye by pristine cotton husk bracts (*Gossypium hirsutum* L.) From agroindustrial waste. *Ind. Crops Prod.* **2024**, *209*, 117947. [CrossRef]
8. Boughrara, L.; Zaoui, F.; Guezzoul, M.; Sebba, F.Z.; Bounaceur, B.; Kada, S.O. New alginic acid derivatives ester for methylene blue dye adsorption: Kinetic, isotherm, thermodynamic, and mechanism study. *Int. J. Biol. Macromol.* **2022**, *205*, 651–663. [CrossRef]
9. Oladoye, P.O.; Ajiboye, T.O.; Omotola, E.O.; Oyewola, O.J. Methylene blue dye: Toxicity and potential elimination technology from wastewater. *Results Eng.* **2022**, *16*, 100678. [CrossRef]

10. Li, H.; Kong, J.; Zhang, H.; Gao, J.; Fang, Y.; Shi, J.; Ge, T.; Fang, T.; Shi, Y.; Zhang, R.; et al. Mechanisms and adsorption capacities of ball milled biomass fly ash/biochar composites for the adsorption of methylene blue dye from aqueous solution. *J. Water Process. Eng.* **2023**, *53*, 103713. [CrossRef]
11. Liu, C.; Wang, W.; Wu, R.; Liu, Y.; Lin, X.; Kan, H.; Zheng, Y. Preparation of acid- and alkali-modified biochar for removal of methylene blue pigment. *ACS Omega* **2020**, *5*, 30906–30922. [CrossRef] [PubMed]
12. Liu, J.; Lin, Q.; Gao, J.; Jia, X.; Cai, M.; Liang, Q. Adsorption properties and mechanisms of methylene blue and tetracycline by nano-silica biochar composites activated by KOH. *Chemosphere* **2023**, *337*, 139395. [CrossRef] [PubMed]
13. Kavitha, E.; Poonguzhali, E.; Nanditha, D.; Kapoor, A.; Arthanareeswaran, G.; Prabhakar, S. Current status and future prospects of membrane separation processes for value recovery from wastewater. *Chemosphere* **2022**, *291*, 132690. [CrossRef]
14. Kanwal, F.; Batool, A.; Aziz, F.; Sandali, Y.; Li, C.; Ullah, H.M.N.; Qasim, M.; Irfan, A.; Sulaman, M. Facile synthesis of silver chloride-polyaniline nanocomposites and its photocatalytic activity for the degradation of methylene blue. *Mater. Sci. Eng. B* **2024**, *299*, 117026. [CrossRef]
15. Akhter, P.; Ali, F.; Ali, A.; Hussain, M. TiO_2 decorated cnts nanocomposite for efficient photocatalytic degradation of methylene blue. *Diam. Relat. Mater.* **2024**, *141*, 110702. [CrossRef]
16. Liu, X.; Chen, Z.; Du, W.; Liu, P.; Zhang, L.; Shi, F. Treatment of wastewater containing methyl orange dye by fluidized three dimensional electrochemical oxidation process integrated with chemical oxidation and adsorption. *J. Environ. Manag.* **2022**, *311*, 114775. [CrossRef]
17. Adeyemo, A.A.; Adeoye, I.O.; Bello, O.S. Adsorption of dyes using different types of clay: A review. *Appl. Water Sci.* **2017**, *7*, 543–568. [CrossRef]
18. Araújo, R.F.; Bezerra, L.C.A.; de Novais, L.M.R.; D'Oca, C.D.R.M.; Avelino, F. Unveiling the mechanistic aspects of methylene blue adsorption onto a novel phosphate-decorated coconut fiber lignin. *Int. J. Biol. Macromol.* **2023**, *253*, 127011. [CrossRef]
19. Ifguis, O.; Ziat, Y.; Belkhanchi, H.; Ammou, F.; Moutcine, A.; Laghlimi, C. Adsorption mechanism of methylene blue from polluted water by *Opuntia ficus indica* of *Beni Mellal* and *Sidi Bou Othmane* areas: A comparative study. *Chem. Phys. Impact* **2023**, *6*, 100235. [CrossRef]
20. Da Rosa, M.P.; Igansi, A.V.; Lütke, S.F.; Cadaval, T.R.S.A.; Santos, A.C.R.D.; de Oliveira Lopes Inacio, A.P.; de Almeida Pinto, L.A.; Beck, P.H. A new approach to convert rice husk waste in a quick and efficient adsorbent to remove cationic dye from water. *J. Environ. Chem. Eng.* **2019**, *7*, 103504. [CrossRef]
21. Natrayan, L.; Niveditha, V.R.; Nadh, V.S.; Srinivas, C.; Dhanraj, J.A.; Saravanan, A. Application of response surface and artificial neural network optimization approaches for exploring methylene blue adsorption using luffa fiber treated with sodium chlorite. *J. Water Process. Eng.* **2024**, *58*, 104778. [CrossRef]
22. Tahir, H.; Sultan, M.; Akhtar, N.; Hameed, U.; Abid, T. Application of natural and modified sugar cane bagasse for the removal of dye from aqueous solution. *J. Saudi Chem. Soc.* **2016**, *20*, S115–S121. [CrossRef]
23. Coupal, B.; Lalancette, J. The treatment of waters with peat moss. *Water Res.* **1976**, *10*, 1071–1076. [CrossRef]
24. Zhang, R.; Leiviskä, T.; Taskila, S.; Tanskanen, J. Iron-loaded sphagnum moss extract residue for phosphate removal. *J. Environ. Manag.* **2018**, *218*, 271–279. [CrossRef] [PubMed]
25. González, A.G.; Jimenez-Villacorta, F.; Beike, A.K.; Reski, R.; Adamo, P.; Pokrovsky, O.S. Chemical and structural characterization of copper adsorbed on mosses (Bryophyta). *J. Hazard. Mater.* **2016**, *308*, 343–354. [CrossRef] [PubMed]
26. Kalmykova, Y.; Strömvall, A.; Steenari, B. Adsorption of Cd, Pb and Zn on *Sphagnum* peat from solutions with low metal concentrations. *J. Hazard. Mater.* **2008**, *152*, 885–891. [CrossRef]
27. Charazińska, S.; Lochyński, P.; Burszta-Adamiak, E. Removal of heavy metal ions form acidic electrolyte for stainless steel electropolishing via adsorption using polish peats. *J. Water Process. Eng.* **2021**, *42*, 102169. [CrossRef]
28. Babel, S.; Kurniawan, T.A. Low-cost adsorbents for heavy metals uptake from contaminated water: A review. *J. Hazard. Mater.* **2003**, *97*, 219–243. [CrossRef]
29. Keskin, N.O.S.; Uyar, G. Methylene blue dye removal using *Sphagnum palustre* L. Bog-moss as a reusable biosorbent. *Anatol. Bryol.* **2019**, *5*, 1–7. [CrossRef]
30. Zhao, S.; Li, Y.; Wang, M.; Chen, B.; Zhang, Y.; Sun, Y.; Chen, K.; Du, Q.; Pi, X.; Wang, Y.; et al. Efficient adsorption of methylene blue in water by nitro-functionalized metal-organic skeleton-calcium alginate composite aerogel. *Int. J. Biol. Macromol.* **2023**, *253*, 126458. [CrossRef]
31. Zhao, S.; Li, Y.; Wang, M.; Chen, B.; Zhang, Y.; Sun, Y.; Chen, K.; Du, Q.; Jing, Z.; Jin, Y. Preparation of MIL-88A micro/nanocrystals with different morphologies in different solvents for efficient removal of Congo red from water: Synthesis, characterization, and adsorption mechanisms. *Microporous Mesoporous Mater.* **2022**, *345*, 112241. [CrossRef]
32. Igwegbe, C.A.; Onukwuli, O.D.; Ighalo, J.O.; Okoye, P.U. Adsorption of cationic dyes on *Dacryodes edulis* seeds activated carbon modified using phosphoric acid and sodium chloride. *Environ. Process.* **2020**, *7*, 1151–1171. [CrossRef]
33. Hachemaoui, M.; Mokhtar, A.; Ismail, I.; Mohamedi, M.W.; Iqbal, J.; Taha, I.; Bennabi, F.; Zaoui, F.; Bengueddach, A.; Hamacha, R.; et al. M (M: Cu, Cr or Fe) nanoparticles-loaded metal-organic framework MIL-101(Cr) material by sonication process: Catalytic activity and antibacterial properties. *Microporous Mesoporous Mater.* **2021**, *323*, 111244. [CrossRef]
34. Lv, P.; Yang, S.; Ma, P. Bio-based oil gelling agent for effective removal of oil spills from the surface of water. *Mater. Chem. Front.* **2018**, *2*, 1784–1790. [CrossRef]

35. Dong, Z.; Wei, Y.; Chen, Y.; Zhang, L.; Xu, D.; Lin, G.; Li, B.; Zhang, S.; Wei, J.; Akhtar, M.A.; et al. Nitrogen migration and transformation in chars and tars during co-pyrolysis of cellulose/xylan/lignin with urea formaldehyde. *J. Anal. Appl. Pyrolysis* **2024**, *178*, 106425. [CrossRef]
36. Hemmati, F.; Norouzbeigi, R.; Sarbisheh, F.; Shayesteh, H. Malachite green removal using modified sphagnum peat moss as a low-cost biosorbent: Kinetic, equilibrium and thermodynamic studies. *J. Taiwan Inst. Chem. Eng.* **2016**, *58*, 482–489. [CrossRef]
37. Karagozoglu, B.; Tasdemir, M.; Demirbas, E.; Kobya, M. The adsorption of basic dye (astrazon blue FGRL) from aqueous solutions onto sepiolite, fly ash and apricot shell activated carbon: Kinetic and equilibrium studies. *J. Hazard. Mater.* **2007**, *147*, 297–306. [CrossRef] [PubMed]
38. Bensedira, A.; Haddaoui, N.; Doufnoune, R.; Meziane, O.; Labidi, N.S. Study of methylene blue dye elimination from water using polyaniline (PANI) and PANI/SiO_2 composite. *Polym. Polym. Compos.* **2022**, *30*, 09673911221141747. [CrossRef]
39. Peretz, S.; Anghel, D.F.; Vasilescu, E.; Florea-Spiroiu, M.; Stoian, C.; Zgherea, G. Synthesis, characterization and adsorption properties of alginate porous beads. *Polym. Bull.* **2015**, *72*, 3169–3182. [CrossRef]
40. Bai, R.; Feng, Y.; Wu, L.; Li, N.; Liu, Q.; Teng, Y.; He, R.; Zhi, K.; Zhou, H.; Qi, X. Adsorption mechanism of methylene blue by magnesium salt-modified lignite-based adsorbents. *J. Environ. Manag.* **2023**, *344*, 118514. [CrossRef]
41. Zhou, L.; Zhou, H.; Hu, Y.; Yan, S.; Yang, J. Adsorption removal of cationic dyes from aqueous solutions using ceramic adsorbents prepared from industrial waste coal gangue. *J. Environ. Manag.* **2019**, *234*, 245–252. [CrossRef] [PubMed]
42. Abdulhameed, A.S.; Jawad, A.H.; Kashi, E.; Radzun, K.A.; ALOthman, Z.A.; Wilson, L.D. Insight into adsorption mechanism, modeling, and desirability function of crystal violet and methylene blue dyes by microalgae: Box-behnken design application. *Algal Res.* **2022**, *67*, 102864. [CrossRef]
43. Kubra, K.T.; Salman, M.S.; Hasan, M.N. Enhanced toxic dye removal from wastewater using biodegradable polymeric natural adsorbent. *J. Mol. Liq.* **2021**, *328*, 115468. [CrossRef]
44. Noori, M.; Tahmasebpoor, M.; Foroutan, R. Enhanced adsorption capacity of low-cost magnetic clinoptilolite powders/beads for the effective removal of methylene blue: Adsorption and desorption studies. *Mater. Chem. Phys.* **2022**, *278*, 125655. [CrossRef]
45. Yardımcı, B.; Kanmaz, N. An effective-green strategy of methylene blue adsorption: Sustainable and low-cost waste cinnamon bark biomass enhanced via MnO_2. *J. Environ. Chem. Eng.* **2023**, *11*, 110254. [CrossRef]
46. Zheng, Y.; Wang, J.; Li, D.; Liu, C.; Lu, Y.; Lin, X.; Zheng, Z. Insight into the KOH/$KMnO_4$ activation mechanism of oxygen-enriched hierarchical porous biochar derived from biomass waste by in-situ pyrolysis for methylene blue enhanced adsorption. *J. Anal. Appl. Pyrolysis* **2021**, *158*, 105269. [CrossRef]
47. Jawad, A.H.; Abdulhameed, A.S.; Wilson, L.D.; Syed-Hassan, S.S.A.; ALOthman, Z.A.; Khan, M.R. High surface area and mesoporous activated carbon from KOH-activated dragon fruit peels for methylene blue dye adsorption: Optimization and mechanism study. *Chin. J. Chem. Eng.* **2021**, *32*, 281–290. [CrossRef]
48. Liao, Y.; Ge, W.; Liu, M.; Bi, W.; Jin, C.; Chen, D.D.Y. Eco-friendly regeneration of lignin with acidic deep eutectic solvent for adsorption of pollutant dyes for water cleanup. *Int. J. Biol. Macromol.* **2024**, *260*, 129677. [CrossRef]
49. Tran, H.N.; Wang, Y.; You, S.; Chao, H. Insights into the mechanism of cationic dye adsorption on activated charcoal: The importance of π–π interactions. *Process Saf. Environ. Protect.* **2017**, *107*, 168–180. [CrossRef]

MDPI AG
Grosspeteranlage 5
4052 Basel
Switzerland
Tel.: +41 61 683 77 34

MDPI Books Editorial Office
E-mail: books@mdpi.com
www.mdpi.com/books

www.ingramcontent.com/pod-product-compliance
Lightning Source LLC
Chambersburg PA
CBHW041352210926
51850CB00013B/3054
* 9 7 8 3 7 2 5 8 2 3 5 1 2 *